Lie Group Mathematics

" The Math of String Theory "

Edited by Paul F. Kisak

Contents

Chapter 1

Lie group

In mathematics, a **Lie group** /'liː/ is a group that is also a differentiable manifold, with the property that the group operations are compatible with the smooth structure. Lie groups are named after Sophus Lie, who laid the foundations of the theory of continuous transformation groups. The term *groupes de Lie* first appeared in French in 1893 in the thesis of Lie's student Arthur Tresse, page 3.[1]

Lie groups represent the best-developed theory of continuous symmetry of mathematical objects and structures, which makes them indispensable tools for many parts of contemporary mathematics, as well as for modern theoretical physics. They provide a natural framework for analysing the continuous symmetries of differential equations (differential Galois theory), in much the same way as permutation groups are used in Galois theory for analysing the discrete symmetries of algebraic equations. An extension of Galois theory to the case of continuous symmetry groups was one of Lie's principal motivations.

1.1   Overview

Lie groups are smooth[Note 1] differentiable manifolds and as such can be studied using differential calculus, in contrast with the case of more general topological groups. One of the key ideas in the theory of Lie groups is to replace the *global* object, the group, with its *local* or linearized version, which Lie himself called its "infinitesimal group" and which has since become known as its Lie algebra.

Lie groups play an enormous role in modern geometry, on several different levels. Felix Klein argued in his Erlangen program that one can consider various "geometries" by specifying an appropriate transformation group that leaves certain geometric properties invariant. Thus Euclidean geometry corresponds to the choice of the group E(3) of distance-preserving transformations of the Euclidean space $\mathbf{R}^3$, conformal geometry corresponds to enlarging the group to the conformal group, whereas in projective geometry one is interested in the properties invariant under the projective group. This idea later led to the notion of a G-structure, where G is a Lie group of "local" symmetries of a manifold. On a "global" level, whenever a Lie group acts on a geometric object, such as a Riemannian or a symplectic manifold, this action provides a measure of rigidity and yields a rich algebraic structure. The presence of continuous symmetries expressed via a Lie group action on a manifold places strong constraints on its geometry and facilitates analysis on the manifold. Linear actions of Lie groups are especially important, and are studied in representation theory.

In the 1940s–1950s, Ellis Kolchin, Armand Borel, and Claude Chevalley realised that many foundational results concerning Lie groups can be developed completely algebraically, giving rise to the theory of algebraic groups defined over an arbitrary field. This insight opened new possibilities in pure algebra, by providing a uniform construction for most finite simple groups, as well as in algebraic geometry. The theory of automorphic forms, an important branch of modern number theory, deals extensively with analogues of Lie groups over adele rings; p-adic Lie groups play an important role, via their connections with Galois representations in number theory.

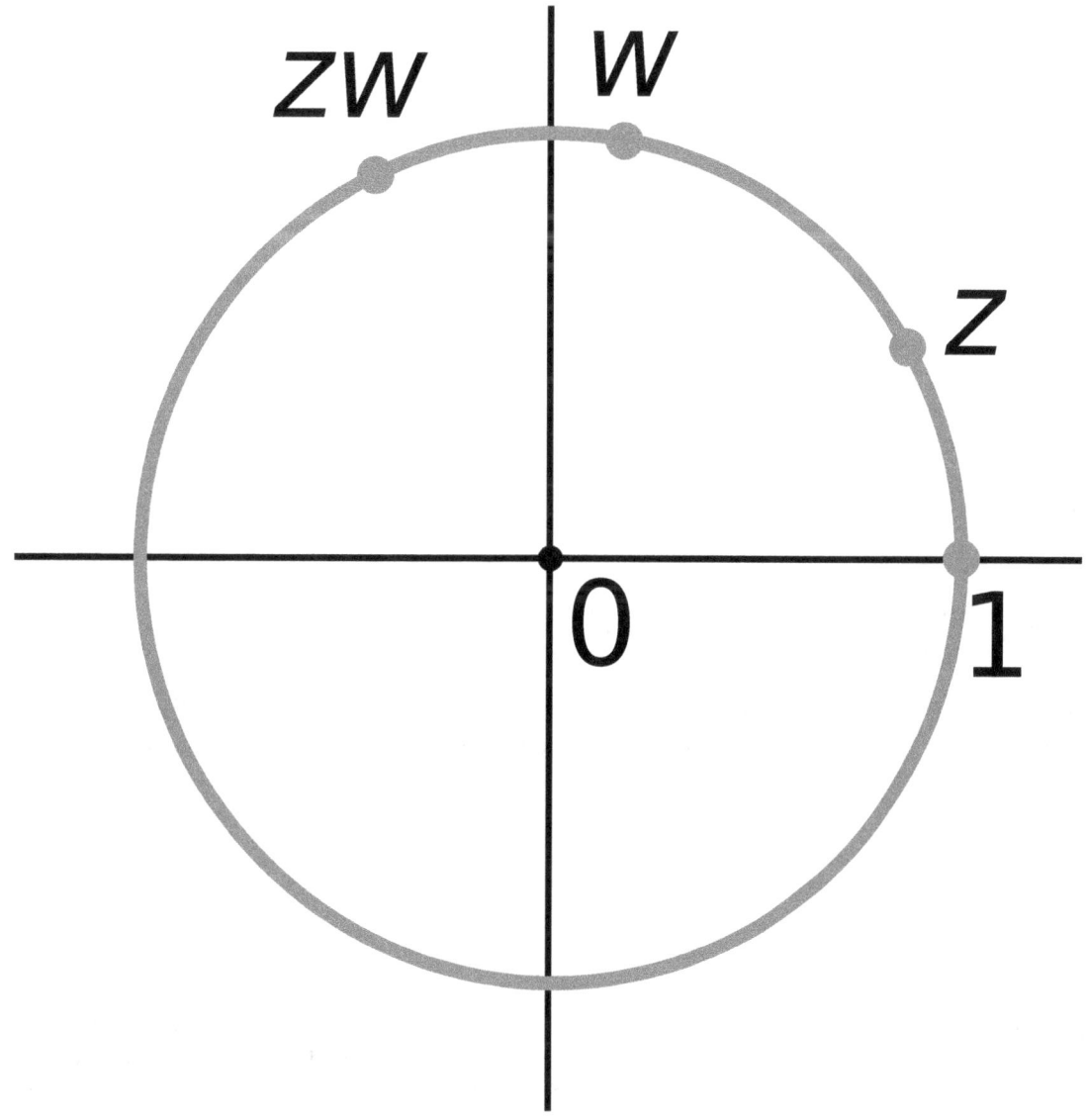

The circle of center 0 and radius 1 in the complex plane is a Lie group with complex multiplication.

1.2   Definitions and examples

A **real Lie group** is a group that is also a finite-dimensional real smooth manifold, in which the group operations of multiplication and inversion are smooth maps. Smoothness of the group multiplication

$$\mu : G \times G \to G \quad \mu(x, y) = xy$$

means that μ is a smooth mapping of the product manifold $G{\times}G$ into G. These two requirements can be combined to the single requirement that the mapping

$$(x, y) \mapsto x^{-1}y$$

be a smooth mapping of the product manifold into *G*.

1.2.1　First examples

- The 2×2 real invertible matrices form a group under multiplication, denoted by GL(2, **R**) or by GL2(**R**):

$$
\mathrm{GL}(2, \mathbf{R}) = \left\{ A = \begin{pmatrix} a & b \\ c & d \end{pmatrix} : \det A = ad - bc \neq 0 \right\}.
$$

This is a four-dimensional noncompact real Lie group. This group is disconnected; it has two connected components corresponding to the positive and negative values of the determinant.

- The rotation matrices form a subgroup of GL(2, **R**), denoted by SO(2, **R**). It is a Lie group in its own right: specifically, a one-dimensional compact connected Lie group which is diffeomorphic to the circle. Using the rotation angle φ as a parameter, this group can be parametrized as follows:

$$
\mathrm{SO}(2, \mathbf{R}) = \left\{ \begin{pmatrix} \cos \varphi & -\sin \varphi \\ \sin \varphi & \cos \varphi \end{pmatrix} : \varphi \in \mathbf{R}/2\pi\mathbf{Z} \right\}.
$$

Addition of the angles corresponds to multiplication of the elements of SO(2, **R**), and taking the opposite angle corresponds to inversion. Thus both multiplication and inversion are differentiable maps.

- The orthogonal group also forms an interesting example of a Lie group.

All of the previous examples of Lie groups fall within the class of classical groups.

1.2.2　Related concepts

A **complex Lie group** is defined in the same way using complex manifolds rather than real ones (example: SL(2, **C**)), and similarly, using an alternate metric completion of **Q**, one can define a ***p*-adic Lie group** over the *p*-adic numbers, a topological group in which each point has a *p*-adic neighborhood. Hilbert's fifth problem asked whether replacing differentiable manifolds with topological or analytic ones can yield new examples. The answer to this question turned out to be negative: in 1952, Gleason, Montgomery and Zippin showed that if *G* is a topological manifold with continuous group operations, then there exists exactly one analytic structure on *G* which turns it into a Lie group (see also Hilbert–Smith conjecture). If the underlying manifold is allowed to be infinite-dimensional (for example, a Hilbert manifold), then one arrives at the notion of an infinite-dimensional Lie group. It is possible to define analogues of many Lie groups over finite fields, and these give most of the examples of finite simple groups.

The language of category theory provides a concise definition for Lie groups: a Lie group is a group object in the category of smooth manifolds. This is important, because it allows generalization of the notion of a Lie group to Lie supergroups.

1.3　More examples of Lie groups

See also: Table of Lie groups and List of simple Lie groups

Lie groups occur in abundance throughout mathematics and physics. Matrix groups or algebraic groups are (roughly) groups of matrices (for example, orthogonal and symplectic groups), and these give most of the more common examples of Lie groups.

1.3.1   Examples with a specific number of dimensions

- The circle group S^1 consisting of angles mod 2π under addition or, alternatively, the complex numbers with absolute value 1 under multiplication. This is a one-dimensional compact connected abelian Lie group.

- The 3-sphere S^3 forms a Lie group by identification with the set of quaternions of unit norm, called versors. The only other spheres that admit the structure of a Lie group are the 0-sphere S^0 (real numbers with absolute value 1) and the circle S^1 (complex numbers with absolute value 1). For example, for even $n > 1$, S^n is not a Lie group because it does not admit a nonvanishing vector field and so *a fortiori* cannot be parallelizable as a differentiable manifold. Of the spheres only S^0, S^1, S^3, and S^7 are parallelizable. The last carries the structure of a Lie quasigroup (a nonassociative group), which can be identified with the set of unit octonions.

- The (3-dimensional) metaplectic group is a double cover of SL(2, $\mathbf{R}$) playing an important role in the theory of modular forms. It is a connected Lie group that cannot be faithfully represented by matrices of finite size, i.e., a nonlinear group.

- The Heisenberg group is a connected nilpotent Lie group of dimension 3, playing a key role in quantum mechanics.

- The Lorentz group is a 6-dimensional Lie group of linear isometries of the Minkowski space.

- The Poincaré group is a 10-dimensional Lie group of affine isometries of the Minkowski space.

- The group U(1)×SU(2)×SU(3) is a Lie group of dimension 1+3+8=12 that is the gauge group of the Standard Model in particle physics. The dimensions of the factors correspond to the 1 photon + 3 vector bosons + 8 gluons of the standard model

- The exceptional Lie groups of types G_2, F_4, E_6, E_7, E_8 have dimensions 14, 52, 78, 133, and 248. Along with the A-B-C-D series of simple Lie groups, the exceptional groups complete the list of simple Lie groups. There is also a Lie group named $E_7\frac{1}{2}$ of dimension 190, but it is not a *simple* Lie group.

1.3.2   Examples with n dimensions

- Euclidean space $\mathbf{R}^n$ with ordinary vector addition as the group operation becomes an n-dimensional noncompact abelian Lie group.

- The Euclidean group E(n, $\mathbf{R}$) is the Lie group of all Euclidean motions, i.e., isometric affine maps, of n-dimensional Euclidean space $\mathbf{R}^n$.

- The orthogonal group O(n, $\mathbf{R}$), consisting of all $n \times n$ orthogonal matrices with real entries is an $n(n-1)/2$-dimensional Lie group. This group is disconnected, but it has a connected subgroup SO(n, $\mathbf{R}$) of the same dimension consisting of orthogonal matrices of determinant 1, called the special orthogonal group (for $n = 3$, the rotation group SO(3)).

- The unitary group U(n) consisting of $n \times n$ unitary matrices (with complex entries) is a compact connected Lie group of dimension n^2. Unitary matrices of determinant 1 form a closed connected subgroup of dimension $n^2 - 1$ denoted SU(n), the special unitary group.

- Spin groups are double covers of the special orthogonal groups, used for studying fermions in quantum field theory (among other things).

- The group GL(n, $\mathbf{R}$) of invertible matrices (under matrix multiplication) is a Lie group of dimension n^2, called the general linear group. It has a closed connected subgroup SL(n, $\mathbf{R}$), the special linear group, consisting of matrices of determinant 1 which is also a Lie group.

- The symplectic group Sp($2n$, $\mathbf{R}$) consists of all $2n \times 2n$ matrices preserving a *symplectic form* on $\mathbf{R}^{2n}$. It is a connected Lie group of dimension $2n^2 + n$.

- The group of invertible upper triangular n by n matrices is a solvable Lie group of dimension $n(n + 1)/2$. (cf. Borel subgroup)

- The A-series, B-series, C-series and D-series, whose elements are denoted by A*n*, B*n*, C*n*, and D*n*, are infinite families of simple Lie groups.

1.3.3 Constructions

There are several standard ways to form new Lie groups from old ones:

- The product of two Lie groups is a Lie group.

- Any topologically closed subgroup of a Lie group is a Lie group. This is known as the Closed subgroup theorem or **Cartan's theorem**.

- The quotient of a Lie group by a closed normal subgroup is a Lie group.

- The universal cover of a connected Lie group is a Lie group. For example, the group $\mathbf{R}$ is the universal cover of the circle group $\mathbf{S}^1$. In fact any covering of a differentiable manifold is also a differentiable manifold, but by specifying *universal* cover, one guarantees a group structure (compatible with its other structures).

1.3.4 Related notions

Some examples of groups that are *not* Lie groups (except in the trivial sense that any group can be viewed as a 0-dimensional Lie group, with the discrete topology), are:

- Infinite-dimensional groups, such as the additive group of an infinite-dimensional real vector space. These are not Lie groups as they are not *finite-dimensional* manifolds.

- Some totally disconnected groups, such as the Galois group of an infinite extension of fields, or the additive group of the *p*-adic numbers. These are not Lie groups because their underlying spaces are not real manifolds. (Some of these groups are "*p*-adic Lie groups".) In general, only topological groups having similar local properties to $\mathbf{R}^n$ for some positive integer n can be Lie groups (of course they must also have a differentiable structure).

1.4 Basic concepts

1.4.1 The Lie algebra associated with a Lie group

Main article: Lie group–Lie algebra correspondence

To every Lie group we can associate a Lie algebra whose underlying vector space is the tangent space of the Lie group at the identity element and which completely captures the local structure of the group. Informally we can think of elements of the Lie algebra as elements of the group that are "infinitesimally close" to the identity, and the Lie bracket of the Lie algebra is related to the commutator of two such infinitesimal elements. Before giving the abstract definition we give a few examples:

- The Lie algebra of the vector space $\mathbf{R}^n$ is just $\mathbf{R}^n$ with the Lie bracket given by
$[A, B] = 0$.
(In general the Lie bracket of a connected Lie group is always 0 if and only if the Lie group is abelian.)

- The Lie algebra of the general linear group $GL(n, \mathbf{R})$ of invertible matrices is the vector space $M(n, \mathbf{R})$ of square matrices with the Lie bracket given by
$[A, B] = AB - BA$.
If G is a closed subgroup of $GL(n, \mathbf{R})$ then the Lie algebra of G can be thought of informally as the matrices m

of M(n, **R**) such that $1 + \varepsilon m$ is in G, where ε is an infinitesimal positive number with $\varepsilon^2 = 0$ (of course, no such real number ε exists). For example, the orthogonal group O(n, **R**) consists of matrices A with $AA^T = 1$, so the Lie algebra consists of the matrices m with $(1 + \varepsilon m)(1 + \varepsilon m)^T = 1$, which is equivalent to $m + m^T = 0$ because $\varepsilon^2 = 0$.

- Formally, when working over the reals, as here, this is accomplished by considering the limit as $\varepsilon \to 0$; but the "infinitesimal" language generalizes directly to Lie groups over general rings.

The concrete definition given above is easy to work with, but has some minor problems: to use it we first need to represent a Lie group as a group of matrices, but not all Lie groups can be represented in this way, and it is not obvious that the Lie algebra is independent of the representation we use. To get around these problems we give the general definition of the Lie algebra of a Lie group (in 4 steps):

1. Vector fields on any smooth manifold M can be thought of as derivations X of the ring of smooth functions on the manifold, and therefore form a Lie algebra under the Lie bracket $[X, Y] = XY - YX$, because the Lie bracket of any two derivations is a derivation.

2. If G is any group acting smoothly on the manifold M, then it acts on the vector fields, and the vector space of vector fields fixed by the group is closed under the Lie bracket and therefore also forms a Lie algebra.

3. We apply this construction to the case when the manifold M is the underlying space of a Lie group G, with G acting on $G = M$ by left translations $Lg(h) = gh$. This shows that the space of left invariant vector fields (vector fields satisfying $Lg*Xh = Xgh$ for every h in G, where $Lg*$ denotes the differential of Lg) on a Lie group is a Lie algebra under the Lie bracket of vector fields.

4. Any tangent vector at the identity of a Lie group can be extended to a left invariant vector field by left translating the tangent vector to other points of the manifold. Specifically, the left invariant extension of an element v of the tangent space at the identity is the vector field defined by $v^{\wedge}g = Lg*v$. This identifies the tangent space TeG at the identity with the space of left invariant vector fields, and therefore makes the tangent space at the identity into a Lie algebra, called the Lie algebra of G, usually denoted by a Fraktur $\mathfrak{g}$. Thus the Lie bracket on $\mathfrak{g}$ is given explicitly by $[v, w] = [v^{\wedge}, w^{\wedge}]e$.

This Lie algebra $\mathfrak{g}$ is finite-dimensional and it has the same dimension as the manifold G. The Lie algebra of G determines G up to "local isomorphism", where two Lie groups are called **locally isomorphic** if they look the same near the identity element. Problems about Lie groups are often solved by first solving the corresponding problem for the Lie algebras, and the result for groups then usually follows easily. For example, simple Lie groups are usually classified by first classifying the corresponding Lie algebras.

We could also define a Lie algebra structure on Te using right invariant vector fields instead of left invariant vector fields. This leads to the same Lie algebra, because the inverse map on G can be used to identify left invariant vector fields with right invariant vector fields, and acts as -1 on the tangent space Te.

The Lie algebra structure on Te can also be described as follows: the commutator operation

$$(x, y) \to xyx^{-1}y^{-1}$$

on $G \times G$ sends (e, e) to e, so its derivative yields a bilinear operation on TeG. This bilinear operation is actually the zero map, but the second derivative, under the proper identification of tangent spaces, yields an operation that satisfies the axioms of a Lie bracket, and it is equal to twice the one defined through left-invariant vector fields.

1.4.2 Homomorphisms and isomorphisms

If G and H are Lie groups, then a Lie group homomorphism $f : G \to H$ is a smooth group homomorphism. In the case of complex Lie groups, such a homomorphism is required to be a holomorphic map. However, these requirements are a bit stringent; over real or complex numbers, every continuous homomorphism between Lie groups turns out to be (real or complex) analytic.

The composition of two Lie homomorphisms is again a homomorphism, and the class of all Lie groups, together with these morphisms, forms a category. Moreover, every Lie group homomorphism induces a homomorphism between the corresponding Lie algebras. Let $\phi: G \to H$ be a Lie group homomorphism and let ϕ_* be its derivative at the identity. If we identify the Lie algebras of G and H with their tangent spaces at the identity elements then ϕ_* is a map between the corresponding Lie algebras:

$$\phi_*: \mathfrak{g} \to \mathfrak{h}$$

One can show that ϕ_* is actually a Lie algebra homomorphism (meaning that it is a linear map which preserves the Lie bracket). In the language of category theory, we then have a covariant functor from the category of Lie groups to the category of Lie algebras which sends a Lie group to its Lie algebra and a Lie group homomorphism to its derivative at the identity.

Two Lie groups are called *isomorphic* if there exists a bijective homomorphism between them whose inverse is also a Lie group homomorphism. Equivalently, it is a diffeomorphism which is also a group homomorphism.

Ado's theorem says every finite-dimensional Lie algebra is isomorphic to a matrix Lie algebra. For every finite-dimensional matrix Lie algebra, there is a linear group (matrix Lie group) with this algebra as its Lie algebra. So every abstract Lie algebra is the Lie algebra of some (linear) Lie group.

The *global structure* of a Lie group is not determined by its Lie algebra; for example, if Z is any discrete subgroup of the center of G then G and G/Z have the same Lie algebra (see the table of Lie groups for examples). A *connected* Lie group is simple, semisimple, solvable, nilpotent, or abelian if and only if its Lie algebra has the corresponding property.

If we require that the Lie group be simply connected, then the global structure is determined by its Lie algebra: for every finite-dimensional Lie algebra $\mathfrak{g}$ over $\mathbf{F}$ there is a simply connected Lie group G with $\mathfrak{g}$ as Lie algebra, unique up to isomorphism. Moreover every homomorphism between Lie algebras lifts to a unique homomorphism between the corresponding simply connected Lie groups.

1.4.3 The exponential map

Main article: Exponential map (Lie theory)

The exponential map from the Lie algebra $M(n, \mathbf{R})$ of the general linear group $GL(n, \mathbf{R})$ to $GL(n, \mathbf{R})$ is defined by the usual power series:

$$\exp(A) = 1 + A + \frac{A^2}{2!} + \frac{A^3}{3!} + \cdots$$

for matrices A. If G is any subgroup of $GL(n, \mathbf{R})$, then the exponential map takes the Lie algebra of G into G, so we have an exponential map for all matrix groups.

The definition above is easy to use, but it is not defined for Lie groups that are not matrix groups, and it is not clear that the exponential map of a Lie group does not depend on its representation as a matrix group. We can solve both problems using a more abstract definition of the exponential map that works for all Lie groups, as follows.

Every vector v in $\mathfrak{g}$ determines a linear map from $\mathbf{R}$ to $\mathfrak{g}$ taking 1 to v, which can be thought of as a Lie algebra homomorphism. Because $\mathbf{R}$ is the Lie algebra of the simply connected Lie group $\mathbf{R}$, this induces a Lie group homomorphism $c : \mathbf{R} \to G$ so that

$$c(s + t) = c(s)c(t)$$

for all s and t. The operation on the right hand side is the group multiplication in G. The formal similarity of this formula with the one valid for the exponential function justifies the definition

$\exp(v) = c(1)$.

This is called the **exponential map**, and it maps the Lie algebra $\mathfrak{g}$ into the Lie group G. It provides a diffeomorphism between a neighborhood of 0 in $\mathfrak{g}$ and a neighborhood of e in G. This exponential map is a generalization of the exponential function for real numbers (because **R** is the Lie algebra of the Lie group of positive real numbers with multiplication), for complex numbers (because **C** is the Lie algebra of the Lie group of non-zero complex numbers with multiplication) and for matrices (because M(n, **R**) with the regular commutator is the Lie algebra of the Lie group GL(n, **R**) of all invertible matrices).

Because the exponential map is surjective on some neighbourhood N of e, it is common to call elements of the Lie algebra **infinitesimal generators** of the group G. The subgroup of G generated by N is the identity component of G.

The exponential map and the Lie algebra determine the *local group structure* of every connected Lie group, because of the Baker–Campbell–Hausdorff formula: there exists a neighborhood U of the zero element of $\mathfrak{g}$, such that for u, v in U we have

$$\exp(u)\, \exp(v) = \exp\left(u + v + \tfrac{1}{2}[u, v] + \tfrac{1}{12}[\,[u, v], v] - \tfrac{1}{12}[\,[u, v], u] - \cdots\right),$$

where the omitted terms are known and involve Lie brackets of four or more elements. In case u and v commute, this formula reduces to the familiar exponential law $\exp(u)\exp(v) = \exp(u + v)$.

The exponential map relates Lie group homomorphisms. That is, if $\phi : G \to H$ is a Lie group homomorphism and $\phi_* : \mathfrak{g} \to \mathfrak{h}$ the induced map on the corresponding Lie algebras, then for all $x \in \mathfrak{g}$ we have

$$\phi(\exp(x)) = \exp(\phi_*(x)).$$

In other words the following diagram commutes,[Note 2]

(In short, exp is a natural transformation from the functor Lie to the identity functor on the category of Lie groups.)

The exponential map from the Lie algebra to the Lie group is not always onto, even if the group is connected (though it does map onto the Lie group for connected groups that are either compact or nilpotent). For example, the exponential map of SL(2, **R**) is not surjective. Also, exponential map is not surjective nor injective for infinite-dimensional (see below) Lie groups modelled on C^∞ Fréchet space, even from arbitrary small neighborhood of 0 to corresponding neighborhood of 1.

See also: derivative of the exponential map and normal coordinates.

1.4.4   Lie subgroup

A **Lie subgroup** H of a Lie group G is a Lie group that is a subset of G and such that the inclusion map from H to G is an injective immersion and group homomorphism. According to Cartan's theorem, a closed subgroup of G admits a unique smooth structure which makes it an embedded Lie subgroup of G—i.e. a Lie subgroup such that the inclusion map is a smooth embedding.

Examples of non-closed subgroups are plentiful; for example take G to be a torus of dimension ≥ 2, and let H be a one-parameter subgroup of *irrational slope*, i.e. one that winds around in G. Then there is a Lie group homomorphism $\varphi : $ **R** $\to G$ with H as its image. The closure of H will be a sub-torus in G.

In terms of the exponential map of G, in general, only some of the Lie subalgebras of the Lie algebra g of G correspond to closed Lie subgroups H of G. There is no criterion solely based on the structure of g which determines which those are.

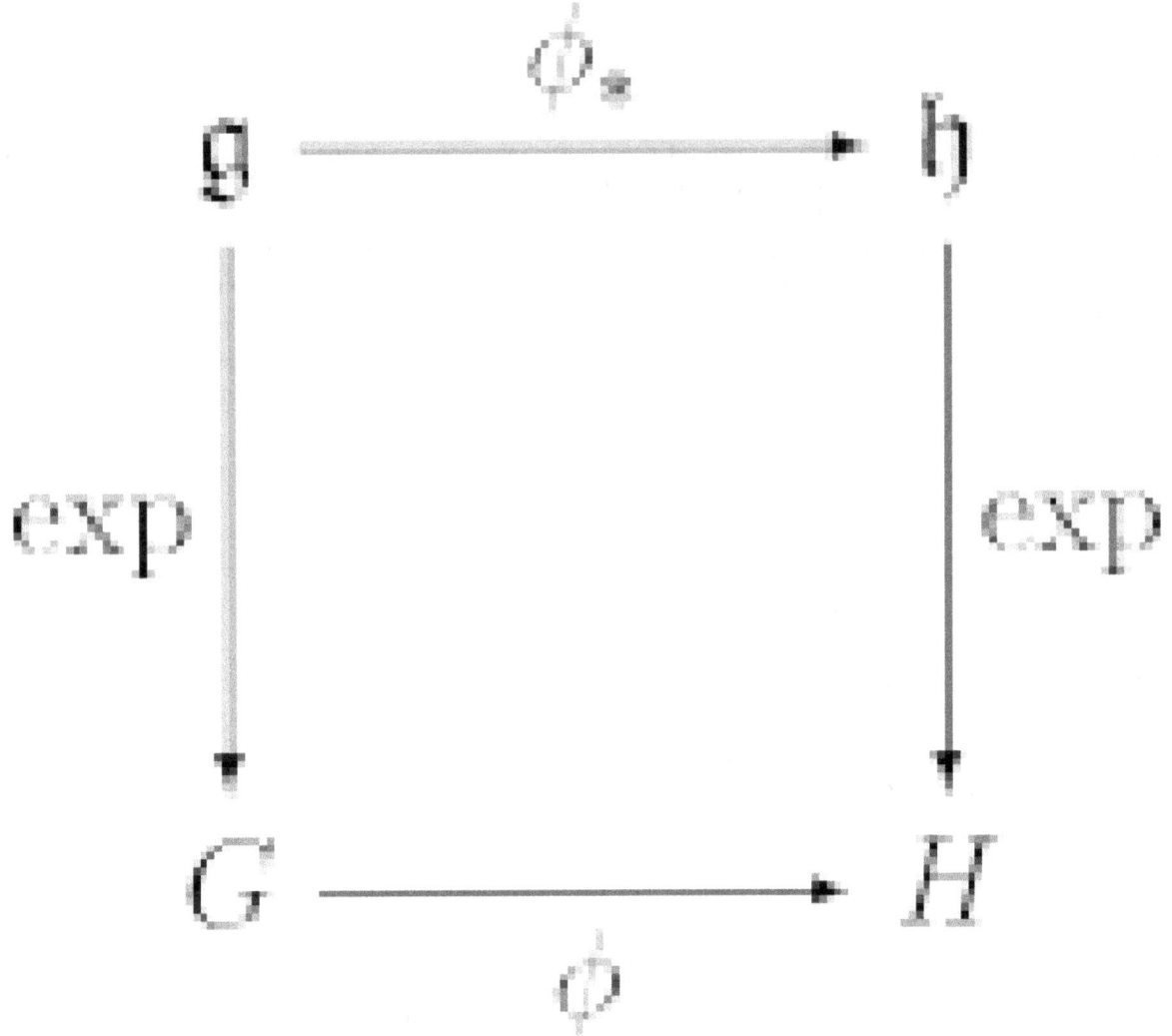

1.5 Early history

According to the most authoritative source on the early history of Lie groups (Hawkins, p. 1), Sophus Lie himself considered the winter of 1873–1874 as the birth date of his theory of continuous groups. Hawkins, however, suggests that it was "Lie's prodigious research activity during the four-year period from the fall of 1869 to the fall of 1873" that led to the theory's creation (*ibid*). Some of Lie's early ideas were developed in close collaboration with Felix Klein. Lie met with Klein every day from October 1869 through 1872: in Berlin from the end of October 1869 to the end of February 1870, and in Paris, Göttingen and Erlangen in the subsequent two years (*ibid*, p. 2). Lie stated that all of the principal results were obtained by 1884. But during the 1870s all his papers (except the very first note) were published in Norwegian journals, which impeded recognition of the work throughout the rest of Europe (*ibid*, p. 76). In 1884 a young German mathematician, Friedrich Engel, came to work with Lie on a systematic treatise to expose his theory of continuous groups. From this effort resulted the three-volume *Theorie der Transformationsgruppen*, published in 1888, 1890, and 1893.

Lie's ideas did not stand in isolation from the rest of mathematics. In fact, his interest in the geometry of differential equations was first motivated by the work of Carl Gustav Jacobi, on the theory of partial differential equations of first order and on the equations of classical mechanics. Much of Jacobi's work was published posthumously in the 1860s, generating enormous interest in France and Germany (Hawkins, p. 43). Lie's *idée fixe* was to develop a theory of symmetries of differential equations that would accomplish for them what Évariste Galois had done for algebraic equations: namely, to classify them in terms of group theory. Lie and other mathematicians showed that the most important equations for special functions and orthogonal polynomials tend to arise from group theoretical symmetries. In Lie's early work, the idea was to construct a theory of *continuous groups*, to complement the theory of discrete groups that had developed in the theory of modular forms, in the hands of Felix Klein and Henri Poincaré. The initial application that Lie had in mind was to the theory of differential equations. On the model of Galois theory and polynomial equations, the driving conception was of a theory capable of unifying, by the study of symmetry, the whole area of ordinary differential equations. However, the hope that Lie Theory would unify the entire field of ordinary differential equations was not fulfilled. Symmetry methods for ODEs continue to be studied, but do not dominate the subject. There is a differential Galois theory, but it was developed by others, such as Picard and Vessiot, and it provides a theory of quadratures, the indefinite integrals required to express solutions.

Additional impetus to consider continuous groups came from ideas of Bernhard Riemann, on the foundations of geometry, and their further development in the hands of Klein. Thus three major themes in 19th century mathematics were combined by Lie in creating his new theory: the idea of symmetry, as exemplified by Galois through the algebraic notion of a group; geometric theory and the explicit solutions of differential equations of mechanics, worked out by Poisson and Jacobi; and the new understanding of geometry that emerged in the works of Plücker, Möbius, Grassmann and others, and culminated in Riemann's revolutionary vision of the subject.

Although today Sophus Lie is rightfully recognized as the creator of the theory of continuous groups, a major stride in the development of their structure theory, which was to have a profound influence on subsequent development of mathematics, was made by Wilhelm Killing, who in 1888 published the first paper in a series entitled *Die Zusammensetzung der stetigen endlichen Transformationsgruppen* (*The composition of continuous finite transformation groups*) (Hawkins, p. 100). The work of Killing, later refined and generalized by Élie Cartan, led to classification of semisimple Lie algebras, Cartan's theory of symmetric spaces, and Hermann Weyl's description of representations of compact and semisimple Lie groups using highest weights.

In 1900 David Hilbert challenged Lie theorists with his Fifth Problem presented at the International Congress of Mathematicians in Paris.

Weyl brought the early period of the development of the theory of Lie groups to fruition, for not only did he classify irreducible representations of semisimple Lie groups and connect the theory of groups with quantum mechanics, but he also put Lie's theory itself on firmer footing by clearly enunciating the distinction between Lie's *infinitesimal groups* (i.e., Lie algebras) and the Lie groups proper, and began investigations of topology of Lie groups.[2] The theory of Lie groups was systematically reworked in modern mathematical language in a monograph by Claude Chevalley.

1.6   The concept of a Lie group, and possibilities of classification

Lie groups may be thought of as smoothly varying families of symmetries. Examples of symmetries include rotation about an axis. What must be understood is the nature of 'small' transformations, e.g., rotations through tiny angles, that link nearby transformations. The mathematical object capturing this structure is called a Lie algebra (Lie himself called them "infinitesimal groups"). It can be defined because Lie groups are manifolds, so have tangent spaces at each point.

The Lie algebra of any compact Lie group (very roughly: one for which the symmetries form a bounded set) can be decomposed as a direct sum of an abelian Lie algebra and some number of simple ones. The structure of an abelian Lie algebra is mathematically uninteresting (since the Lie bracket is identically zero); the interest is in the simple summands. Hence the question arises: what are the simple Lie algebras of compact groups? It turns out that they mostly fall into four infinite families, the "classical Lie algebras" A_n, B_n, C_n and D_n, which have simple descriptions in terms of symmetries of Euclidean space. But there are also just five "exceptional Lie algebras" that do not fall into any of these families. E_8 is the largest of these.

Lie groups are classified according to their algebraic properties (simple, semisimple, solvable, nilpotent, abelian), their connectedness (connected or simply connected) and their compactness.

- Compact Lie groups are all known: they are finite central quotients of a product of copies of the circle group S^1 and simple compact Lie groups (which correspond to connected Dynkin diagrams).

- Any simply connected solvable Lie group is isomorphic to a closed subgroup of the group of invertible upper triangular matrices of some rank, and any finite-dimensional irreducible representation of such a group is 1-dimensional. Solvable groups are too messy to classify except in a few small dimensions.

- Any simply connected nilpotent Lie group is isomorphic to a closed subgroup of the group of invertible upper triangular matrices with 1's on the diagonal of some rank, and any finite-dimensional irreducible representation of such a group is 1-dimensional. Like solvable groups, nilpotent groups are too messy to classify except in a few small dimensions.

- Simple Lie groups are sometimes defined to be those that are simple as abstract groups, and sometimes defined to be connected Lie groups with a simple Lie algebra. For example, SL(2, **R**) is simple according to the second definition but not according to the first. They have all been classified (for either definition).

- Semisimple Lie groups are Lie groups whose Lie algebra is a product of simple Lie algebras.[3] They are central extensions of products of simple Lie groups.

The identity component of any Lie group is an open normal subgroup, and the quotient group is a discrete group. The universal cover of any connected Lie group is a simply connected Lie group, and conversely any connected Lie group is a quotient of a simply connected Lie group by a discrete normal subgroup of the center. Any Lie group G can be decomposed into discrete, simple, and abelian groups in a canonical way as follows. Write

G_{con} for the connected component of the identity

G_{sol} for the largest connected normal solvable subgroup

G_{nil} for the largest connected normal nilpotent subgroup

so that we have a sequence of normal subgroups

$1 \subseteq G_{\mathrm{nil}} \subseteq G_{\mathrm{sol}} \subseteq G_{\mathrm{con}} \subseteq G.$

Then

G/G_{con} is discrete

$G_{\mathrm{con}}/G_{\mathrm{sol}}$ is a central extension of a product of simple connected Lie groups.

$G_{\mathrm{sol}}/G_{\mathrm{nil}}$ is abelian. A connected abelian Lie group is isomorphic to a product of copies of **R** and the circle group S^1.

$G_{\mathrm{nil}}/1$ is nilpotent, and therefore its ascending central series has all quotients abelian.

This can be used to reduce some problems about Lie groups (such as finding their unitary representations) to the same problems for connected simple groups and nilpotent and solvable subgroups of smaller dimension.

- The diffeomorphism group of a Lie group acts transitively on the Lie group

- Every Lie group is parallelizable, and hence an orientable manifold (there is a bundle isomorphism between its tangent bundle and the product of itself with the tangent space at the identity)

1.7 Infinite-dimensional Lie groups

Lie groups are often defined to be finite-dimensional, but there are many groups that resemble Lie groups, except for being infinite-dimensional. The simplest way to define infinite-dimensional Lie groups is to model them on Banach spaces, and in this case much of the basic theory is similar to that of finite-dimensional Lie groups. However this is inadequate for many applications, because many natural examples of infinite-dimensional Lie groups are not Banach manifolds. Instead one needs to define Lie groups modeled on more general locally convex topological vector spaces. In this case the relation between the Lie algebra and the Lie group becomes rather subtle, and several results about finite-dimensional Lie groups no longer hold.

Some of the examples that have been studied include:

- The group of diffeomorphisms of a manifold. Quite a lot is known about the group of diffeomorphisms of the circle. Its Lie algebra is (more or less) the Witt algebra, which has a central extension called the Virasoro algebra, used in string theory and conformal field theory. Diffeomorphism groups of compact manifolds of larger dimension are regular Fréchet Lie groups; very little about their structure is known.

The diffeomorphism group of spacetime sometimes appears in attempts to quantize gravity.

- The group of smooth maps from a manifold to a finite-dimensional Lie group is an example of a gauge group (with operation of pointwise multiplication), and is used in quantum field theory and Donaldson theory. If the manifold is a circle these are called loop groups, and have central extensions whose Lie algebras are (more or less) Kac–Moody algebras.

- There are infinite-dimensional analogues of general linear groups, orthogonal groups, and so on. One important aspect is that these may have *simpler* topological properties: see for example Kuiper's theorem. In M-Theory theory, for example, a 10 dimensional SU(N) gauge theory becomes an 11 dimensional theory when N becomes infinite.

- A specific example is that $SU(\infty)$ is equal to the group of area preserving diffeomorphisms of a torus.

1.8   See also

- Lie subgroup

- E_8

- Adjoint representation of a Lie group

- Adjoint endomorphism

- Haar measure

- Homogeneous space

- List of Lie group topics

- List of simple Lie groups

- Moufang polygon

- Riemannian manifold

- Representations of Lie groups

- Table of Lie groups

- Lie algebra

- Symmetry in quantum mechanics

- Lie group action

1.9 Notes

1.9.1 Explanatory notes

[1] having derivatives of all orders

[2] http://www.math.sunysb.edu/~{}vkiritch/MAT552/ProblemSet1.pdf

1.9.2 Citations

[1] Arthur Tresse (1893). "Sur les invariants différentiels des groupes continus de transformations". *Acta Mathematica* **18**: 1–88. doi:10.1007/bf02418270.

[2] Borel (2001).

[3] Helgason, Sigurdur (1978). *Differential Geometry, Lie Groups, and Symmetric Spaces.* New York: Academic Press. p. 131. ISBN 0-12-338460-5.

1.10 References

- Adams, John Frank (1969), *Lectures on Lie Groups*, Chicago Lectures in Mathematics, Chicago: Univ. of Chicago Press, ISBN 0-226-00527-5.

- Borel, Armand (2001), *Essays in the history of Lie groups and algebraic groups*, History of Mathematics **21**, Providence, R.I.: American Mathematical Society, ISBN 978-0-8218-0288-5, MR 1847105

- Bourbaki, Nicolas, *Elements of mathematics: Lie groups and Lie algebras.* Chapters 1–3 ISBN 3-540-64242-0, Chapters 4–6 ISBN 3-540-42650-7, Chapters 7–9 ISBN 3-540-43405-4

- Chevalley, Claude (1946), *Theory of Lie groups*, Princeton: Princeton University Press, ISBN 0-691-04990-4.

- P. M. Cohn (1957) *Lie Groups*, Cambridge Tracts in Mathematical Physics.

- J. L. Coolidge (1940) *A History of Geometrical Methods*, pp 304–17, Oxford University Press (Dover Publications 2003).

- Fulton, William; Harris, Joe (1991), *Representation theory. A first course*, Graduate Texts in Mathematics, Readings in Mathematics **129**, New York: Springer-Verlag, ISBN 978-0-387-97495-8, MR 1153249, ISBN 978-0-387-97527-6

- Robert Gilmore (2008) *Lie groups, physics, and geometry: an introduction for physicists, engineers and chemists*, Cambridge University Press ISBN 9780521884006 .

- Hall, Brian C. (2003), *Lie Groups, Lie Algebras, and Representations: An Elementary Introduction*, Springer, ISBN 0-387-40122-9.

- F. Reese Harvey (1990) *Spinors and calibrations*, Academic Press, ISBN 0-12-329650-1 .

- Hawkins, Thomas (2000), *Emergence of the theory of Lie groups*, Sources and Studies in the History of Mathematics and Physical Sciences, Berlin, New York: Springer-Verlag, ISBN 978-0-387-98963-1, MR 1771134 Borel's review

- Helgason, Sigurdur (2001), *Differential geometry, Lie groups, and symmetric spaces*, Graduate Studies in Mathematics **34**, Providence, R.I.: American Mathematical Society, ISBN 978-0-8218-2848-9, MR 1834454

- Knapp, Anthony W. (2002), *Lie Groups Beyond an Introduction*, Progress in Mathematics **140** (2nd ed.), Boston: Birkhäuser, ISBN 0-8176-4259-5.

- Nijenhuis, Albert (1959). "Review: *Lie groups*, by P. M. Cohn". *Bulletin of the American Mathematical Society* **65** (6): 338–341. doi:10.1090/s0002-9904-1959-10358-x.

- Rossmann, Wulf (2001), *Lie Groups: An Introduction Through Linear Groups*, Oxford Graduate Texts in Mathematics, Oxford University Press, ISBN 978-0-19-859683-7. The 2003 reprint corrects several typographical mistakes.

- Sattinger, David H.; Weaver, O. L. (1986). *Lie groups and algebras with applications to physics, geometry, and mechanics.* Springer-Verlag. ISBN 3-540-96240-9.

- Serre, Jean-Pierre (1965), *Lie Algebras and Lie Groups: 1964 Lectures given at Harvard University*, Lecture notes in mathematics **1500**, Springer, ISBN 3-540-55008-9.

- Stillwell, John (2008). *Naive Lie Theory*. Springer. ISBN 0-387-98289-2.

- Heldermann Verlag Journal of Lie Theory

- Warner, Frank W. (1983), *Foundations of differentiable manifolds and Lie groups*, Graduate Texts in Mathematics **94**, New York Berlin Heidelberg: Springer-Verlag, ISBN 978-0-387-90894-6, MR 0722297

- Steeb, Willi-Hans (2007), *Continuous Symmetries, Lie algebras, Differential Equations and Computer Algebra: second edition*, World Scientific Publishing, ISBN 981-270-809-X.

- Lie Groups. Representation Theory and Symmetric Spaces Wolfgang Ziller, Vorlesung 2010

Chapter 2

Table of Lie groups

This article gives a table of some common Lie groups and their associated Lie algebras.

The following are noted: the topological properties of the group (dimension; connectedness; compactness; the nature of the fundamental group; and whether or not they are simply connected) as well as on their algebraic properties (abelian; simple; semisimple).

For more examples of Lie groups and other related topics see the list of simple Lie groups; the Bianchi classification of groups of up to three dimensions; and the list of Lie group topics.

2.1 Real Lie groups and their algebras

Column legend

- **CM**: Is this group G compact? (Yes or No)

- π_0 : Gives the group of components of G. The order of the component group gives the number of connected components. The group is connected if and only if the component group is trivial (denoted by 0).

- π_1 : Gives the fundamental group of G whenever G is connected. The group is simply connected if and only if the fundamental group is trivial (denoted by 0).

- **UC**: If G is not simply connected, gives the universal cover of G.

2.2 Real Lie algebras

Table legend:

- **S**: Is this algebra simple? (Yes or No)

- **SS**: Is this algebra semi-simple? (Yes or No)

2.3 Complex Lie groups and their algebras

The dimensions given are dimensions over **C**. Note that every complex Lie group/algebra can also be viewed as a real Lie group/algebra of twice the dimension.

2.4   Complex Lie algebras

The dimensions given are dimensions over **C**. Note that every complex Lie algebra can also be viewed as a real Lie algebra of twice the dimension.

2.5   References

- Fulton, William; Harris, Joe (1991), *Representation theory. A first course*, Graduate Texts in Mathematics, Readings in Mathematics **129**, New York: Springer-Verlag, ISBN 978-0-387-97495-8, MR 1153249, ISBN 978-0-387-97527-6

Chapter 3

Simple Lie group

In group theory, a **simple Lie group** is a connected non-abelian Lie group G which does not have nontrivial connected normal subgroups.

A **simple Lie algebra** is a non-abelian Lie algebra whose only ideals are 0 and itself. A direct sum of simple Lie algebras is called a semisimple Lie algebra.

An equivalent definition of a simple Lie group follows from the Lie correspondence: a connected Lie group is simple if its Lie algebra is simple. An important technical point is that a simple Lie group may contain *discrete* normal subgroups, hence being a simple Lie group is different from being simple as an abstract group.

Simple Lie groups include many classical Lie groups, which provide a group-theoretic underpinning for spherical geometry, projective geometry and related geometries in the sense of Felix Klein's Erlangen programme. It emerged in the course of classification of simple Lie groups that there exist also several exceptional possibilities not corresponding to any familiar geometry. These *exceptional groups* account for many special examples and configurations in other branches of mathematics, as well as contemporary theoretical physics.

While the notion of a simple Lie group is satisfying from the axiomatic perspective, in applications of Lie theory, such as the theory of Riemannian symmetric spaces, somewhat more general notions of semisimple and reductive Lie groups proved to be even more useful. In particular, every connected compact Lie group is reductive, and the study of representations of general reductive groups is a major branch of representation theory.

3.1 Comments on the definition

Unfortunately there is no single standard definition of a simple Lie group. The definition given above is sometimes varied in the following ways:

- Connectedness: Usually simple Lie groups are connected by definition. This excludes discrete simple groups (these are zero-dimensional Lie groups that are simple as abstract groups) as well as disconnected orthogonal groups.

- Center: Usually simple Lie groups are allowed to have a discrete center; for example, SL(2, **R**) has a center of order 2, but is still counted as a simple Lie group. If the center is non-trivial (and not the whole group) then the simple Lie group is not simple as an abstract group. Some authors require that the center of a simple Lie group be finite (or trivial); the universal cover of SL(2, **R**) is an example of a simple Lie group with infinite center.

- **R**: Usually the group **R** of real numbers under addition (and its quotient **R/Z**) are not counted as simple Lie groups, even though they are connected and have a Lie algebra with no proper non-zero ideals. Occasionally authors define simple Lie groups in such a way that **R** is simple, though this sometimes seems to be an accident caused by overlooking this case.

- Matrix groups: Some authors restrict themselves to Lie groups that can be represented as groups of finite matrices. The metaplectic group is an example of a simple Lie group that cannot be represented in this way.

- Complex Lie algebras: The definition of a simple Lie algebra is not stable under the *extension of scalars*. The complexification of a complex simple Lie algebra, such as **sl**(n, **C**) is semisimple, but not simple.

The most common definition is the one above: simple Lie groups have to be connected, they are allowed to have non-trivial centers (possibly infinite), they need not be representable by finite matrices, and they must be non-abelian.

3.2   Method of classification

Main article: list of simple Lie groups

Such groups are classified using the prior classification of the complex simple Lie algebras: for which see the page on root systems. It is shown that a simple Lie group has a simple Lie algebra that will occur on the list given there, once it is complexified (that is, made into a complex vector space rather than a real one). This reduces the classification to two further matters.

3.3   Real forms

The groups SO(p,q,**R**) and SO($p+q$,**R**), for example, give rise to different real Lie algebras, but having the same Dynkin diagram. In general there may be different *real forms* of the same complex Lie algebra.

3.4   Relationship of simple Lie algebras to groups

Secondly the Lie algebra only determines uniquely the simply connected (universal) cover G^* of the component containing the identity of a Lie group G. It may well happen that G^* isn't actually a simple group, for example having a non-trivial center. We have therefore to worry about the global topology, by computing the fundamental group of G (an abelian group: a Lie group is an H-space). This was done by Élie Cartan.

For an example, take the special orthogonal groups in even dimension. With the non-identity matrix $-I$ in the center, these aren't actually simple groups; and having a twofold spin cover, they aren't simply-connected either. They lie 'between' G^* and G, in the notation above.

3.5   Classification by Dynkin diagram

Main article: root system

According to Dynkin's classification, we have as possibilities these only, where n is the number of nodes:

3.6 Infinite series

3.6.1 A series

A_1, A_2, ...

A_r corresponds to the special unitary group, $SU(r + 1)$.

3.6.2 B series

B_2, B_3, ...

B_r corresponds to the special orthogonal group, $SO(2r + 1)$.

3.6.3 C series

C_3, C_4, ...

C_r corresponds to the symplectic group, $Sp(2r)$.

3.6.4 D series

D_4, D_5, ...

D_r corresponds to the special orthogonal group, $SO(2r)$. Note that $SO(4)$ is not a simple group, though. The Dynkin diagram has two nodes that are not connected. There is a surjective homomorphism from $SO(3)^* \times SO(3)^*$ to $SO(4)$ given by quaternion multiplication; see quaternions and spatial rotation. Therefore the simple groups here start with D_3, which as a diagram straightens out to A_3. With D_4 there is an 'exotic' symmetry of the diagram, corresponding to so-called triality.

3.7 Exceptional cases

For the so-called exceptional cases see G_2, F_4, E_6, E_7, and E_8. These cases are deemed 'exceptional' because they do not fall into infinite series of groups of increasing dimension. From the point of view of each group taken separately, there is nothing so unusual about them. These exceptional groups were discovered around 1890 in the classification of the simple Lie algebras, over the complex numbers (Wilhelm Killing, re-done by Élie Cartan). For some time it was a research issue to find concrete ways in which they arise, for example as a symmetry group of a differential system.

See also $E_7\frac{1}{2}$.

3.8   Simply laced groups

A **simply laced group** is a Lie group whose Dynkin diagram only contain simple links, and therefore all the nonzero roots of the corresponding Lie algebra have the same length. The A, D and E series groups are all simply laced, but no group of type B, C, F, or G is simply laced.

3.9   See also

- Cartan matrix

- Coxeter matrix

- Weyl group

- Coxeter group

- Kac–Moody algebra

- Catastrophe theory

3.10   References

- Jacobson, Nathan (1971-06-01). *Exceptional Lie Algebras* (1 ed.). CRC Press. ISBN 0-8247-1326-5.

Chapter 4

Lie algebra

"Lie bracket" redirects here. For the operation on vector fields, see Lie bracket of vector fields.

In mathematics, a **Lie algebra** (/ˈliː/, not /ˈlaɪ/) is a vector space together with a non-associative multiplication called "Lie bracket" $[x, y]$. It was introduced to study the concept of infinitesimal transformations. Hermann Weyl introduced the term "Lie algebra" (after Sophus Lie) in the 1930s. In older texts, the name "**infinitesimal group**" is used.

Lie algebras are closely related to Lie groups which are groups that are also smooth manifolds, with the property that the group operations of multiplication and inversion are smooth maps. Any Lie group gives rise to a Lie algebra. Conversely, to any finite-dimensional Lie algebra over real or complex numbers, there is a corresponding connected Lie group unique up to covering (Lie's third theorem). This correspondence between Lie groups and Lie algebras allows one to study Lie groups in terms of Lie algebras.

4.1 Definitions

A **Lie algebra** is a vector space $\mathfrak{g}$ over some field F together with a binary operation $[\cdot, \cdot] : \mathfrak{g} \times \mathfrak{g} \to \mathfrak{g}$ called the **Lie bracket**, which satisfies the following axioms:

- Bilinearity,

$$[ax + by, z] = a[x, z] + b[y, z], \quad [z, ax + by] = a[z, x] + b[z, y]$$

for all scalars a, b in F and all elements x, y, z in $\mathfrak{g}$.

- Alternativity on $\mathfrak{g}$,

$$[x, x] = 0$$

for all x in $\mathfrak{g}$.

- The Jacobi identity,

$$[x, [y, z]] + [z, [x, y]] + [y, [z, x]] = 0$$

for all x, y, z in $\mathfrak{g}$.

Note that the bilinearity and alternating properties imply

- Anticommutativity,

  $$[x,y] = -[y,x],$$

  for all elements x, y in $\mathfrak{g}$, while anticommutativity only implies the alternating property if the field's characteristic is not 2.[1]

It is customary to express a Lie algebra in lower-case fraktur, like $\mathfrak{g}$. If a Lie algebra is associated with a Lie group, then the spelling of the Lie algebra is the same as that Lie group. For example, the Lie algebra of SU(n) is written as $\mathfrak{su}(n)$.

4.1.1   Generators and dimension

Elements of a Lie algebra $\mathfrak{g}$ are said to be **generators** of the Lie algebra if the smallest subalgebra of $\mathfrak{g}$ containing them is $\mathfrak{g}$ itself. The **dimension** of a Lie algebra is its dimension as a vector space over F. The cardinality of a minimal generating set of a Lie algebra is always less than or equal to its dimension.

4.1.2   Subalgebras, ideals, and homomorphisms

The Lie bracket is not associative in general, meaning that $[[x, y], z]$ need not equal $[x, [y, z]]$. Nonetheless, much of the terminology that was developed in the theory of associative rings or associative algebras is commonly applied to Lie algebras. A subspace $\mathfrak{h} \subseteq \mathfrak{g}$ that is closed under the Lie bracket is called a **Lie subalgebra**. If a subspace $I \subseteq \mathfrak{g}$ satisfies a stronger condition that

$$[\mathfrak{g}, I] \subseteq I,$$

then I is called an **ideal** in the Lie algebra $\mathfrak{g}$.[2] A **homomorphism** between two Lie algebras (over the same base field) is a linear map that is compatible with the respective Lie brackets:

$$f : \mathfrak{g} \to \mathfrak{g}', \quad f([x,y]) = [f(x), f(y)],$$

for all elements x and y in $\mathfrak{g}$. As in the theory of associative rings, ideals are precisely the kernels of homomorphisms, given a Lie algebra $\mathfrak{g}$ and an ideal I in it, one constructs the **factor algebra** $\mathfrak{g}/I$, and the first isomorphism theorem holds for Lie algebras.

Let S be a subset of $\mathfrak{g}$. The set of elements x such that $[x, s] = 0$ for all s in S forms a subalgebra called the centralizer of S. The centralizer of $\mathfrak{g}$ itself is called the center of $\mathfrak{g}$. Similar to centralizers, if S is a subspace,[3] then the set of x such that $[x, s]$ is in S for all s in S forms a subalgebra called the normalizer of S.

4.1.3   Direct sum and semidirect product

Given two Lie algebras $\mathfrak{g}$ and $\mathfrak{g}'$, their direct sum is the Lie algebra consisting of the vector space $\mathfrak{g} \oplus \mathfrak{g}'$, of the pairs (x, x'), $x \in \mathfrak{g}, x' \in \mathfrak{g}'$, with the operation

$$[(x, x'), (y, y')] = ([x, y], [x', y']), \quad x, y \in \mathfrak{g}, \, x', y' \in \mathfrak{g}'.$$

Let $\mathfrak{g}$ be a Lie algebra and $\mathfrak{i}$ its ideal. If the canonical map $\mathfrak{g} \to \mathfrak{g}/\mathfrak{i}$ splits (i.e., admits a section), then $\mathfrak{g}$ is said to be a semidirect product of $\mathfrak{i}$ and $\mathfrak{g}/\mathfrak{i}$.

Levi's theorem says that a finite-dimensional Lie algebra is a semidirect product of its radical and the complementary subalgebra (Levi subalgebra).

4.2 Properties

4.2.1 Admits an enveloping algebra

See also: Universal enveloping algebra

For any associative algebra A with multiplication $*$, one can construct a Lie algebra $L(A)$. As a vector space, $L(A)$ is the same as A. The Lie bracket of two elements of $L(A)$ is defined to be their commutator in A:

$$[a, b] = a * b - b * a.$$

The associativity of the multiplication * in A implies the Jacobi identity of the commutator in $L(A)$. For example, the associative algebra of $n \times n$ matrices over a field F gives rise to the general linear Lie algebra $\mathfrak{gl}_n(F)$. The associative algebra A is called an **enveloping algebra** of the Lie algebra $L(A)$. Every Lie algebra can be embedded into one that arises from an associative algebra in this fashion; see universal enveloping algebra.

4.2.2 Representation

Given a vector space V, let $\mathfrak{gl}(V)$ denote the Lie algebra enveloped by the associative algebra of all linear endomorphisms of V. A representation of a Lie algebra $\mathfrak{g}$ on V is a Lie algebra homomorphism

$$\pi : \mathfrak{g} \to \mathfrak{gl}(V).$$

A representation is said to be faithful if its kernel is trivial. Every finite-dimensional Lie algebra has a faithful representation on a finite-dimensional vector space (Ado's theorem).[4]

For example,

$$\mathrm{ad} : \mathfrak{g} \to \mathfrak{gl}(\mathfrak{g})$$

given by $\mathrm{ad}(x)(y) = [x, y]$ is a representation of $\mathfrak{g}$ on the vector space $\mathfrak{g}$ called the adjoint representation. A derivation on the Lie algebra $\mathfrak{g}$ (in fact on any non-associative algebra) is a linear map $\delta : \mathfrak{g} \to \mathfrak{g}$ that obeys the Leibniz' law, that is,

$$\delta([x, y]) = [\delta(x), y] + [x, \delta(y)]$$

for all x and y in the algebra. For any x, $\mathrm{ad}(x)$ is a derivation; a consequence of the Jacobi identity. Thus, the image of ad lies in the subalgebra of $\mathfrak{gl}(\mathfrak{g})$ consisting of derivations on $\mathfrak{g}$. A derivation that happens to be in the image of ad is called an inner derivation. If $\mathfrak{g}$ is semisimple, every derivation on $\mathfrak{g}$ is inner.

4.3 Examples

4.3.1 Vector spaces

- Any vector space V endowed with the identically zero Lie bracket becomes a Lie algebra. Such Lie algebras are called abelian, cf. below. Any one-dimensional Lie algebra over a field is abelian, by the antisymmetry of the Lie bracket.

- The real vector space of all $n \times n$ skew-hermitian matrices is closed under the commutator and forms a real Lie algebra denoted $\mathfrak{u}(n)$. This is the Lie algebra of the unitary group $U(n)$.

4.3.2   Subspaces

- The subspace of the general linear Lie algebra $\mathfrak{gl}_n(F)$ consisting of matrices of trace zero is a subalgebra,[5] the special linear Lie algebra, denoted $\mathfrak{sl}_n(F)$.

4.3.3   Real matrix groups

- Any Lie group G defines an associated real Lie algebra $\mathfrak{g}$ =Lie(G). The definition in general is somewhat technical, but in the case of real matrix groups, it can be formulated via the exponential map, or the matrix exponent. The Lie algebra $\mathfrak{g}$ consists of those matrices X for which $\exp(tX) \in G$, $\forall$ real numbers t.

  The Lie bracket of $\mathfrak{g}$ is given by the commutator of matrices. As a concrete example, consider the special linear group SL(n,**R**), consisting of all $n \times n$ matrices with real entries and determinant 1. This is a matrix Lie group, and its Lie algebra consists of all $n \times n$ matrices with real entries and trace 0.

4.3.4   Three dimensions

- The three-dimensional Euclidean space $\mathbf{R}^3$ with the Lie bracket given by the cross product of vectors becomes a three-dimensional Lie algebra.

- The Heisenberg algebra $H_3(\mathrm{R})$ is a three-dimensional Lie algebra generated by elements x, y and z with Lie brackets

$$[x,y] = z, \quad [x,z] = 0, \quad [y,z] = 0$$

  It is explicitly realized as the space of 3×3 strictly upper-triangular matrices, with the Lie bracket given by the matrix commutator,

$$x = \begin{pmatrix} 0 & 1 & 0 \\ 0 & 0 & 0 \\ 0 & 0 & 0 \end{pmatrix}, \quad y = \begin{pmatrix} 0 & 0 & 0 \\ 0 & 0 & 1 \\ 0 & 0 & 0 \end{pmatrix}, \quad z = \begin{pmatrix} 0 & 0 & 1 \\ 0 & 0 & 0 \\ 0 & 0 & 0 \end{pmatrix}.$$

  Any element of the Heisenberg group is thus representable as a product of group generators, i.e., matrix exponentials of these Lie algebra generators,

$$\begin{pmatrix} 1 & a & c \\ 0 & 1 & b \\ 0 & 0 & 1 \end{pmatrix} = e^{by}e^{cz}e^{ax}.$$

- The commutation relations between the x, y, and z components of the angular momentum operator in quantum mechanics are the same as those of $\mathfrak{su}(2)$ and $\mathfrak{so}(3)$,

$$[L_x, L_y] = i\hbar L_z$$

$$[L_y, L_z] = i\hbar L_x$$

$$[L_z, L_x] = i\hbar L_y$$

(The physicist convention for Lie algebras is used in the above equations, hence the factor of i.) The Lie algebra formed by these operators have, in fact, representations of all finite dimensions.

4.3.5 Infinite dimensions

- An important class of infinite-dimensional real Lie algebras arises in differential topology. The space of smooth vector fields on a differentiable manifold M forms a Lie algebra, where the Lie bracket is defined to be the commutator of vector fields. One way of expressing the Lie bracket is through the formalism of Lie derivatives, which identifies a vector field X with a first order partial differential operator LX acting on smooth functions by letting $LX(f)$ be the directional derivative of the function f in the direction of X. The Lie bracket $[X,Y]$ of two vector fields is the vector field defined through its action on functions by the formula:

$$L_{[X,Y]}f = L_X(L_Y f) - L_Y(L_X f).$$

- A Kac–Moody algebra is an example of an infinite-dimensional Lie algebra.

- The Moyal algebra is an infinite-dimensional Lie algebra which contains all classical Lie algebras as subalgebras.

4.4   Structure theory and classification

Lie algebras can be classified to some extent. In particular, this has an application to the classification of Lie groups.

4.4.1   Abelian, nilpotent, and solvable

Analogously to abelian, nilpotent, and solvable groups, defined in terms of the derived subgroups, one can define abelian, nilpotent, and solvable Lie algebras.

A Lie algebra $\mathfrak{g}$ is **abelian** if the Lie bracket vanishes, i.e. $[x,y] = 0$, for all x and y in $\mathfrak{g}$. Abelian Lie algebras correspond to commutative (or abelian) connected Lie groups such as vector spaces K^n or tori T^n, and are all of the form $\mathfrak{k}^n$, meaning an n-dimensional vector space with the trivial Lie bracket.

A more general class of Lie algebras is defined by the vanishing of all commutators of given length. A Lie algebra $\mathfrak{g}$ is **nilpotent** if the lower central series

$$\mathfrak{g} > [\mathfrak{g},\mathfrak{g}] > [[\mathfrak{g},\mathfrak{g}],\mathfrak{g}] > [[[\mathfrak{g},\mathfrak{g}],\mathfrak{g}],\mathfrak{g}] > \cdots$$

becomes zero eventually. By Engel's theorem, a Lie algebra is nilpotent if and only if for every u in $\mathfrak{g}$ the adjoint endomorphism

$$\mathrm{ad}(u) : \mathfrak{g} \to \mathfrak{g}, \quad \mathrm{ad}(u)v = [u,v]$$

is nilpotent.

More generally still, a Lie algebra $\mathfrak{g}$ is said to be **solvable** if the derived series:

$$\mathfrak{g} > [\mathfrak{g},\mathfrak{g}] > [[\mathfrak{g},\mathfrak{g}],[\mathfrak{g},\mathfrak{g}]] > [[[\mathfrak{g},\mathfrak{g}],[\mathfrak{g},\mathfrak{g}]],[[\mathfrak{g},\mathfrak{g}],[\mathfrak{g},\mathfrak{g}]]] > \cdots$$

becomes zero eventually.

Every finite-dimensional Lie algebra has a unique maximal solvable ideal, called its radical. Under the Lie correspondence, nilpotent (respectively, solvable) connected Lie groups correspond to nilpotent (respectively, solvable) Lie algebras.

4.4.2   Simple and semisimple

A Lie algebra is "simple" if it has no non-trivial ideals and is not abelian. A Lie algebra $\mathfrak{g}$ is called **semisimple** if its radical is zero. Equivalently, $\mathfrak{g}$ is semisimple if it does not contain any non-zero abelian ideals. In particular, a simple Lie algebra is semisimple. Conversely, it can be proven that any semisimple Lie algebra is the direct sum of its minimal ideals, which are canonically determined simple Lie algebras.

The concept of semisimplicity for Lie algebras is closely related with the complete reducibility (semisimplicity) of their representations. When the ground field F has characteristic zero, any finite-dimensional representation of a semisimple Lie algebra is semisimple (i.e., direct sum of irreducible representations.) In general, a Lie algebra is called reductive if the adjoint representation is semisimple. Thus, a semisimple Lie algebra is reductive.

4.4.3   Cartan's criterion

Cartan's criterion gives conditions for a Lie algebra to be nilpotent, solvable, or semisimple. It is based on the notion of the Killing form, a symmetric bilinear form on $\mathfrak{g}$ defined by the formula

$$K(u, v) = \operatorname{tr}(\operatorname{ad}(u)\operatorname{ad}(v)),$$

where tr denotes the trace of a linear operator. A Lie algebra $\mathfrak{g}$ is semisimple if and only if the Killing form is nondegenerate. A Lie algebra $\mathfrak{g}$ is solvable if and only if $K(\mathfrak{g}, [\mathfrak{g}, \mathfrak{g}]) = 0$.

4.4.4   Classification

The Levi decomposition expresses an arbitrary Lie algebra as a semidirect sum of its solvable radical and a semisimple Lie algebra, almost in a canonical way. Furthermore, semisimple Lie algebras over an algebraically closed field have been completely classified through their root systems. However, the classification of solvable Lie algebras is a 'wild' problem, and cannot be accomplished in general.

4.5   Relation to Lie groups

See also: Lie group–Lie algebra correspondence

Although Lie algebras are often studied in their own right, historically they arose as a means to study Lie groups.

Lie's fundamental theorems describe a relation between Lie groups and Lie algebras. In particular, any Lie group gives rise to a canonically determined Lie algebra (concretely, *the tangent space at the identity*); and, conversely, for any Lie algebra there is a corresponding connected Lie group (Lie's third theorem; see the Baker–Campbell–Hausdorff formula). This Lie group is not determined uniquely; however, any two connected Lie groups with the same Lie algebra are *locally isomorphic*, and in particular, have the same universal cover. For instance, the special orthogonal group SO(3) and the special unitary group SU(2) give rise to the same Lie algebra, which is isomorphic to $\mathbf{R}^3$ with the cross-product, while SU(2) is a simply-connected twofold cover of SO(3).

Given a Lie group, a Lie algebra can be associated to it either by endowing the tangent space to the identity with the differential of the adjoint map, or by considering the left-invariant vector fields as mentioned in the examples. In the case of real matrix groups, the Lie algebra $\mathfrak{g}$ consists of those matrices X for which $\exp(tX) \in G$ for all real numbers t, where exp is the exponential map.

Some examples of Lie algebras corresponding to Lie groups are the following:

- The Lie algebra $\mathfrak{gl}_n(\mathbb{C})$ for the group $\mathrm{GL}_n(\mathbb{C})$ is the algebra of complex $n{\times}n$ matrices

- The Lie algebra $\mathfrak{sl}_n(\mathbb{C})$ for the group $\mathrm{SL}_n(\mathbb{C})$ is the algebra of complex $n{\times}n$ matrices with trace 0

- The Lie algebras $\mathfrak{o}(n)$ for the group $\mathrm{O}(n)$ and $\mathfrak{so}(n)$ for $\mathrm{SO}(n)$ are both the algebra of real anti-symmetric $n{\times}n$ matrices (See Antisymmetric matrix: Infinitesimal rotations for a discussion)

- The Lie algebra $\mathfrak{u}(n)$ for the group $\mathrm{U}(n)$ is the algebra of skew-Hermitian complex $n{\times}n$ matrices while the Lie algebra $\mathfrak{su}(n)$ for $\mathrm{SU}(n)$ is the algebra of skew-Hermitian, traceless complex $n{\times}n$ matrices.

In the above examples, the Lie bracket $[X, Y]$ (for X and Y matrices in the Lie algebra) is defined as $[X, Y] = XY - YX$.

Given a set of generators T^a, the **structure constants** $f^{\,abc}$ express the Lie brackets of pairs of generators as linear combinations of generators from the set, i.e., $[T^a,\ T^b] = f^{\,abc}\ T^c$. The structure constants determine the Lie brackets of elements of the Lie algebra, and consequently nearly completely determine the group structure of the Lie group. The structure of the Lie group near the identity element is displayed explicitly by the Baker–Campbell–Hausdorff formula, an expansion in Lie algebra elements X, Y and their Lie brackets, all nested together within a single exponent, $\exp(tX)\,\exp(tY) = \exp(tX+tY+\tfrac{1}{2}\,t^2[X,Y] + \mathrm{O}(t^3)\,)$).

The mapping from Lie groups to Lie algebras is functorial, which implies that homomorphisms of Lie groups lift to homomorphisms of Lie algebras, and various properties are satisfied by this lifting: it commutes with composition, it maps Lie subgroups, kernels, quotients and cokernels of Lie groups to subalgebras, kernels, quotients and cokernels of Lie algebras, respectively.

The functor **L** which takes each Lie group to its Lie algebra and each homomorphism to its differential is faithful and exact. It is however not an equivalence of categories: different Lie groups may have isomorphic Lie algebras (for example SO(3) and SU(2)), and there are (infinite dimensional) Lie algebras that are not associated to any Lie group.[6]

However, when the Lie algebra $\mathfrak{g}$ is finite-dimensional, one can associate to it a simply connected Lie group having $\mathfrak{g}$ as its Lie algebra. More precisely, the Lie algebra functor **L** has a left adjoint functor $\boldsymbol{\Gamma}$ from finite-dimensional (real) Lie algebras to Lie groups, factoring through the full subcategory of simply connected Lie groups.[7] In other words, there is a natural isomorphism of bifunctors

$$\mathrm{Hom}(\Gamma(\mathfrak{g}), H) \cong \mathrm{Hom}(\mathfrak{g}, \mathrm{L}(H)).$$

The adjunction $\mathfrak{g} \to \mathrm{L}(\Gamma(\mathfrak{g}))$ (corresponding to the identity on $\Gamma(\mathfrak{g})$) is an isomorphism, and the other adjunction $\Gamma(\mathrm{L}(H)) \to H$ is the projection homomorphism from the universal cover group of the identity component of H to H. It follows immediately that if G is simply connected, then the Lie algebra functor establishes a bijective correspondence between Lie group homomorphisms $G{\to}H$ and Lie algebra homomorphisms $\mathrm{L}(G){\to}\mathrm{L}(H)$.

The universal cover group above can be constructed as the image of the Lie algebra under the exponential map. More generally, we have that the Lie algebra is homeomorphic to a neighborhood of the identity. But globally, if the Lie group is compact, the exponential will not be injective, and if the Lie group is not connected, simply connected or compact, the exponential map need not be surjective.

If the Lie algebra is infinite-dimensional, the issue is more subtle. In many instances, the exponential map is not even locally a homeomorphism (for example, in $\mathrm{Diff}(\mathbf{S}^1)$, one may find diffeomorphisms arbitrarily close to the identity that are not in the image of exp). Furthermore, some infinite-dimensional Lie algebras are not the Lie algebra of any group.

The correspondence between Lie algebras and Lie groups is used in several ways, including in the classification of Lie groups and the related matter of the representation theory of Lie groups. Every representation of a Lie algebra lifts uniquely to a representation of the corresponding connected, simply connected Lie group, and conversely every representation of any Lie group induces a representation of the group's Lie algebra; the representations are in one to one correspondence. Therefore, knowing the representations of a Lie algebra settles the question of representations of the group.

As for classification, it can be shown that any connected Lie group with a given Lie algebra is isomorphic to the universal cover mod a discrete central subgroup. So classifying Lie groups becomes simply a matter of counting the discrete subgroups of the center, once the classification of Lie algebras is known (solved by Cartan et al. in the semisimple case).

4.6   Category theoretic definition

Using the language of category theory, a **Lie algebra** can be defined as an object A in **Vec**k, the category of vector spaces over a field k of characteristic not 2, together with a morphism $[.,.]: A \otimes A \to A$, where $\otimes$ refers to the monoidal product of **Vec**k, such that

- $[\cdot, \cdot] \circ (\mathrm{id} + \tau_{A,A}) = 0$
- $[\cdot, \cdot] \circ ([\cdot, \cdot] \otimes \mathrm{id}) \circ (\mathrm{id} + \sigma + \sigma^2) = 0$

where $\tau (a \otimes b) := b \otimes a$ and σ is the cyclic permutation braiding $(\mathrm{id} \otimes \tau A,A) \circ (\tau A,A \otimes \mathrm{id})$. In diagrammatic form:

4.7   See also

4.8   Notes

[1] Humphreys p. 1

[2] Due to the anticommutativity of the commutator, the notions of a left and right ideal in a Lie algebra coincide.

[3] Jacobson 1962, pg. 28

[4] Jacobson 1962, Ch. VI

[5] Humphreys p.2

[6] Beltita 2005, pg. 75

[7] Adjoint property is discussed in more general context in Hofman & Morris (2007) (e.g., page 130) but is a straightforward consequence of, e.g., Bourbaki (1989) Theorem 1 of page 305 and Theorem 3 of page 310.

4.9   References

- Beltita, Daniel. *Smooth Homogeneous Structures in Operator Theory*, CRC Press, 2005. ISBN 978-1-4200-3480-6

- Boza, Luis; Fedriani, Eugenio M. & Núñez, Juan. *A new method for classifying complex filiform Lie algebras*, Applied Mathematics and Computation, 121 (2-3): 169–175, 2001

- Bourbaki, Nicolas. "Lie Groups and Lie Algebras - Chapters 1-3", Springer, 1989, ISBN 3-540-64242-0

- Erdmann, Karin & Wildon, Mark. *Introduction to Lie Algebras*, 1st edition, Springer, 2006. ISBN 1-84628-040-0

- Hall, Brian C. *Lie Groups, Lie Algebras, and Representations: An Elementary Introduction*, Springer, 2003. ISBN 0-387-40122-9

- Hofman, Karl & Morris, Sidney. "The Lie Theory of Connected Pro-Lie Groups", European Mathematical Society, 2007, ISBN 978-3-03719-032-6

- Humphreys, James E. *Introduction to Lie Algebras and Representation Theory*, Second printing, revised. Graduate Texts in Mathematics, 9. Springer-Verlag, New York, 1978. ISBN 0-387-90053-5

- Jacobson, Nathan, *Lie algebras*, Republication of the 1962 original. Dover Publications, Inc., New York, 1979. ISBN 0-486-63832-4

- Kac, Victor G. et al. *Course notes for MIT 18.745: Introduction to Lie Algebras*, math.mit.edu

- O'Connor, J.J. & Robertson, E.F. Biography of Sophus Lie, MacTutor History of Mathematics Archive, www-history.mcs.st-andrews.ac.uk

- O'Connor, J.J. & Robertson, E.F. Biography of Wilhelm Killing, MacTutor History of Mathematics Archive, www-history.mcs.st-andrews.ac.uk

- Serre, Jean-Pierre. "Lie Algebras and Lie Groups", 2nd edition, Springer, 2006. ISBN 3-540-55008-9

- Steeb, W.-H. *Continuous Symmetries, Lie Algebras, Differential Equations and Computer Algebra*, second edition, World Scientific, 2007, ISBN 978-981-270-809-0

- Varadarajan, V.S. *Lie Groups, Lie Algebras, and Their Representations*, 1st edition, Springer, 2004. ISBN 0-387-90969-9.

4.10 External links

- Hazewinkel, Michiel, ed. (2001), "Lie algebra", *Encyclopedia of Mathematics*, Springer, ISBN 978-1-55608-010-4

Chapter 5

Differentiable manifold

"Smooth curve" redirects here. For the equivalent concept in algebraic geometry, see singular point of an algebraic variety.
In mathematics, a **differentiable manifold** is a type of manifold that is locally similar enough to a linear space to allow

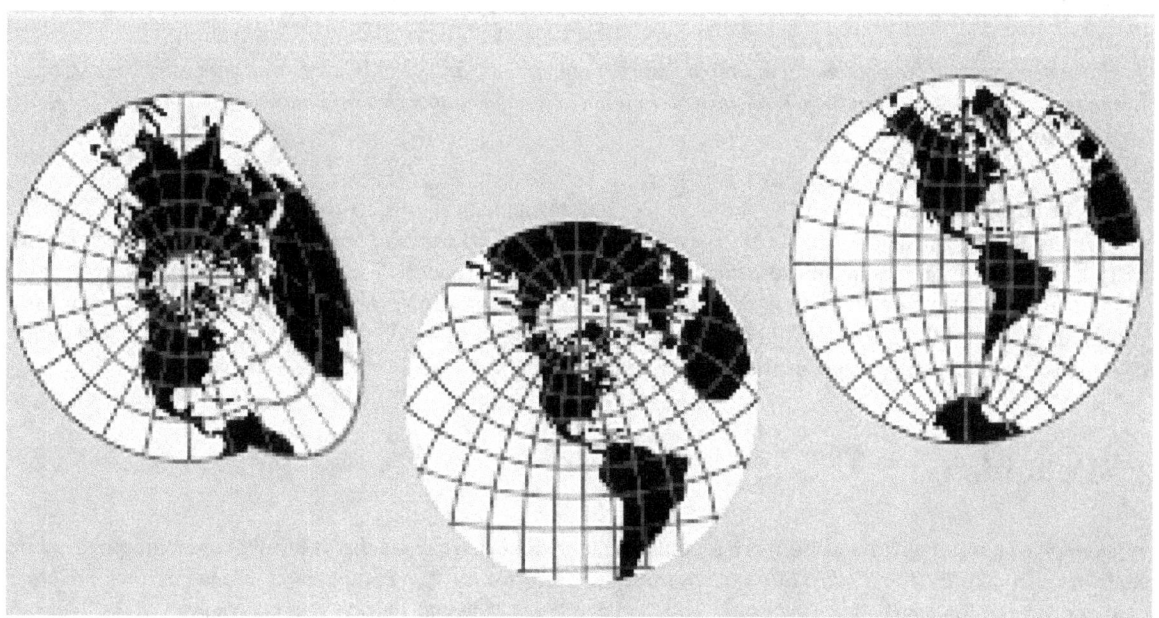

A nondifferentiable atlas of charts for the globe. The results of calculus may not be compatible between charts if the atlas is not differentiable. In the center and right charts the Tropic of Cancer is a smooth curve, whereas in the left chart it has a sharp corner. The notion of a differentiable manifold refines that of a manifold by requiring the functions that transform between charts to be differentiable.

one to do calculus. Any manifold can be described by a collection of charts, also known as an atlas. One may then apply ideas from calculus while working within the individual charts, since each chart lies within a linear space to which the usual rules of calculus apply. If the charts are suitably compatible (namely, the transition from one chart to another is differentiable), then computations done in one chart are valid in any other differentiable chart.

In formal terms, a **differentiable manifold** is a topological manifold with a globally defined differential structure. Any topological manifold can be given a differential structure *locally* by using the homeomorphisms in its atlas and the standard differential structure on a linear space. To induce a global differential structure on the local coordinate systems induced by the homeomorphisms, their composition on chart intersections in the atlas must be differentiable functions on the corresponding linear space. In other words, where the domains of charts overlap, the coordinates defined by each chart are required to be differentiable with respect to the coordinates defined by every chart in the atlas. The maps that relate the coordinates defined by the various charts to one another are called *transition maps*.

Differentiability means different things in different contexts including: continuously differentiable, k times differentiable, smooth, and holomorphic. Furthermore, the ability to induce such a differential structure on an abstract space allows one to extend the definition of differentiability to spaces without global coordinate systems. A differential structure allows one to define the globally differentiable tangent space, differentiable functions, and differentiable tensor and vector fields. Differentiable manifolds are very important in physics. Special kinds of differentiable manifolds form the basis for physical theories such as classical mechanics, general relativity, and Yang–Mills theory. It is possible to develop a calculus for differentiable manifolds. This leads to such mathematical machinery as the exterior calculus. The study of calculus on differentiable manifolds is known as differential geometry.

5.1 History

Main article: History of manifolds and varieties

The emergence of differential geometry as a distinct discipline is generally credited to Carl Friedrich Gauss and Bernhard Riemann. Riemann first described manifolds in his famous habilitation lecture[1] before the faculty at Göttingen. He motivated the idea of a manifold by an intuitive process of varying a given object in a new direction, and presciently described the role of coordinate systems and charts in subsequent formal developments:

> *Having constructed the notion of a manifoldness of n dimensions, and found that its true character consists in the property that the determination of position in it may be reduced to n determinations of magnitude, ...–* B. Riemann

The works of physicists such as James Clerk Maxwell,[2] and mathematicians Gregorio Ricci-Curbastro and Tullio Levi-Civita[3] led to the development of tensor analysis and the notion of covariance, which identifies an intrinsic geometric property as one that is invariant with respect to coordinate transformations. These ideas found a key application in Einstein's theory of general relativity and its underlying equivalence principle. A modern definition of a 2-dimensional manifold was given by Hermann Weyl in his 1913 book on Riemann surfaces.[4] The widely accepted general definition of a manifold in terms of an atlas is due to Hassler Whitney.[5]

5.2 Definition

A *presentation* of a **topological manifold** is a second countable Hausdorff space that is locally homeomorphic to a linear space, by a collection (called an *atlas*) of homeomorphisms called *charts*. The composition of one chart with the inverse of another chart is a function called a *transition map*, and defines a homeomorphism of an open subset of the linear space onto another open subset of the linear space. This formalizes the notion of "patching together pieces of a space to make a manifold" – the manifold produced also contains the data of how it has been patched together. However, different atlases (patchings) may produce "the same" manifold; a manifold does not come with a preferred atlas. And, thus, one defines a **topological manifold** to be a space as above with an *equivalence class* of atlases, where one defines equivalence of atlases below.

There are a number of different types of differentiable manifolds, depending on the precise differentiability requirements on the transition functions. Some common examples include the following.

- A **differentiable manifold** is a topological manifold equipped with an equivalence class of atlases whose transition maps are all differentiable. In broader terms, a C^k-manifold is a topological manifold with an atlas whose transition maps are all k-times continuously differentiable.

- A **smooth manifold** or C^∞-manifold is a differentiable manifold for which all the transition maps are smooth. That is, derivatives of all orders exist; so it is a C^k-manifold for all k. An equivalence class of such atlases is said to be a smooth structure.

- An **analytic manifold**, or C^ω-manifold is a smooth manifold with the additional condition that each transition map is analytic: the Taylor expansion is absolutely convergent and equals the function on some open ball.

- A **complex manifold** is a topological space modeled on a Euclidean space over the complex field and for which all the transition maps are holomorphic.

While there is a meaningful notion of a C^k *atlas,* there is no distinct notion of a C^k *manifold* other than C^0 (continuous maps: a topological manifold) and C^∞ (smooth maps: a smooth manifold), because for every C^k-structure with $k > 0$, there is a unique C^k-equivalent C^∞-structure (every C^k-structure is *uniquely smoothable* to a C^∞-structure) – a result of Whitney.[6] In fact, every C^k-structure is uniquely smoothable to a C^ω-structure. Furthermore, two C^k atlases that are equivalent to a single C^∞ atlas are equivalent as C^k atlases, so two distinct C^k atlases do not collide. See Differential structure: Existence and uniqueness theorems for details. Thus one uses the terms "differentiable manifold" and "smooth manifold" interchangeably; this is in stark contrast to C^k *maps,* where there are meaningful differences for different k. For example, the Nash embedding theorem states that any manifold can be C^k isometrically embedded in Euclidean space $\mathbf{R}^N$ – for any $1 \le k \le \infty$ there is a sufficiently large N, but N depends on k.

On the other hand, complex manifolds are significantly more restrictive. As an example, Chow's theorem states that any projective complex manifold is in fact a projective variety – it has an algebraic structure.

5.2.1   Atlases

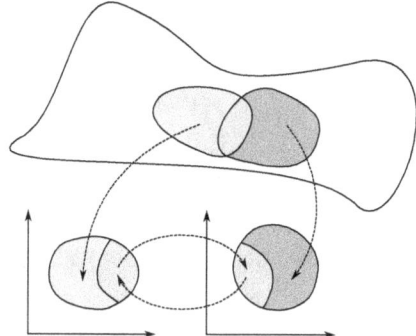

X
U_α
U_β
φ_α
φ_β
$\varphi_{\alpha\beta}$
$\varphi_{\beta\alpha}$
$\mathbf{R}^n$
$\mathbf{R}^n$
Charts on a manifold

An atlas on a topological space X is a collection of pairs $\{(U\alpha, \varphi\alpha)\}$ called *charts*, where the $U\alpha$ are open sets that cover X, and for each index α

$$\varphi_\alpha : U_\alpha \to \mathbf{R}^n$$

is a homeomorphism of $U\alpha$ onto an open subset of n-dimensional real space. The **transition maps** of the atlas are the functions

$$\varphi_{\alpha\beta} = \varphi_\beta \circ \varphi_\alpha^{-1}|_{\varphi_\alpha(U_\alpha \cap U_\beta)} : \varphi_\alpha(U_\alpha \cap U_\beta) \to \varphi_\beta(U_\alpha \cap U_\beta).$$

Every topological manifold has an atlas. A C^k-atlas is an atlas whose transition maps are C^k. A topological manifold has a C^0-atlas and in general a C^k-manifold has a C^k-atlas. A continuous atlas is a C^0 atlas, a smooth atlas is a C^∞ atlas and an analytic atlas is a C^ω atlas. If the atlas is at least C^1, it is also called a *differential structure* or *differentiable structure*. A *holomorphic atlas* is an atlas whose underlying Euclidean space is defined on the complex field and whose transition maps are biholomorphic.

5.2.2   Compatible atlases

Different atlases can give rise to, in essence, the same manifold. The circle can be mapped by two coordinate charts, but if the domains of these charts are changed slightly a different atlas for the same manifold is obtained. These different atlases can be combined into a bigger atlas. It can happen that the transition maps of such a combined atlas are not as smooth as those of the constituent atlases. If C^k atlases can be combined to form a C^k atlas, then they are called compatible. Compatibility of atlases is an equivalence relation; by combining all the atlases in an equivalence class, a **maximal atlas** can be constructed. Each C^k atlas belongs to a unique maximal C^k atlas.

5.3   Alternative definitions

5.3.1   Pseudogroups

The notion of a pseudogroup[7] provides a flexible generalization of atlases in order to allow a variety of different structures to be defined on manifolds in a uniform way. A *pseudogroup* consists of a topological space S and a collection Γ consisting of homeomorphisms from open subsets of S to other open subsets of S such that

1. If $f \in \Gamma$, and U is an open subset of the domain of f, then the restriction $f|U$ is also in Γ.

2. If f is a homeomorphism from a union of open subsets of S, $\cup_i U_i$, to an open subset of S, then $f \in \Gamma$ provided $f|_{U_i} \in \Gamma$ for every i.

3. For every open $U \subset S$, the identity transformation of U is in Γ.

4. If $f \in \Gamma$, then $f^{-1} \in \Gamma$.

5. The composition of two elements of Γ is in Γ.

These last three conditions are analogous to the definition of a group. Note that Γ need not be a group, however, since the functions are not globally defined on S. For example, the collection of all local C^k diffeomorphisms on $\mathbf{R}^n$ form a pseudogroup. All biholomorphisms between open sets in $\mathbf{C}^n$ form a pseudogroup. More examples include: orientation preserving maps of $\mathbf{R}^n$, symplectomorphisms, Möbius transformations, affine transformations, and so on. Thus a wide variety of function classes determine pseudogroups.

An atlas $(Ui, \varphi i)$ of homeomorphisms φi from $Ui \subset M$ to open subsets of a topological space S is said to be *compatible* with a pseudogroup Γ provided that the transition functions $\varphi j \circ \varphi i^{-1} : \varphi i(Ui \cap Uj) \to \varphi j(Ui \cap Uj)$ are all in Γ.

A differentiable manifold is then an atlas compatible with the pseudogroup of C^k functions on $\mathbf{R}^n$. A complex manifold is an atlas compatible with the biholomorphic functions on open sets in $\mathbf{C}^n$. And so forth. Thus pseudogroups provide a single framework in which to describe many structures on manifolds of importance to differential geometry and topology.

5.3.2   Structure sheaf

Sometimes it can be useful to use an alternative approach to endow a manifold with a C^k-structure. Here $k = 1, 2, ..., \infty$, or ω for real analytic manifolds. Instead of considering coordinate charts, it is possible to start with functions defined on the manifold itself. The structure sheaf of M, denoted $\mathbf{C}^k$, is a sort of functor that defines, for each open set $U \subset M$, an algebra $\mathbf{C}^k(U)$ of continuous functions $U \to \mathbf{R}$. A structure sheaf $\mathbf{C}^k$ is said to give M the structure of a C^k manifold of dimension n provided that, for any $p \in M$, there exists a neighborhood U of p and n functions $x^1, ..., x^n \in \mathbf{C}^k(U)$ such that the map $f = (x^1, ..., x^n): U \to \mathbf{R}^n$ is a homeomorphism onto an open set in $\mathbf{R}^n$, and such that $\mathbf{C}^k|U$ is the pullback of the sheaf of k-times continuously differentiable functions on $\mathbf{R}^n$.[8]

In particular, this latter condition means that any function h in $\mathbf{C}^k(V)$, for V, can be written uniquely as $h(x) = H(x^1(x),...,x^n(x))$, where H is a k-times differentiable function on $f(V)$ (an open set in $\mathbf{R}^n$). Thus, the sheaf-theoretic viewpoint is that the functions on a differentiable manifold can be expressed in local coordinates as differentiable functions on $\mathbf{R}^n$, and *a fortiori* this is sufficient to characterize the differential structure on the manifold.

Sheaves of local rings

A similar, but more technical, approach to defining differentiable manifolds can be formulated using the notion of a ringed space. This approach is strongly influenced by the theory of schemes in algebraic geometry, but uses local rings of the germs of differentiable functions. It is especially popular in the context of *complex* manifolds.

We begin by describing the basic structure sheaf on $\mathbf{R}^n$. If U is an open set in $\mathbf{R}^n$, let

$$\mathbf{O}(U) = C^k(U, \mathbf{R})$$

consist of all real-valued k-times continuously differentiable functions on U. As U varies, this determines a sheaf of rings on $\mathbf{R}^n$. The stalk $\mathbf{O}p$ for $p \in \mathbf{R}^n$ consists of germs of functions near p, and is an algebra over $\mathbf{R}$. In particular, this is a local ring whose unique maximal ideal consists of those functions that vanish at p. The pair $(\mathbf{R}^n, \mathbf{O})$ is an example of a locally ringed space: it is a topological space equipped with a sheaf whose stalks are each local rings.

A differentiable manifold (of class C^k) consists of a pair $(M, \mathbf{O}M)$ where M is a second countable Hausdorff space, and $\mathbf{O}M$ is a sheaf of local $\mathbf{R}$-algebras defined on M, such that the locally ringed space $(M, \mathbf{O}M)$ is locally isomorphic to $(\mathbf{R}^n, \mathbf{O})$. In this way, differentiable manifolds can be thought of as schemes modelled on $\mathbf{R}^n$. This means that,[9] for each point $p \in M$, there is a neighborhood U of p, and a pair of functions $(f, f^{\#})$ where

1. $f: U \to f(U) \subset \mathbf{R}^n$ is a homeomorphism onto an open set in $\mathbf{R}^n$.

2. $f^{\#}: \mathbf{O}|f_{(U)} \to f^* (\mathbf{O}M|U)$ is an isomorphism of sheaves.

3. The localization of $f^{\#}$ is an isomorphism of local rings

$$f^{\#}{}_{f(p)}: \mathbf{O}f_{(p)} \to \mathbf{O}M, p.$$

There are a number of important motivations for studying differentiable manifolds within this abstract framework. First, there is no *a priori* reason that the model space needs to be $\mathbf{R}^n$. For example (in particular in algebraic geometry), one could take this to be the space of complex numbers $\mathbf{C}^n$ equipped with the sheaf of holomorphic functions (thus arriving at the spaces of complex analytic geometry), or the sheaf of polynomials (thus arriving at the spaces of interest in complex *algebraic* geometry). In broad terms, this concept can be adapted for any suitable notion of a scheme (see topos theory). Second, coordinates are no longer explicitly necessary to the construction. The analog of a coordinate system is the pair $(f, f^{\#})$, but these merely quantify the idea of *local isomorphism* rather than being central to the discussion (as in the case of charts and atlases). Third, the sheaf $\mathbf{O}M$ is not manifestly a sheaf of functions at all. Rather, it emerges as a sheaf of functions as a *consequence* of the construction (via the quotients of local rings by their maximal ideals). Hence it is a more primitive definition of the structure (see synthetic differential geometry).

A final advantage of this approach is that it allows for natural direct descriptions of many of the fundamental objects of study to differential geometry and topology.

- The cotangent space at a point is Ip/Ip^2, where Ip is the maximal ideal of the stalk $\mathbf{O}M, p$.

- In general, the entire cotangent bundle can be obtained by a related technique (see cotangent bundle for details).

- Taylor series (and jets) can be approached in a coordinate-independent manner using the Ip-adic filtration on $\mathbf{O}M$, p.

- The tangent bundle (or more precisely its sheaf of sections) can be identified with the sheaf of morphisms of $\mathbf{O}M$ into the ring of dual numbers.

5.4   Differentiable functions

A real valued function f on an n-dimensional differentiable manifold M is called **differentiable** at a point $p \in M$ if it is differentiable in any coordinate chart defined around p. In more precise terms, if (U, φ) is a chart where U is an open set in M containing p and $\varphi: U \to \mathbf{R}^n$ is the map defining the chart, then f is differentiable if and only if

$$f \circ \phi^{-1}: \phi(U) \subset \mathbf{R}^n \to \mathbf{R}$$

is differentiable at $\varphi(p)$. In general there will be many available charts; however, the definition of differentiability does not depend on the choice of chart at p. It follows from the chain rule applied to the transition functions between one chart and another that if f is differentiable in any particular chart at p, then it is differentiable in all charts at p. Analogous considerations apply to defining C^k functions, smooth functions, and analytic functions.

5.4.1   Differentiation of functions

There are various ways to define the derivative of a function on a differentiable manifold, the most fundamental of which is the directional derivative. The definition of the directional derivative is complicated by the fact that a manifold will lack a suitable affine structure with which to define vectors. The directional derivative therefore looks at curves in the manifold instead of vectors.

Directional differentiation

Given a real valued function f on an m dimensional differentiable manifold M, the directional derivative of f at a point p in M is defined as follows. Suppose that $\gamma(t)$ is a curve in M with $\gamma(0) = p$, which is *differentiable* in the sense that its composition with any chart is a differentiable curve in $\mathbf{R}^m$. Then the **directional derivative** of f at p along γ is

$$\frac{d}{dt} f(\gamma(t)) \bigg|_{t=0}.$$

If γ_1 and γ_2 are two curves such that $\gamma_1(0) = \gamma_2(0) = p$, and in any coordinate chart φ,

$$\frac{d}{dt} \phi \circ \gamma_1(t) \bigg|_{t=0} = \frac{d}{dt} \phi \circ \gamma_2(t) \bigg|_{t=0}$$

then, by the chain rule, f has the same directional derivative at p along γ_1 as along γ_2. This means that the directional derivative depends only on the tangent vector of the curve at p. Thus the more abstract definition of directional differentiation adapted to the case of differentiable manifolds ultimately captures the intuitive features of directional differentiation in an affine space.

Tangent vectors and the differential

A **tangent vector** at $p \in M$ is an equivalence class of differentiable curves γ with $\gamma(0) = p$, modulo the equivalence relation of first-order contact between the curves. Therefore,

$$\gamma_1 \equiv \gamma_2 \iff \frac{d}{dt}\phi \circ \gamma_1(t)\bigg|_{t=0} = \frac{d}{dt}\phi \circ \gamma_2(t)\bigg|_{t=0}$$

in every coordinate chart φ. Therefore, the equivalence classes are curves through p with a prescribed velocity vector at p. The collection of all tangent vectors at p forms a vector space: the tangent space to M at p, denoted T_pM.

If X is a tangent vector at p and f a differentiable function defined near p, then differentiating f along any curve in the equivalence class defining X gives a well-defined directional derivative along X:

$$Xf(p) := \frac{d}{dt}f(\gamma(t))\bigg|_{t=0}.$$

Once again, the chain rule establishes that this is independent of the freedom in selecting γ from the equivalence class, since any curve with the same first order contact will yield the same directional derivative.

If the function f is fixed, then the mapping

$$X \mapsto Xf(p)$$

is a linear functional on the tangent space. This linear functional is often denoted by $df(p)$ and is called the **differential** of f at p:

$$df(p) : T_pM \to \mathbf{R}.$$

5.4.2   Partitions of unity

One of the topological features of the sheaf of differentiable functions on a differentiable manifold is that it admits partitions of unity. This distinguishes the differential structure on a manifold from stronger structures (such as analytic and holomorphic structures) that in general fail to have partitions of unity.

Suppose that M is a manifold of class C^k, where $0 \le k \le \infty$. Let $\{U\alpha\}$ be an open covering of M. Then a **partition of unity** subordinate to the cover $\{U\alpha\}$ is a collection of real-valued C^k functions φi on M satisfying the following conditions:

- The supports of the φi are compact and locally finite;
- The support of φi is completely contained in $U\alpha$ for some α;
- The φi sum to one at each point of M:

$$\sum_i \phi_i(x) = 1.$$

(Note that this last condition is actually a finite sum at each point because of the local finiteness of the supports of the φi.)

Every open covering of a C^k manifold M has a C^k partition of unity. This allows for certain constructions from the topology of C^k functions on $\mathbf{R}^n$ to be carried over to the category of differentiable manifolds. In particular, it is possible to discuss integration by choosing a partition of unity subordinate to a particular coordinate atlas, and carrying out the integration in each chart of $\mathbf{R}^n$. Partitions of unity therefore allow for certain other kinds of function spaces to be considered: for instance L^p spaces, Sobolev spaces, and other kinds of spaces that require integration.

5.4.3 Differentiability of mappings between manifolds

Suppose M and N are two differentiable manifolds with dimensions m and n, respectively, and f is a function from M to N. Since differentiable manifolds are topological spaces we know what it means for f to be continuous. But what does "f is $C^k(M, N)$" mean for $k \geq 1$? We know what that means when f is a function between Euclidean spaces, so if we compose f with a chart of M and a chart of N such that we get a map that goes from Euclidean space to M to N to Euclidean space we know what it means for that map to be $C^k(\mathbf{R}^m, \mathbf{R}^n)$. We define "$f$ is $C^k(M, N)$" to mean that all such compositions of f with charts are $C^k(\mathbf{R}^m, \mathbf{R}^n)$. Once again the chain rule guarantees that the idea of differentiability does not depend on which charts of the atlases on M and N are selected. However, defining the derivative itself is more subtle. If M or N is itself already a Euclidean space, then we don't need a chart to map it to one.

5.4.4 Algebra of scalars

For a C^k manifold M, the set of real-valued C^k functions on the manifold forms an algebra under pointwise addition and multiplication, called the *algebra of scalar fields* or simply the *algebra of scalars*. This algebra has the constant function 1 as the multiplicative identity, and is a differentiable analog of the ring of regular functions in algebraic geometry.

It is possible to reconstruct a manifold from its algebra of scalars, first as a set, but also as a topological space – this is an application of the Banach–Stone theorem, and is more formally known as the spectrum of a C*-algebra. First, there is a one-to-one correspondence between the points of M and the algebra homomorphisms $\varphi \colon C^k(M) \to \mathbf{R}$, as such a homomorphism φ corresponds a codimension one ideal in $C^k(M)$ (namely the kernel of φ), which is necessarily a maximal ideal. On the converse, every maximal ideal in this algebra is an ideal of functions vanishing at a single point, which demonstrates that MSpec (the Max Spec) of $C^k(M)$ recovers M as a point set, though in fact it recovers M as a topological space.

One can define various geometric structures algebraically in terms of the algebra of scalars, and these definitions often generalize to algebraic geometry (interpreting rings geometrically) and operator theory (interpreting Banach spaces geometrically). For example, the tangent bundle to M can be defined as the derivations of the algebra of smooth functions on M.

This "algebraization" of a manifold (replacing a geometric object with an algebra) leads to the notion of a C*-algebra – a commutative C*-algebra being precisely the ring of scalars of a manifold, by Banach–Stone, and allows one to consider *non*commutative C*-algebras as non-commutative generalizations of manifolds. This is the basis of the field of noncommutative geometry.

5.5 Bundles

5.5.1 Tangent bundle

For more details on this topic, see tangent bundle.

The tangent space of a point consists of the possible directional derivatives at that point, and has the same dimension n as does the manifold. For a set of (non-singular) coordinates xk local to the point, the coordinate derivatives $\partial_k = \frac{\partial}{\partial x_k}$ typically define a basis of the tangent space. The collection of tangent spaces at all points can in turn be made into a manifold, the tangent bundle, whose dimension is $2n$. The tangent bundle is where tangent vectors lie, and is itself a differentiable manifold. The Lagrangian is a function on the tangent bundle. One can also define the tangent bundle as the bundle of 1-jets from $\mathbf{R}$ (the real line) to M.

One may construct an atlas for the tangent bundle consisting of charts based on $U\alpha \times \mathbf{R}^n$, where $U\alpha$ denotes one of the charts in the atlas for M. Each of these new charts is the tangent bundle for the charts $U\alpha$. The transition maps on this atlas are defined from the transition maps on the original manifold, and retain the original differentiability class.

5.5.2   Cotangent bundle

For more details on this topic, see cotangent bundle.

The dual space of a vector space is the set of real valued linear functions on the vector space. The cotangent space at a point is the dual of the tangent space at that point, and the cotangent bundle is the collection of all cotangent spaces.

Like the tangent bundle the cotangent bundle is again a differentiable manifold. The Hamiltonian is a scalar on the cotangent bundle. The total space of a cotangent bundle has the structure of a symplectic manifold. Cotangent vectors are sometimes called *covectors*. One can also define the cotangent bundle as the bundle of 1-jets of functions from M to **R**.

Elements of the cotangent space can be thought of as infinitesimal displacements: if f is a differentiable function we can define at each point p a cotangent vector dfp, which sends a tangent vector Xp to the derivative of f associated with Xp. However, not every covector field can be expressed this way. Those that can are referred to as exact differentials. For a given set of local coordinates x^k the differentials dxk
p form a basis of the cotangent space at p.

5.5.3   Tensor bundle

For more details on this topic, see tensor bundle.

The tensor bundle is the direct sum of all tensor products of the tangent bundle and the cotangent bundle. Each element of the bundle is a tensor field, which can act as a multilinear operator on vector fields, or on other tensor fields.

The tensor bundle cannot be a differentiable manifold, since it is infinite dimensional. It is however an algebra over the ring of scalar functions. Each tensor is characterized by its ranks, which indicate how many tangent and cotangent factors it has. Sometimes these ranks are referred to as *covariant* and *contravariant* ranks, signifying tangent and cotangent ranks, respectively.

5.5.4   Frame bundle

For more details on this topic, see frame bundle.

A frame (or, in more precise terms, a tangent frame) is an ordered basis of particular tangent space. Likewise, a tangent frame is a linear isomorphism of $\mathbf{R}^n$ to this tangent space. A moving tangent frame is an ordered list of vector fields that give a basis at every point of their domain. One may also regard a moving frame as a section of the frame bundle F(M), a GL(n, **R**) principal bundle made up of the set of all frames over M. The frame bundle is useful because tensor fields on M can be regarded as equivariant vector-valued functions on F(M).

5.5.5   Jet bundles

For more details on this topic, see jet bundle.

On a manifold that is sufficiently smooth, various kinds of jet bundles can also be considered. The (first-order) tangent bundle of a manifold is the collection of curves in the manifold modulo the equivalence relation of first-order contact. By analogy, the k-th order tangent bundle is the collection of curves modulo the relation of k-th order contact. Likewise, the cotangent bundle is the bundle of 1-jets of functions on the manifold: the k-jet bundle is the bundle of their k-jets. These and other examples of the general idea of jet bundles play a significant role in the study of differential operators on manifolds.

The notion of a frame also generalizes to the case of higher-order jets. Define a k-th order frame to be the k-jet of a diffeomorphism from $\mathbf{R}^n$ to M.[10] The collection of all k-th order frames, $F^k(M)$, is a principal G^k bundle over M, where G^k is the group of k-jets; i.e., the group made up of k-jets of diffeomorphisms of $\mathbf{R}^n$ that fix the origin. Note that GL(n, $\mathbf{R}$) is naturally isomorphic to G^1, and a subgroup of every G^k, $k \geq 2$. In particular, a section of $F^2(M)$ gives the frame components of a connection on M. Thus, the quotient bundle $F^2(M)/$ GL(n, $\mathbf{R}$) is the bundle of linear connections over M.

5.6 Calculus on manifolds

Many of the techniques from multivariate calculus also apply, *mutatis mutandis*, to differentiable manifolds. One can define the directional derivative of a differentiable function along a tangent vector to the manifold, for instance, and this leads to a means of generalizing the total derivative of a function: the differential. From the perspective of calculus, the derivative of a function on a manifold behaves in much the same way as the ordinary derivative of a function defined on a Euclidean space, at least locally. For example, there are versions of the implicit and inverse function theorems for such functions.

There are, however, important differences in the calculus of vector fields (and tensor fields in general). In brief, the directional derivative of a vector field is not well-defined, or at least not defined in a straightforward manner. Several generalizations of the derivative of a vector field (or tensor field) do exist, and capture certain formal features of differentiation in Euclidean spaces. The chief among these are:

- The Lie derivative, which is uniquely defined by the differential structure, but fails to satisfy some of the usual features of directional differentiation.

- An affine connection, which is not uniquely defined, but generalizes in a more complete manner the features of ordinary directional differentiation. Because an affine connection is not unique, it is an additional piece of data that must be specified on the manifold.

Ideas from integral calculus also carry over to differential manifolds. These are naturally expressed in the language of exterior calculus and differential forms. The fundamental theorems of integral calculus in several variables — namely Green's theorem, the divergence theorem, and Stokes' theorem — generalize to a theorem (also called Stokes' theorem) relating the exterior derivative and integration over submanifolds.

5.6.1 Differential calculus of functions

Differentiable functions between two manifolds are needed in order to formulate suitable notions of submanifolds, and other related concepts. If $f: M \to N$ is a differentiable function from a differentiable manifold M of dimension m to another differentiable manifold N of dimension n, then the differential of f is a mapping $df: \mathrm{T}M \to \mathrm{T}N$. It also denoted by Tf and called the **tangent map**. At each point of M, this is a linear transformation from one tangent space to another:

$$df(p): T_p M \to T_{f(p)} N.$$

The **rank** of f at p is the rank of this linear transformation.

Usually the rank of a function is a pointwise property. However, if the function has maximal rank, then the rank will remain constant in a neighborhood of a point. A differentiable function "usually" has maximal rank, in a precise sense given by Sard's theorem. Functions of maximal rank at a point are called immersions and submersions:

- If $m \leq n$, and $f: M \to N$ has rank m at $p \in M$, then f is called an **immersion** at p. If f is an immersion at all points of M and is a homeomorphism onto its image, then f is an **embedding**. Embeddings formalize the notion of M being a submanifold of N. In general, an embedding is an immersion without self-intersections and other sorts of non-local topological irregularities.

- If $m \geq n$, and $f \colon M \to N$ has rank n at $p \in M$, then f is called a **submersion** at p. The implicit function theorem states that if f is a submersion at p, then M is locally a product of N and $\mathbf{R}^{m-n}$ near p. In formal terms, there exist coordinates $(y_1, ..., yn)$ in a neighborhood of $f(p)$ in N, and $m-n$ functions $x_1,...,xm\_n$ defined in a neighborhood of p in M such that

$$(y_1 \circ f, \dots, y_n \circ f, x_1, \dots, x_{m-n})$$

is a system of local coordinates of M in a neighborhood of p. Submersions form the foundation of the theory of fibrations and fibre bundles.

5.6.2 Lie derivative

A Lie derivative, named after Sophus Lie, is a derivation on the algebra of tensor fields over a manifold M. The vector space of all Lie derivatives on M forms an infinite dimensional Lie algebra with respect to the Lie bracket defined by

$$[A, B] := \mathcal{L}_A B = -\mathcal{L}_B A.$$

The Lie derivatives are represented by vector fields, as infinitesimal generators of flows (active diffeomorphisms) on M. Looking at it the other way round, the group of diffeomorphisms of M has the associated Lie algebra structure, of Lie derivatives, in a way directly analogous to the Lie group theory.

5.6.3 Exterior calculus

For more details on this topic, see differential form.

The exterior calculus allows for a generalization of the gradient, divergence and curl operators.

The bundle of differential forms, at each point, consists of all totally antisymmetric multilinear maps on the tangent space at that point. It is naturally divided into n-forms for each n at most equal to the dimension of the manifold; an n-form is an n-variable form, also called a form of degree n. The 1-forms are the cotangent vectors, while the 0-forms are just scalar functions. In general, an n-form is a tensor with cotangent rank n and tangent rank 0. But not every such tensor is a form, as a form must be antisymmetric.

Exterior derivative

There is a map from scalars to covectors called the exterior derivative

$$\mathrm{d} \colon \mathcal{C}(M) \to \mathrm{T}^*(M) : f \mapsto \mathrm{d}f$$

such that

$$\mathrm{d}f \colon \mathrm{T}(M) \to \mathcal{C}(M) : V \mapsto V(f).$$

This map is the one that relates covectors to infinitesimal displacements, mentioned above; some covectors are the exterior derivatives of scalar functions. It can be generalized into a map from the n-forms onto the $(n+1)$-forms. Applying this derivative twice will produce a zero form. Forms with zero derivative are called closed forms, while forms that are themselves exterior derivatives are known as exact forms.

The space of differential forms at a point is the archetypal example of an exterior algebra; thus it possesses a wedge product, mapping a *k*-form and *l*-form to a (*k+l*)-form. The exterior derivative extends to this algebra, and satisfies a version of the product rule:

$$d(\omega \wedge \eta) = d\omega \wedge \eta + (-1)^{\deg \omega}(\omega \wedge d\eta).$$

From the differential forms and the exterior derivative, one can define the de Rham cohomology of the manifold. The rank *n* cohomology group is the quotient group of the closed forms by the exact forms.

5.7 Topology of differentiable manifolds

5.7.1 Relationship with topological manifolds

Every topological manifold in dimension 1, 2, or 3 has a unique differential structure (up to diffeomorphism); thus the concepts of topological and differentiable manifold are distinct only in higher dimensions. It is known that in each higher dimension, there are some topological manifolds with no smooth structure, and some with multiple non-diffeomorphic structures.

The existence of non-smoothable manifolds was proven by Kervaire (1960), see Kervaire manifold, and later explained in the context of Donaldson's theorem (compare Hilbert's fifth problem);[11] a good example of a non-smoothable manifold is the E_8 manifold.

The classic example of manifolds with multiple incompatible structures are the exotic 7-spheres of John Milnor.[12]

5.7.2 Classification

Every second-countable 1-manifold without boundary is homeomorphic to a disjoint union of countably many copies of **R** (the real line) and **S** (the circle); the only connected examples are **R** and **S**, and of these only **S** is compact. In higher dimensions, classification theory normally focuses only on compact connected manifolds.

For a classification of 2-manifolds, see surface: in particular compact connected oriented 2-manifolds are classified by their genus, which is a nonnegative integer.

A classification of 3-manifolds follows *in principle* from the geometrization of 3-manifolds and various recognition results for geometrizable 3-manifolds, such as Mostow rigidity and Sela's algorithm for the isomorphism problem for hyperbolic groups.[13]

The classification of *n*-manifolds for *n* greater than three is known to be impossible, even up to homotopy equivalence. Given any finitely presented group, one can construct a closed 4-manifold having that group as fundamental group. Since there is no algorithm to decide the isomorphism problem for finitely presented groups, there is no algorithm to decide whether two 4-manifolds have the same fundamental group. Since the previously described construction results in a class of 4-manifolds that are homeomorphic if and only if their groups are isomorphic, the homeomorphism problem for 4-manifolds is undecidable. In addition, since even recognizing the trivial group is undecidable, it is not even possible in general to decide whether a manifold has trivial fundamental group, i.e. is simply connected.

Simply connected 4-manifolds have been classified up to homeomorphism by Freedman using the intersection form and Kirby–Siebenmann invariant. Smooth 4-manifold theory is known to be much more complicated, as the exotic smooth structures on $\mathbf{R}^4$ demonstrate.

However, the situation becomes more tractable for simply connected smooth manifolds of dimension ≥ 5, where the h-cobordism theorem can be used to reduce the classification to a classification up to homotopy equivalence, and surgery theory can be applied.[14] This has been carried out to provide an explicit classification of simply connected 5-manifolds by Dennis Barden.

5.8   Structures on manifolds

5.8.1   (Pseudo-)Riemannian manifolds

A Riemannian manifold is a differentiable manifold on which the tangent spaces are equipped with inner products in a differentiable fashion. The inner product structure is given in the form of a symmetric 2-tensor called the Riemannian metric. This metric can be used to interconvert vectors and covectors, and to define a rank 4 Riemann curvature tensor. On a Riemannian manifold one has notions of length, volume, and angle. Any differentiable manifold can be given a Riemannian structure.

A pseudo-Riemannian manifold is a variant of Riemannian manifold where the metric tensor is allowed to have an indefinite signature (as opposed to a positive-definite one). Pseudo-Riemannian manifolds of signature (3, 1) are important in general relativity. Not every differentiable manifold can be given a pseudo-Riemannian structure; there are topological restrictions on doing so.

A Finsler manifold is a generalization of a Riemannian manifold, in which the inner product is replaced with a vector norm; this allows the definition of length, but not angle.

5.8.2   Symplectic manifolds

For more details on this topic, see symplectic manifold.

A symplectic manifold is a manifold equipped with a closed, nondegenerate 2-form. This condition forces symplectic manifolds to be even-dimensional. Cotangent bundles, which arise as phase spaces in Hamiltonian mechanics, are the motivating example, but many compact manifolds also have symplectic structure. All orientable surfaces embedded in Euclidean space have a symplectic structure, the signed area form on each tangent space induced by the ambient Euclidean inner product.[note 1] Every Riemann surface is an example of such a surface, and hence a symplectic manifold, when considered as a real manifold.

5.8.3   Lie groups

For more details on this topic, see Lie group.

A Lie group is C^∞ manifold that also carries a group structure whose product and inversion operations are smooth as maps of manifolds. These objects arise naturally in describing symmetries.

5.9   Generalizations

The category of smooth manifolds with smooth maps lacks certain desirable properties, and people have tried to generalize smooth manifolds in order to rectify this. Diffeological spaces use a different notion of chart known as a "plot". Frölicher spaces and orbifolds are other attempts.

A rectifiable set generalizes the idea of a piece-wise smooth or rectifiable curve to higher dimensions; however, rectifiable sets are not in general manifolds.

Banach manifolds and Fréchet manifolds, in particular manifolds of mappings are infinite dimensional differentiable manifolds.

5.10   See also

- Affine connection

- Atlas (topology)

- Christoffel symbols

- Differential geometry

- Introduction to mathematics of general relativity

- List of formulas in Riemannian geometry

- Riemannian geometry

- Space (mathematics)

5.11 Notes

[1] This form is clearly nondegenerate, and it must be closed because it is top-dimensional with respect to the surface; this reflects the exceptional isomorphism of Lie groups $Sp(2, \mathbf{R}) \cong SL(2, \mathbf{R})$ between the symplectic group (corresponding to symplectic structure) and the special linear group (corresponding to orientable structure). Note that a symplectic structure requires an additional integrability condition, beyond this isomorphism of groups: it is not just a G-structure.

5.12 References

[1] B. Riemann (1867).

[2] Maxwell himself worked with quaternions rather than tensors, but his equations for electromagnetism were used as an early example of the tensor formalism; see Dimitrienko, Yuriy I. (2002), *Tensor Analysis and Nonlinear Tensor Functions*, Springer, p. xi, ISBN 9781402010156.

[3] See G. Ricci (1888), G. Ricci and T. Levi-Civita (1901), T. Levi-Civita (1927).

[4] See H. Weyl (1955).

[5] H. Whitney (1936).

[6] H. Whitney (1936).

[7] Kobayashi and Nomizu (1963), Volume 1.

[8] This definition can be found in MacLane and Moerdijk (1992). For an equivalent, *ad hoc* definition, see Sternberg (1964) Chapter II.

[9] Hartshorne (1997)

[10] See S. Kobayashi (1972).

[11] S. Donaldson (1983).

[12] J. Milnor (1956). These are the first examples of exotic spheres.

[13] Z. Sela (1995). However, 3-manifolds are only classified in the sense that there is an (impractical) algorithm for generating a non-redundant list of all compact 3-manifolds.

[14] See A. Ranicki (2002).

5.13   Bibliography

- Donaldson, Simon (1983). "An application of gauge theory to four-dimensional topology". *Journal of Differential Geometry* **18** (2): 279–315.

- Hartshorne, Robin (1977). *Algebraic Geometry*. Springer-Verlag. ISBN 0-387-90244-9.

- Hazewinkel, Michiel, ed. (2001), "Differentiable manifold", *Encyclopedia of Mathematics*, Springer, ISBN 978-1-55608-010-4

- Kervaire, Michel A. (1960). "A manifold which does not admit any differentiable structure". *Coment. Math. Helv.* **34** (1): 257–270. doi:10.1007/BF02565940..

- Kobayashi, S. (1972). *Transformation groups in differential geometry*. Springer.

- Lee, Jeffrey M. (2009), *Manifolds and Differential Geometry*, Graduate Studies in Mathematics, Vol. 107, Providence: American Mathematical Society .

- Levi-Civita, Tullio (1927). *The absolute differential calculus (calculus of tensors)*.

- MacLane, S.; Moerdijk, I. (1992). *Sheaves in Geometry and Logic*. Springer. ISBN 0-387-97710-4.

- Milnor, John (1956). "On Manifolds Homeomorphic to the 7-Sphere". *Annals of Mathematics* **64**: 399–405. doi:10.2307/1969983. JSTOR 1969983.

- Ranicki, Andrew (2002). *Algebraic and Geometric Surgery*. Oxford Mathematical Monographs, Clarendon Press. ISBN 0-19-850924-3.

- Ricci-Curbastro, Gregorio; Levi-Civita, Tullio (1901). *Die Methoden des absoluten Differentialkalkuls*.

- Ricci-Curbastro, Gregorio (1888). "Delle derivazioni covarianti e controvarianti e del loro uso nella analisi applicata (Italian)".

- Riemann, Bernhard (1867). "Ueber die Hypothesen, welche der Geometrie zu Grunde liegen (On the Hypotheses which lie at the Bases of Geometry)". *Abhandlungen der Königlichen Gesellschaft der Wissenschaften zu Göttingen* **13**. Available online at Trinity College Dublin

- Sela, Zlil (1995). "The isomorphism problem for hyperbolic groups. I". *Annals of Mathematics* (Annals of Mathematics) **141** (2): 217–283. doi:10.2307/2118520. JSTOR 2118520.

- Sternberg, Shlomo (1964). *Lectures on Differential Geometry*. Prentice-Hall.

- Weisstein, Eric W. "Smooth Manifold". Retrieved 2008-03-04.

- Weyl, Hermann (1955). *Die Idee der Riemannschen Fläche*. Teubner.

- Whitney, Hassler (1936). "Differentiable Manifolds". *Annals of Mathematics* (Annals of Mathematics) **37** (3): 645–680. doi:10.2307/1968482. JSTOR 1968482.

Chapter 6

Differential structure

In mathematics, an n-dimensional **differential structure** (or **differentiable structure**) on a set M makes M into an n-dimensional differential manifold, which is a topological manifold with some additional structure that allows for differential calculus on the manifold. If M is already a topological manifold, it is required that the new topology be identical to the existing one.

6.1   Definition

For a natural number n and some k which may be a non-negative integer or infinity, an **n-dimensional C^k differential structure** [1] is defined using a **C^k-atlas**, which is a set of bijections called **charts** between a collection of subsets of M (whose union is the whole of M), and a set of open subsets of $\mathbb{R}^n$:

$$\varphi_i : M \supset W_i \to U_i \subset \mathbb{R}^n$$

which are **C^k-compatible** (in the sense defined below):

Each such map provides a way in which certain subsets of the manifold may be viewed as being like open subsets of $\mathbb{R}^n$ but the usefulness of this notion depends on to what extent these notions agree when the domains of two such maps overlap.

Consider two charts:

$$\varphi_i : W_i \to U_i,$$

$$\varphi_j : W_j \to U_j.$$

The intersection of the domains of these two functions is:

$$W_{ij} = W_i \cap W_j$$

and its map by the two chart maps to the two images:

$$U_{ij} = \varphi_i \left(W_{ij} \right),$$

$$U_{ji} = \varphi_j \left(W_{ij} \right)$$

The transition map between the two charts is the map between the two images of this intersection under the two chart maps.

$$\varphi_{ij} : U_{ij} \to U_{ji}$$

$$\varphi_{ij}(x) = \varphi_j \left(\varphi_i^{-1}(x) \right).$$

Two charts φ_i, φ_j are **C^k-compatible** if

$$U_{ij}, \ U_{ji}$$

are open, and the transition maps

$$\varphi_{ij}, \ \varphi_{ji}$$

have continuous derivatives of order k. If $k = 0$, we only require that the transition maps are continuous, consequently a C^0-atlas is simply another way to define a topological manifold. If $k = \infty$, derivatives of all orders must be continuous. A family of C^k-compatible charts covering the whole manifold is a C^k-atlas defining a C^k differential manifold. Two atlases are **C^k-equivalent** if the union of their sets of charts forms a C^k-atlas. In particular, a C^k-atlas that is C^0-compatible with a C^0-atlas that defines a topological manifold is said to determine a C^k differential structure on the topological manifold. The C^k equivalence classes of such atlases are the **distinct C^k differential structures** of the manifold. Each distinct differential structure is determined by a unique maximal atlas, which is simply the union of all atlases in the equivalence class.

For simplification of language, without any loss of precision, one might just call a maximal C^k-atlas on a given set a C^k-manifold. This maximal atlas then uniquely determines both the topology and the underlying set, the latter being the union of the domains of all charts, and the former having the set of all these domains as a basis.

6.2   Existence and uniqueness theorems

For $0 < k < \infty$ and any n-dimensional C^k-manifold, the maximal atlas contains a C^∞-atlas on the same underlying set by a theorem due to Whitney. However, a given maximal C^k-atlas contains *distinct* maximal C^∞-atlases whenever $n > 0$ but there is a C^∞-diffeomorphism between any two of these distinct C^∞-atlases. Thus there is only one class of pairwise smoothly diffeomorphic smooth, i.e. C^∞-structures in a C^k-manifold. A bit loosely, one might express this by saying that the smooth structure is (essentially) unique. The case for $k = 0$ is different. Namely, there exist topological manifolds which admit no C^1-structure, a result proved by Kervaire (1960),[2] and later explained in the context of Donaldson's theorem (compare Hilbert's fifth problem).

Smooth structures on an orientable manifold are usually counted modulo orientation-preserving smooth homeomorphisms. There then arises the question whether orientation-reversing diffeomorphisms exist. There is an "essentially unique" smooth structure for any topological manifold of dimension smaller than 4. For compact manifolds of dimension greater than 4, there is a finite number of "smooth types", i.e. equivalence classes of pairwise smoothly diffeomorphic smooth structures. In the case of $\mathbf{R}^n$ with $n \neq 4$, the number of these types is one, whereas for $n = 4$, there are uncountably many such types. One refers to these by exotic $\mathbf{R}^4$.

6.3   Differential structures on spheres of dimension 1 to 20

The following table lists the number of smooth types of the topological m-sphere $\mathbf{S}^m$ for the values of the dimension m from 1 up to 20. Spheres with a smooth, i.e. C^∞-differential structure not smoothly diffeomorphic to the usual one are known as exotic spheres.

It is not currently known how many smooth types the topological 4-sphere $\mathbf{S}^4$ has, except that there is at least one. There may be one, a finite number, or an infinite number. The claim that there is just one is known as the *smooth* Poincaré conjecture (see generalized Poincaré conjecture). Most mathematicians believe that this conjecture is false, i.e. that $\mathbf{S}^4$ has more than one smooth type. The problem is connected with the existence of more than one smooth type of the topological 4-disk (or 4-ball).

6.4   Differential structures on topological manifolds

As mentioned above, in dimensions smaller than 4, there is only one differential structure for each topological manifold. That was proved by Johann Radon for dimension 1 and 2, and by Edwin E. Moise in dimension 3.[3] By using obstruction theory, Robion Kirby and Laurent Siebenmann [4] were able to show that the number of PL structures for compact topological manifolds of dimension greater than 4 is finite. John Milnor, Michel Kervaire, and Morris Hirsch proved that the number of smooth structures on a compact PL manifold is finite and agrees with the number of differential structures on the sphere for the same dimension (see the book Asselmeyer-Maluga, Brans chapter 7) By combining these results, the number of smooth structures on a compact topological manifold of dimension not equal to 4 is finite.

Dimension 4 is more complicated. For compact manifolds, results depend on the complexity of the manifold as measured by the second Betti number b_2. For large Betti numbers $b_2 > 18$ in a simply connected 4-manifold, one can use a surgery along a knot or link to produce a new differential structure. With the help of this procedure one can produce countably infinite many differential structures. But even for simple spaces like $S^4, \mathbb{C}P^2, \ldots$ one doesn't know the construction of other differential structures. For non-compact 4-manifolds there are many examples like $\mathbb{R}^4, S^3 \times \mathbb{R}, M^4 \setminus \{*\}, \ldots$ having uncountably many differential structures.

6.5   See also

- Mathematical structure

- Atlas

- Exotic R^4

- Exotic sphere

6.6   References

[1] Hirsch, Morris, *Differential Topology*, Springer (1997), ISBN 0-387-90148-5. for a general mathematical account of differential structures

[2] Kervaire (1960), "A manifold which does not admit any differentiable structure", *Coment. Math. Helv.* **34**: 257–270, doi:10.1007/BF02565940

[3] Moise, Edwin E., *Affine structures in 3-manifolds. V. The triangulation theorem and Hauptvermutung*. Annals of Mathematics. Second Series, Vol. 56 pg 96-114 (1952)

[4] Kirby, Robion C. and Siebenmann, Laurence C., *Foundational Essays on Topological Manifolds. Smoothings, and Triangulations*. Princeton, New Jersey: Princeton University Press (1977), ISBN 0-691-08190-5.

Chapter 7

Glossary of group theory

In **group theory**, a group $(G, \bullet)$ is a set G closed under a binary operation $\bullet$ satisfying the following 3 axioms:

- *Associativity*: For all a, b and c in G, $(a \bullet b) \bullet c = a \bullet (b \bullet c)$.

- *Identity element*: There exists an $e \in G$ such that for all a in G, $e \bullet a = a \bullet e = a$.

- *Inverse element*: For each a in G, there is an element b in G such that $a \bullet b = b \bullet a = e$, where e is an identity element.

Basic examples for groups are the integers $\mathbf{Z}$ with addition operation, or rational numbers without zero $\mathbf{Q} \setminus \{0\}$ with multiplication. More generally, for any ring R, the units in R form a multiplicative group. See the group article for an illustration of this definition and for further examples. Groups include, however, much more general structures than the above. Group theory is concerned with proving abstract statements about groups, regardless of the actual nature of element and the operation of the groups in question.

This glossary provides short explanations of some basic notions used throughout group theory. Please refer to group theory for a general description of the topic. See also list of group theory topics.

7.1 Basic definitions

A subset $H \subset G$ is a *subgroup* if the restriction of $\bullet$ to H is a group operation on H. It is called *normal*, if left and right cosets agree, i.e. $gH = Hg$ for all g in G. Normal subgroups play a distinguished role by virtue of the fact that the collection of cosets of a normal subgroup N in a group G naturally inherits a group structure, enabling the formation of the quotient group, usually denoted G/N (also called a *factor group*). The *butterfly lemma* is a technical result on the lattice of subgroups of a group.

Given a subset S of a group G, the smallest subgroup of G containing S is called the subgroup *generated by S*. It is often denoted $<S>$.

Both subgroups and normal subgroups of a given group form a complete lattice under inclusion of subsets; this property and some related results are described by the lattice theorem.

Given any set A, one can define a group as the smallest group containing the free semigroup of A. This group consists of the finite strings called words that can be composed by elements from A and their inverses. Multiplication of strings is defined by concatenation, for instance $(abb) * (bca) = abbbca$.

Every group G is basically a factor group of a free group generated by the set of its elements. This phenomenon is made formal with group presentations.

The *direct product*, *direct sum*, and *semidirect product* of groups glue several groups together, in different ways. The direct product of a family of groups Gi, for example, is the cartesian product of the sets underlying the various Gi, and the group operation is performed component-wise.

A *group homomorphism* is a map $f : G \rightarrow H$ between two groups that preserves the structure imposed by the operation, i.e.

$$f(a \bullet b) = f(a) \bullet f(b).$$

Bijective (in-, surjective) maps are isomorphisms of groups (mono-, epimorphisms, respectively). The kernel ker(f) is always a normal subgroup of the group. For f as above, the *fundamental theorem on homomorphisms* relates the structure of G and H, and of the kernel and image of the homomorphism, namely

$$G / \ker(f) \cong \mathrm{im}(f).$$

One of the fundamental problems of group theory is the *classification of groups* up to isomorphism.

Groups together with group homomorphisms form a category.

In universal algebra, groups are generally treated as algebraic structures of the form $(G, \bullet, e, {}^{-1})$, i.e. the identity element e and the map that takes every element a of the group to its inverse a^{-1} are treated as integral parts of the formal definition of a group.

7.2 Finiteness conditions

The *order* |G| (or o(G)) of a group is the cardinality of G. If the order |G| is (in-)finite, then G itself is called (in-)finite. An important class is the *group of permutations* or symmetric groups of N letters, denoted SN. *Cayley's theorem* exhibits any finite group G as a subgroup of the symmetric group on G. The theory of finite groups is very rich. *Lagrange's theorem* states that the order of any subgroup H of a finite group G divides the order of G. A partial converse is given by the *Sylow theorems*: if p^n is the greatest power of a prime p dividing the order of a finite group G, then there exists a subgroup of order p^n, and the number of these subgroups is also known. A projective limit of finite groups is called profinite.[1] An important profinite group, fundamental for p-adic analysis, class field theory, and l-adic cohomology is the ring of p-adic integers and the profinite completion of **Z**, respectively

$$\mathbb{Z}_p := \varprojlim_n \mathbb{Z}/p^n \text{ and } \hat{\mathbb{Z}} := \varprojlim_n \mathbb{Z}/n. \text{ [2]}$$

Most of the facts from finite groups can be generalized directly to the profinite case.[3]

Certain conditions on chains of subgroups, parallel to the notion of Noetherian and Artinian rings, allow to deduce further properties. For example the *Krull-Schmidt theorem* states that a group satisfying certain finiteness conditions for chains of its subgroups, can be uniquely written as a finite direct product of indecomposable subgroups.

Another, yet slightly weaker, level of finiteness is the following: a subset A of G is said to generate the group if any element h can be written as the product of elements of A. A group is said to be finitely generated if it is possible to find a finite subset A generating the group. Finitely generated groups are in many respects as well-treatable as finite groups.

7.3 Abelian groups

The category of groups can be subdivided in several ways. A particularly well-understood class of groups are the so-called abelian (in honor of Niels Abel) or commutative groups, i.e. the ones satisfying

$$\forall a, b \in G, \ a * b = b * a.$$

Another way of saying this is that the commutator

$$[a, b] =: a^{-1}b^{-1}ab$$

equals the identity element for all a and b. A *non-abelian* group is a group that is not abelian. Even more particular, cyclic groups are the groups generated by a single element. Being either isomorphic to $\mathbf{Z}$ or to $\mathbf{Z}n$, the integers modulo n, they are always abelian. Any finitely generated abelian group is known to be a direct sum of groups of these two types. The category of abelian groups is an abelian category. In fact, abelian groups serve as the prototype of abelian categories. A converse is given by Mitchell's embedding theorem.

7.4   Normal series

Most of the notions developed in group theory are designed to tackle non-abelian groups. There are several notions designed to measure how far a group is from being abelian. The commutator subgroup (or derived group) is the subgroup generated by commutators $[a, b]$, whereas the center is the subgroup of elements that commute with every other group element.

Given a group G and a normal subgroup N of G, denoted $N \lhd G$, there is an exact sequence:

$$1 \to N \to G \to H \to 1,$$

where 1 denotes the trivial group and H is the quotient G/N. This permits the decomposition of G into two smaller pieces. The other way round, given two groups N and H, a group G fitting into an exact sequence as above is called an *extension* of H by N. Given H and N there are many different group extensions G, which leads to the extension problem. There is always at least one extension, called the trivial extension, namely the direct sum $G = N \oplus H$, but usually there are more. For example, the Klein four-group is a non-trivial extension of $\mathbf{Z}_2$ by $\mathbf{Z}_2$. This is a first glimpse of homological algebra and *Ext* functors.[4]

Many properties for groups, for example being a finite group or a p-group (i.e. the order of every element is a power of p) are stable under extensions and sub- and quotient groups, i.e. if N and H have the property, then so does G and vice versa. This kind of information is therefore preserved while breaking it into pieces by means of exact sequences. If this process has come to an end, i.e. if a group G does not have any (non-trivial) normal subgroups, G is called simple. The name is misleading because a simple group can in fact be very complex. An example is the monster group, whose order is about 10^{54}. The finite simple groups are known and classified.

Repeatedly taking normal subgroups (if they exist) leads to *normal series*:

$$1 = G_0 \lhd G_1 \lhd \ldots \lhd G_n = G,$$

i.e. any Gi is a normal subgroup of the next one Gi_{+1}. A group is *solvable* (or *soluble*) if it has a normal series all of whose quotients are abelian. Imposing further commutativity constraints on the quotients Gi_{+1} / Gi, one obtains central series which lead to *nilpotent groups*. They are an approximation of abelian groups in the sense that

$$[\ldots[[g_1, g_2], g_3] \ldots, gn]=1$$

for all choices of group elements gi.

There may be distinct normal series for a group G. If it is impossible to refine a given series by inserting further normal subgroups, it is called *composition series*. By the *Jordan–Hölder theorem* any two composition series of a given group are equivalent.[5]

7.5   Other notions

General linear group, denoted by GL(n, F), is the group of n -by- n invertible matrices, where the elements of the matrices are taken from a field F such as the real numbers or the complex numbers.

Group representation (not to be confused with the *presentation* of a group). A *group representation* is a homomorphism from a group to a general linear group. One basically tries to "represent" a given abstract group as a concrete group of invertible matrices which is much easier to study.

7.6 Notes

[1] Shatz 1972

[2] These two groups play a central role for maximal abelian extension of number fields, see Kronecker–Weber theorem

[3] For example the Sylow theorems.

[4] Weibel 1994

[5] This can be shown using the Schreier refinement theorem.

7.7 References

- Rotman, Joseph (1994). *An introduction to the theory of groups.* New York: Springer-Verlag. ISBN 0-387-94285-8. A standard contemporary reference.

- Weibel, Charles A. (1994), *An introduction to homological algebra*, Cambridge Studies in Advanced Mathematics **38**, Cambridge University Press, ISBN 978-0-521-55987-4, OCLC 36131259, MR 1269324

- Shatz, Stephen S. (1972). *Profinite groups, arithmetic, and geometry.* Annals of Mathematics Studies **67**. Princeton, NJ: Princeton University Press. ISBN 0-691-08017-8. MR 0347778. Zbl 0236.12002.

Chapter 8

Group theory

This article covers advanced notions. For basic topics, see Group (mathematics).
For group theory in social sciences, see social group.

In mathematics and abstract algebra, **group theory** studies the algebraic structures known as groups. The concept of a group is central to abstract algebra: other well-known algebraic structures, such as rings, fields, and vector spaces, can all be seen as groups endowed with additional operations and axioms. Groups recur throughout mathematics, and the methods of group theory have influenced many parts of algebra. Linear algebraic groups and Lie groups are two branches of group theory that have experienced advances and have become subject areas in their own right.

Various physical systems, such as crystals and the hydrogen atom, can be modelled by symmetry groups. Thus group theory and the closely related representation theory have many important applications in physics, chemistry, and materials science. Group theory is also central to public key cryptography.

One of the most important mathematical achievements of the 20th century[1] was the collaborative effort, taking up more than 10,000 journal pages and mostly published between 1960 and 1980, that culminated in a complete classification of finite simple groups.

8.1 History

Main article: History of group theory

Group theory has three main historical sources: number theory, the theory of algebraic equations, and geometry. The number-theoretic strand was begun by Leonhard Euler, and developed by Gauss's work on modular arithmetic and additive and multiplicative groups related to quadratic fields. Early results about permutation groups were obtained by Lagrange, Ruffini, and Abel in their quest for general solutions of polynomial equations of high degree. Évariste Galois coined the term "group" and established a connection, now known as Galois theory, between the nascent theory of groups and field theory. In geometry, groups first became important in projective geometry and, later, non-Euclidean geometry. Felix Klein's Erlangen program proclaimed group theory to be the organizing principle of geometry.

Galois, in the 1830s, was the first to employ groups to determine the solvability of polynomial equations. Arthur Cayley and Augustin Louis Cauchy pushed these investigations further by creating the theory of permutation groups. The second historical source for groups stems from geometrical situations. In an attempt to come to grips with possible geometries (such as euclidean, hyperbolic or projective geometry) using group theory, Felix Klein initiated the Erlangen programme. Sophus Lie, in 1884, started using groups (now called Lie groups) attached to analytic problems. Thirdly, groups were, at first implicitly and later explicitly, used in algebraic number theory.

The different scope of these early sources resulted in different notions of groups. The theory of groups was unified starting around 1880. Since then, the impact of group theory has been ever growing, giving rise to the birth of abstract algebra in the early 20th century, representation theory, and many more influential spin-off domains. The classification of finite

The popular puzzle Rubik's cube invented in 1974 by Ernő Rubik has been used as an illustration of permutation groups.

simple groups is a vast body of work from the mid 20th century, classifying all the finite simple groups.

8.2 Main classes of groups

Main articles: Group (mathematics) and Glossary of group theory

The range of groups being considered has gradually expanded from finite permutation groups and special examples of matrix groups to abstract groups that may be specified through a presentation by generators and relations.

8.2.1 Permutation groups

The first class of groups to undergo a systematic study was permutation groups. Given any set X and a collection G of bijections of X into itself (known as *permutations*) that is closed under compositions and inverses, G is a group acting on X. If X consists of n elements and G consists of *all* permutations, G is the symmetric group Sn; in general, any permutation group G is a subgroup of the symmetric group of X. An early construction due to Cayley exhibited any group as a permutation group, acting on itself ($X = G$) by means of the left regular representation.

In many cases, the structure of a permutation group can be studied using the properties of its action on the corresponding set. For example, in this way one proves that for $n \geq 5$, the alternating group An is simple, i.e. does not admit any proper normal subgroups. This fact plays a key role in the impossibility of solving a general algebraic equation of degree $n' \geq 5$ *in radicals*.

8.2.2 Matrix groups

The next important class of groups is given by *matrix groups*, or linear groups. Here G is a set consisting of invertible matrices of given order n over a field K that is closed under the products and inverses. Such a group acts on the n-dimensional vector space K^n by linear transformations. This action makes matrix groups conceptually similar to permutation groups, and the geometry of the action may be usefully exploited to establish properties of the group G.

8.2.3 Transformation groups

Permutation groups and matrix groups are special cases of transformation groups: groups that act on a certain space X preserving its inherent structure. In the case of permutation groups, X is a set; for matrix groups, X is a vector space. The concept of a transformation group is closely related with the concept of a symmetry group: transformation groups frequently consist of *all* transformations that preserve a certain structure.

The theory of transformation groups forms a bridge connecting group theory with differential geometry. A long line of research, originating with Lie and Klein, considers group actions on manifolds by homeomorphisms or diffeomorphisms. The groups themselves may be discrete or continuous.

8.2.4 Abstract groups

Most groups considered in the first stage of the development of group theory were "concrete", having been realized through numbers, permutations, or matrices. It was not until the late nineteenth century that the idea of an abstract group as a set with operations satisfying a certain system of axioms began to take hold. A typical way of specifying an abstract group is through a presentation by *generators and relations*,

$$G = \langle S | R \rangle.$$

A significant source of abstract groups is given by the construction of a *factor group*, or quotient group, G/H, of a group G by a normal subgroup H. Class groups of algebraic number fields were among the earliest examples of factor groups, of much interest in number theory. If a group G is a permutation group on a set X, the factor group G/H is no longer acting on X; but the idea of an abstract group permits one not to worry about this discrepancy.

The change of perspective from concrete to abstract groups makes it natural to consider properties of groups that are independent of a particular realization, or in modern language, invariant under isomorphism, as well as the classes of group with a given such property: finite groups, periodic groups, simple groups, solvable groups, and so on. Rather than exploring properties of an individual group, one seeks to establish results that apply to a whole class of groups. The new paradigm was of paramount importance for the development of mathematics: it foreshadowed the creation of abstract algebra in the works of Hilbert, Emil Artin, Emmy Noether, and mathematicians of their school.

8.2.5 Topological and algebraic groups

An important elaboration of the concept of a group occurs if G is endowed with additional structure, notably, of a topological space, differentiable manifold, or algebraic variety. If the group operations m (multiplication) and i (inversion),

$$m : G \times G \to G, (g, h) \mapsto gh, \quad i : G \to G, g \mapsto g^{-1},$$

are compatible with this structure, i.e. are continuous, smooth or regular (in the sense of algebraic geometry) maps, then G becomes a topological group, a Lie group, or an algebraic group.[2]

The presence of extra structure relates these types of groups with other mathematical disciplines and means that more tools are available in their study. Topological groups form a natural domain for abstract harmonic analysis, whereas Lie groups (frequently realized as transformation groups) are the mainstays of differential geometry and unitary representation theory. Certain classification questions that cannot be solved in general can be approached and resolved for special subclasses of groups. Thus, compact connected Lie groups have been completely classified. There is a fruitful relation between infinite abstract groups and topological groups: whenever a group Γ can be realized as a lattice in a topological group G, the geometry and analysis pertaining to G yield important results about Γ. A comparatively recent trend in the theory of finite groups exploits their connections with compact topological groups (profinite groups): for example, a single p-adic analytic group G has a family of quotients which are finite p-groups of various orders, and properties of G translate into the properties of its finite quotients.

8.3 Branches of group theory

8.3.1 Finite group theory

Main article: Finite group

During the twentieth century, mathematicians investigated some aspects of the theory of finite groups in great depth, especially the local theory of finite groups and the theory of solvable and nilpotent groups. As a consequence, the complete classification of finite simple groups was achieved, meaning that all those simple groups from which all finite groups can be built are now known.

During the second half of the twentieth century, mathematicians such as Chevalley and Steinberg also increased our understanding of finite analogs of classical groups, and other related groups. One such family of groups is the family of general linear groups over finite fields. Finite groups often occur when considering symmetry of mathematical or physical objects, when those objects admit just a finite number of structure-preserving transformations. The theory of Lie groups, which may be viewed as dealing with "continuous symmetry", is strongly influenced by the associated Weyl groups. These are finite groups generated by reflections which act on a finite-dimensional Euclidean space. The properties of finite groups can thus play a role in subjects such as theoretical physics and chemistry.

8.3.2 Representation of groups

Main article: Representation theory

Saying that a group G *acts* on a set X means that every element of G defines a bijective map on the set X in a way compatible with the group structure. When X has more structure, it is useful to restrict this notion further: a representation of G on a vector space V is a group homomorphism:

$$\varrho : G \to \mathrm{GL}(V),$$

where $\mathrm{GL}(V)$ consists of the invertible linear transformations of V. In other words, to every group element g is assigned an automorphism $\varrho(g)$ such that $\varrho(g) \circ \varrho(h) = \varrho(gh)$ for any h in G.

This definition can be understood in two directions, both of which give rise to whole new domains of mathematics.[3] On the one hand, it may yield new information about the group G: often, the group operation in G is abstractly given, but via ϱ, it corresponds to the multiplication of matrices, which is very explicit.[4] On the other hand, given a well-understood group acting on a complicated object, this simplifies the study of the object in question. For example, if G is finite, it is known that V above decomposes into irreducible parts. These parts in turn are much more easily manageable than the whole V (via Schur's lemma).

Given a group G, representation theory then asks what representations of G exist. There are several settings, and the employed methods and obtained results are rather different in every case: representation theory of finite groups and representations of Lie groups are two main subdomains of the theory. The totality of representations is governed by the group's characters. For example, Fourier polynomials can be interpreted as the characters of U(1), the group of complex numbers of absolute value 1, acting on the L^2-space of periodic functions.

8.3.3   Lie theory

Main article: Lie group

A Lie group is a group that is also a differentiable manifold, with the property that the group operations are compatible with the smooth structure. Lie groups are named after Sophus Lie, who laid the foundations of the theory of continuous transformation groups. The term *groupes de Lie* first appeared in French in 1893 in the thesis of Lie's student Arthur Tresse, page 3.[5]

Lie groups represent the best-developed theory of continuous symmetry of mathematical objects and structures, which makes them indispensable tools for many parts of contemporary mathematics, as well as for modern theoretical physics. They provide a natural framework for analysing the continuous symmetries of differential equations (differential Galois theory), in much the same way as permutation groups are used in Galois theory for analysing the discrete symmetries of algebraic equations. An extension of Galois theory to the case of continuous symmetry groups was one of Lie's principal motivations.

8.3.4   Combinatorial and geometric group theory

Main article: Geometric group theory

Groups can be described in different ways. Finite groups can be described by writing down the group table consisting of all possible multiplications $g \cdot h$. A more compact way of defining a group is by *generators and relations*, also called the *presentation* of a group. Given any set F of generators $\{g_i\}_{i \in I}$, the free group generated by F subjects onto the group G. The kernel of this map is called subgroup of relations, generated by some subset D. The presentation is usually denoted by $\langle F \mid D \rangle$. For example, the group $\mathbf{Z} = \langle a \mid \rangle$ can be generated by one element a (equal to +1 or −1) and no relations, because $n \cdot 1$ never equals 0 unless n is zero. A string consisting of generator symbols and their inverses is called a *word*.

Combinatorial group theory studies groups from the perspective of generators and relations.[6] It is particularly useful where finiteness assumptions are satisfied, for example finitely generated groups, or finitely presented groups (i.e. in addition the relations are finite). The area makes use of the connection of graphs via their fundamental groups. For example, one can show that every subgroup of a free group is free.

There are several natural questions arising from giving a group by its presentation. The *word problem* asks whether two words are effectively the same group element. By relating the problem to Turing machines, one can show that there is in general no algorithm solving this task. Another, generally harder, algorithmically insoluble problem is the group isomorphism problem, which asks whether two groups given by different presentations are actually isomorphic. For example the additive group $\mathbf{Z}$ of integers can also be presented by

$$\langle x, y \mid xyxyx = e \rangle;$$

it may not be obvious that these groups are isomorphic.[7]

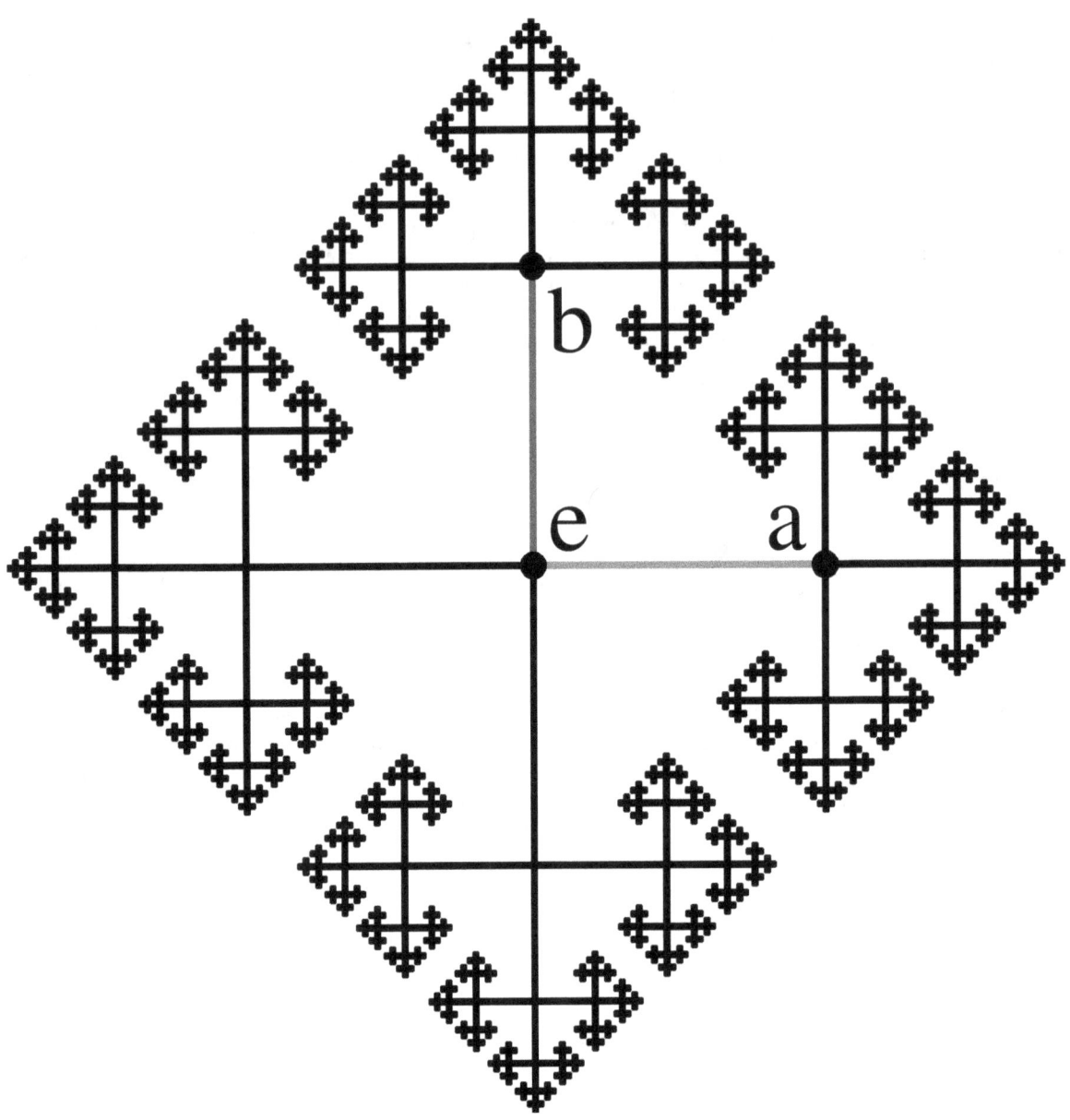

The Cayley graph of ⟨x, y | ⟩, the free group of rank 2.

Geometric group theory attacks these problems from a geometric viewpoint, either by viewing groups as geometric objects, or by finding suitable geometric objects a group acts on.[8] The first idea is made precise by means of the Cayley graph, whose vertices correspond to group elements and edges correspond to right multiplication in the group. Given two elements, one constructs the word metric given by the length of the minimal path between the elements. A theorem of Milnor and Svarc then says that given a group G acting in a reasonable manner on a metric space X, for example a compact manifold, then G is quasi-isometric (i.e. looks similar from a distance) to the space X.

8.4 Connection of groups and symmetry

Main article: Symmetry group

Given a structured object X of any sort, a symmetry is a mapping of the object onto itself which preserves the structure. This occurs in many cases, for example

1. If X is a set with no additional structure, a symmetry is a bijective map from the set to itself, giving rise to permutation groups.

2. If the object X is a set of points in the plane with its metric structure or any other metric space, a symmetry is a bijection of the set to itself which preserves the distance between each pair of points (an isometry). The corresponding group is called isometry group of X.

3. If instead angles are preserved, one speaks of conformal maps. Conformal maps give rise to Kleinian groups, for example.

4. Symmetries are not restricted to geometrical objects, but include algebraic objects as well. For instance, the equation

$$x^2 - 3 = 0$$

has the two solutions $+\sqrt{3}$, and $-\sqrt{3}$. In this case, the group that exchanges the two roots is the Galois group belonging to the equation. Every polynomial equation in one variable has a Galois group, that is a certain permutation group on its roots.

The axioms of a group formalize the essential aspects of symmetry. Symmetries form a group: they are closed because if you take a symmetry of an object, and then apply another symmetry, the result will still be a symmetry. The identity keeping the object fixed is always a symmetry of an object. Existence of inverses is guaranteed by undoing the symmetry and the associativity comes from the fact that symmetries are functions on a space, and composition of functions are associative.

Frucht's theorem says that every group is the symmetry group of some graph. So every abstract group is actually the symmetries of some explicit object.

The saying of "preserving the structure" of an object can be made precise by working in a category. Maps preserving the structure are then the morphisms, and the symmetry group is the automorphism group of the object in question.

8.5   Applications of group theory

Applications of group theory abound. Almost all structures in abstract algebra are special cases of groups. Rings, for example, can be viewed as abelian groups (corresponding to addition) together with a second operation (corresponding to multiplication). Therefore, group theoretic arguments underlie large parts of the theory of those entities.

8.5.1   Galois theory

Main article: Galois theory

Galois theory uses groups to describe the symmetries of the roots of a polynomial (or more precisely the automorphisms of the algebras generated by these roots). The fundamental theorem of Galois theory provides a link between algebraic field extensions and group theory. It gives an effective criterion for the solvability of polynomial equations in terms of the solvability of the corresponding Galois group. For example, S_5, the symmetric group in 5 elements, is not solvable which implies that the general quintic equation cannot be solved by radicals in the way equations of lower degree can. The theory, being one of the historical roots of group theory, is still fruitfully applied to yield new results in areas such as class field theory.

8.5.2 Algebraic topology

Main article: Algebraic topology

Algebraic topology is another domain which prominently associates groups to the objects the theory is interested in. There, groups are used to describe certain invariants of topological spaces. They are called "invariants" because they are defined in such a way that they do not change if the space is subjected to some deformation. For example, the fundamental group "counts" how many paths in the space are essentially different. The Poincaré conjecture, proved in 2002/2003 by Grigori Perelman, is a prominent application of this idea. The influence is not unidirectional, though. For example, algebraic topology makes use of Eilenberg–MacLane spaces which are spaces with prescribed homotopy groups. Similarly algebraic K-theory relies in a way on classifying spaces of groups. Finally, the name of the torsion subgroup of an infinite group shows the legacy of topology in group theory.

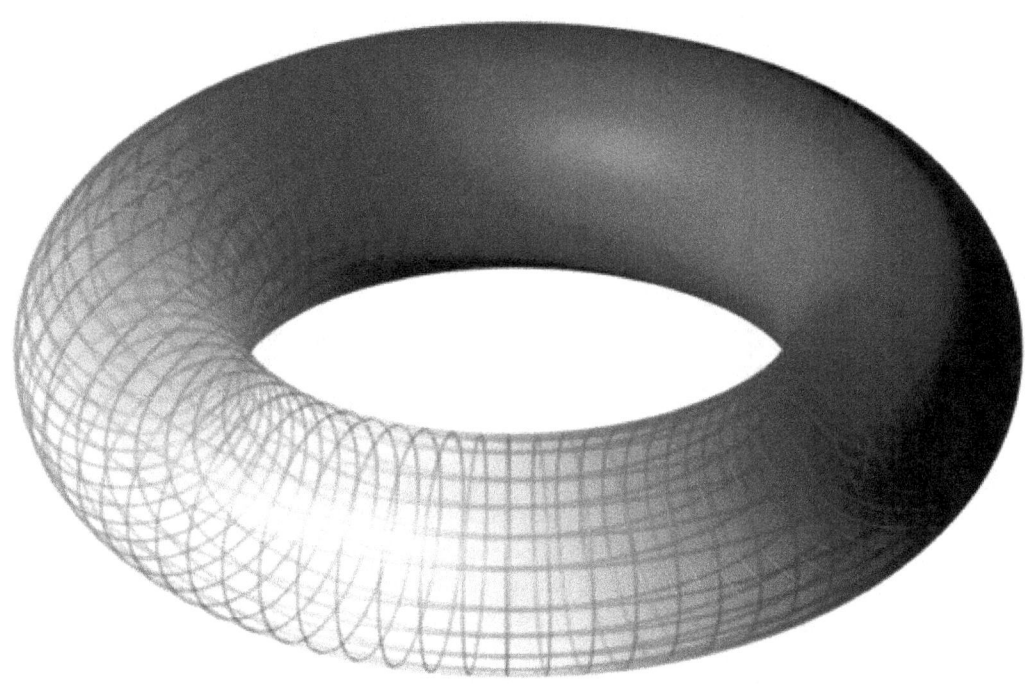

A torus. Its abelian group structure is induced from the map $C \to C/Z + \tau Z$, where τ is a parameter living in the upper half plane.

8.5.3 Algebraic geometry and cryptography

Main articles: Algebraic geometry and Cryptography

Algebraic geometry and cryptography likewise uses group theory in many ways. Abelian varieties have been introduced above. The presence of the group operation yields additional information which makes these varieties particularly accessible. They also often serve as a test for new conjectures.[9] The one-dimensional case, namely elliptic curves is studied in particular detail. They are both theoretically and practically intriguing.[10] Very large groups of prime order constructed in Elliptic-Curve Cryptography serve for public key cryptography. Cryptographical methods of this kind benefit from the flexibility of the geometric objects, hence their group structures, together with the complicated structure of these groups, which make the discrete logarithm very hard to calculate. One of the earliest encryption protocols, Caesar's cipher, may also be interpreted as a (very easy) group operation. In another direction, toric varieties are algebraic varieties

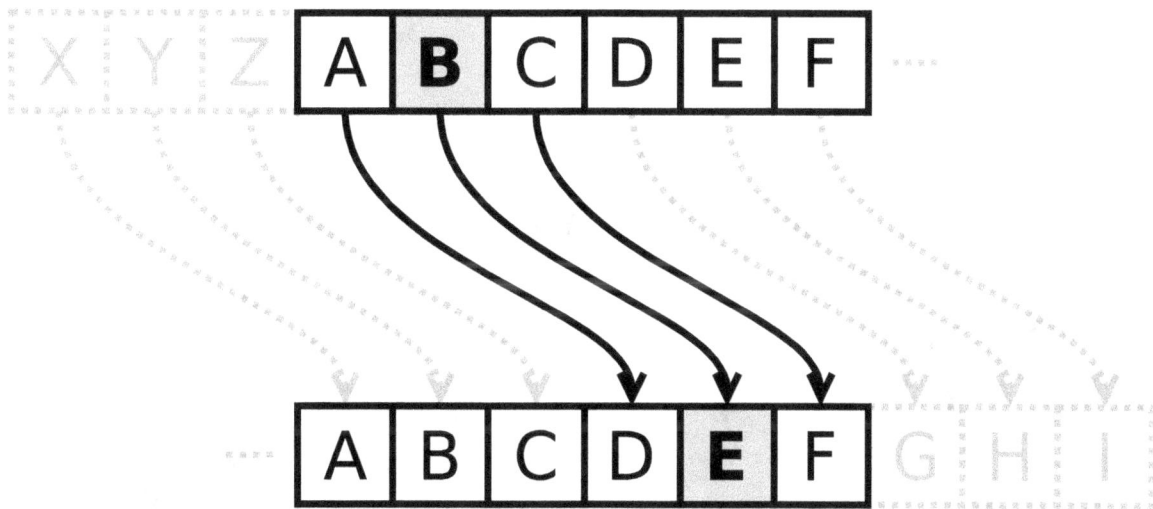

The cyclic group $\mathbf{Z}_{26}$ underlies Caesar's cipher.

acted on by a torus. Toroidal embeddings have recently led to advances in algebraic geometry, in particular resolution of singularities.[11]

8.5.4 Algebraic number theory

Main article: Algebraic number theory

Algebraic number theory is a special case of group theory, thereby following the rules of the latter. For example, Euler's product formula

$$\sum_{n \geq 1} \frac{1}{n^s} = \prod_{p\,prime} \frac{1}{1 - p^{-s}}$$

captures the fact that any integer decomposes in a unique way into primes. The failure of this statement for more general rings gives rise to class groups and regular primes, which feature in Kummer's treatment of Fermat's Last Theorem.

8.5.5 Harmonic analysis

Main article: Harmonic analysis

Analysis on Lie groups and certain other groups is called harmonic analysis. Haar measures, that is, integrals invariant under the translation in a Lie group, are used for pattern recognition and other image processing techniques.[12]

8.5.6 Combinatorics

In combinatorics, the notion of permutation group and the concept of group action are often used to simplify the counting of a set of objects; see in particular Burnside's lemma.

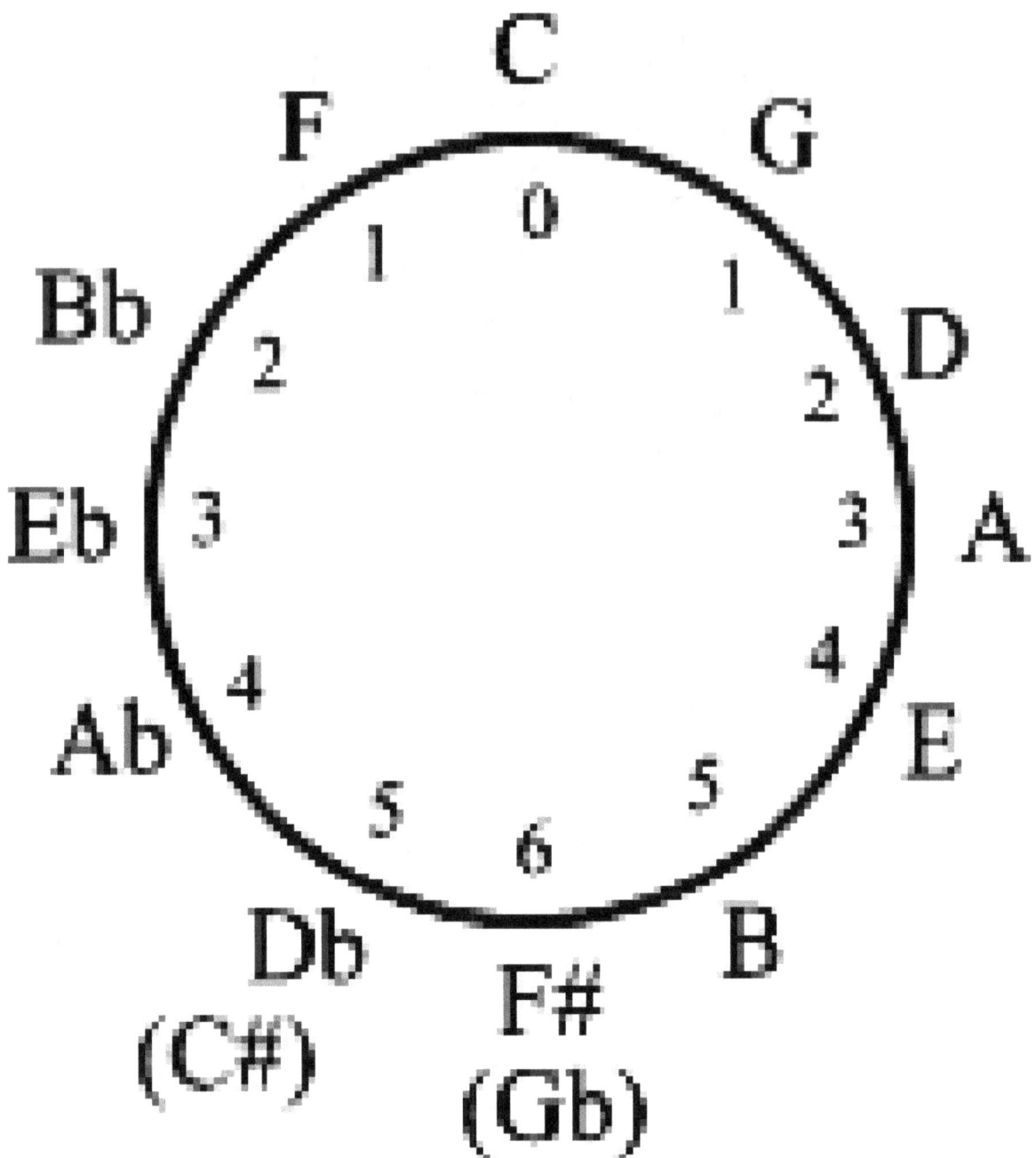

The circle of fifths may be endowed with a cyclic group structure

8.5.7 Music

The presence of the 12-periodicity in the circle of fifths yields applications of elementary group theory in musical set theory.

8.5.8 Physics

In physics, groups are important because they describe the symmetries which the laws of physics seem to obey. According to Noether's theorem, every continuous symmetry of a physical system corresponds to a conservation law of the system.

Physicists are very interested in group representations, especially of Lie groups, since these representations often point the way to the "possible" physical theories. Examples of the use of groups in physics include the Standard Model, gauge theory, the Lorentz group, and the Poincaré group.

8.5.9   Chemistry and materials science

In chemistry and materials science, groups are used to classify crystal structures, regular polyhedra, and the symmetries of molecules. The assigned point groups can then be used to determine physical properties (such as polarity and chirality), spectroscopic properties (particularly useful for Raman spectroscopy and infrared spectroscopy), and to construct molecular orbitals.

Molecular symmetry is responsible for many physical and spectroscopic properties of compounds and provides relevant information about how chemical reactions occur. In order to assign a point group for any given molecule, it is necessary to find the set of symmetry operations present on it. The symmetry operation is an action, such as a rotation around an axis or a reflection through a mirror plane. In other words, it is an operation that moves the molecule such that it is indistinguishable from the original configuration. In group theory, the rotation axes and mirror planes are called "symmetry elements". These elements can be a point, line or plane with respect to which the symmetry operation is carried out. The symmetry operations of a molecule determine the specific point group for this molecule.

In chemistry, there are five important symmetry operations. The identity operation (E) consists of leaving the molecule as it is. This is equivalent to any number of full rotations around any axis. This is a symmetry of all molecules, whereas the symmetry group of a chiral molecule consists of only the identity operation. Rotation around an axis (Cn) consists of rotating the molecule around a specific axis by a specific angle. For example, if a water molecule rotates 180° around the axis that passes through the oxygen atom and between the hydrogen atoms, it is in the same configuration as it started. In this case, $n = 2$, since applying it twice produces the identity operation. Other symmetry operations are: reflection, inversion and improper rotation (rotation followed by reflection).[13]

8.6   See also

- Group (mathematics)

- Glossary of group theory

- List of group theory topics

8.7   Notes

[1]   - Elwes, Richard, "An enormous theorem: the classification of finite simple groups," *Plus Magazine*, Issue 41, December 2006.

[2]   This process of imposing extra structure has been formalized through the notion of a group object in a suitable category. Thus Lie groups are group objects in the category of differentiable manifolds and affine algebraic groups are group objects in the category of affine algebraic varieties.

[3]   Such as group cohomology or equivariant K-theory.

[4]   In particular, if the representation is faithful.

[5]   Arthur Tresse (1893). "Sur les invariants différentiels des groupes continus de transformations". *Acta Mathematica* **18**: 1–88. doi:10.1007/bf02418270.

[6]   Schupp & Lyndon 2001

[7]   Writing $z = xy$, one has $G = \langle z, y \mid z^3 = y\square = \square z\square\rangle$.

[8]   La Harpe 2000

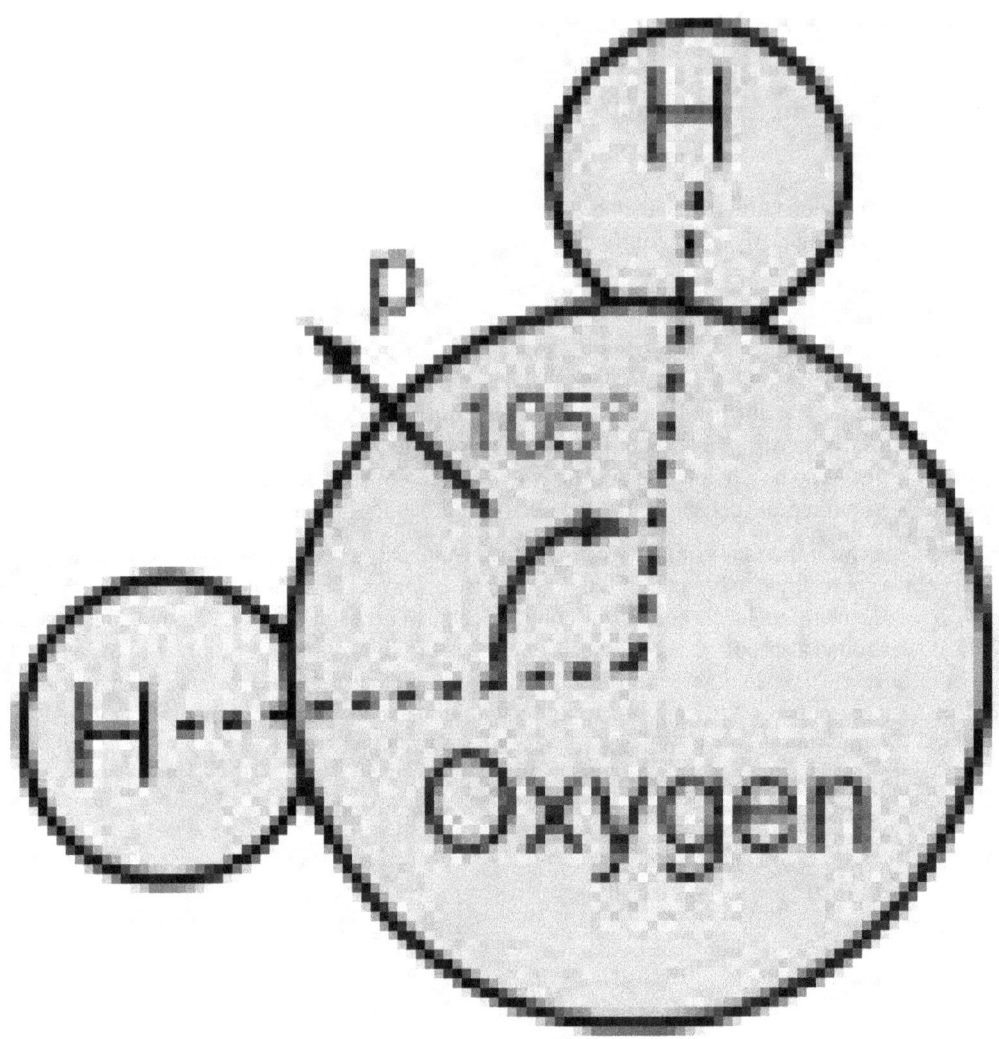

Water molecule with symmetry axis

[9] For example the Hodge conjecture (in certain cases).

[10] See the Birch-Swinnerton-Dyer conjecture, one of the millennium problems

[11] Abramovich, Dan; Karu, Kalle; Matsuki, Kenji; Wlodarczyk, Jaroslaw (2002), "Torification and factorization of birational maps", *Journal of the American Mathematical Society* **15** (3): 531–572, doi:10.1090/S0894-0347-02-00396-X, MR 1896232

[12] Lenz, Reiner (1990), *Group theoretical methods in image processing*, Lecture Notes in Computer Science **413**, Berlin, New York: Springer-Verlag, doi:10.1007/3-540-52290-5, ISBN 978-0-387-52290-6

[13] Shriver, D.F.; Atkins, P.W. Química Inorgânica, 3ª ed., Porto Alegre, Bookman, 2003.

8.8 References

- Borel, Armand (1991), *Linear algebraic groups*, Graduate Texts in Mathematics **126** (2nd ed.), Berlin, New York:

Springer-Verlag, ISBN 978-0-387-97370-8, MR 1102012

- Carter, Nathan C. (2009), *Visual group theory*, Classroom Resource Materials Series, Mathematical Association of America, ISBN 978-0-88385-757-1, MR 2504193

- Cannon, John J. (1969), "Computers in group theory: A survey", *Communications of the Association for Computing Machinery* **12**: 3–12, doi:10.1145/362835.362837, MR 0290613

- Frucht, R. (1939), "Herstellung von Graphen mit vorgegebener abstrakter Gruppe", *Compositio Mathematica* **6**: 239–50, ISSN 0010-437X

- Golubitsky, Martin; Stewart, Ian (2006), "Nonlinear dynamics of networks: the groupoid formalism", *Bull. Amer. Math. Soc. (N.S.)* **43** (03): 305–364, doi:10.1090/S0273-0979-06-01108-6, MR 2223010 Shows the advantage of generalising from group to groupoid.

- Judson, Thomas W. (1997), *Abstract Algebra: Theory and Applications* An introductory undergraduate text in the spirit of texts by Gallian or Herstein, covering groups, rings, integral domains, fields and Galois theory. Free downloadable PDF with open-source GFDL license.

- Kleiner, Israel (1986), "The evolution of group theory: a brief survey", *Mathematics Magazine* **59** (4): 195–215, doi:10.2307/2690312, ISSN 0025-570X, JSTOR 2690312, MR 863090

- La Harpe, Pierre de (2000), *Topics in geometric group theory*, University of Chicago Press, ISBN 978-0-226-31721-2

- Livio, M. (2005), *The Equation That Couldn't Be Solved: How Mathematical Genius Discovered the Language of Symmetry*, Simon & Schuster, ISBN 0-7432-5820-7 Conveys the practical value of group theory by explaining how it points to symmetries in physics and other sciences.

- Mumford, David (1970), *Abelian varieties*, Oxford University Press, ISBN 978-0-19-560528-0, OCLC 138290

- Ronan M., 2006. *Symmetry and the Monster*. Oxford University Press. ISBN 0-19-280722-6. For lay readers. Describes the quest to find the basic building blocks for finite groups.

- Rotman, Joseph (1994), *An introduction to the theory of groups*, New York: Springer-Verlag, ISBN 0-387-94285-8 A standard contemporary reference.

- Schupp, Paul E.; Lyndon, Roger C. (2001), *Combinatorial group theory*, Berlin, New York: Springer-Verlag, ISBN 978-3-540-41158-1

- Scott, W. R. (1987) [1964], *Group Theory*, New York: Dover, ISBN 0-486-65377-3 Inexpensive and fairly readable, but somewhat dated in emphasis, style, and notation.

- Shatz, Stephen S. (1972), *Profinite groups, arithmetic, and geometry*, Princeton University Press, ISBN 978-0-691-08017-8, MR 0347778

- Weibel, Charles A. (1994), *An introduction to homological algebra*, Cambridge Studies in Advanced Mathematics **38**, Cambridge University Press, ISBN 978-0-521-55987-4, OCLC 36131259, MR 1269324

8.9   External links

- History of the abstract group concept

- Higher dimensional group theory This presents a view of group theory as level one of a theory which extends in all dimensions, and has applications in homotopy theory and to higher dimensional nonabelian methods for local-to-global problems.

- Plus teacher and student package: Group Theory This package brings together all the articles on group theory from *Plus*, the online mathematics magazine produced by the Millennium Mathematics Project at the University of Cambridge, exploring applications and recent breakthroughs, and giving explicit definitions and examples of groups.

- US Naval Academy group theory guide A general introduction to group theory with exercises written by Tony Gaglione.

Chapter 9

General linear group

For other uses of "GLN", see GLN.

In mathematics, the **general linear group** (GLN) of degree n is the set of $n \times n$ invertible matrices, together with the operation of ordinary matrix multiplication. This forms a group, because the product of two invertible matrices is again invertible, and the inverse of an invertible matrix is invertible. The group is so named because the columns of an invertible matrix are linearly independent, hence the vectors/points they define are in general linear position, and matrices in the general linear group take points in general linear position to points in general linear position.

To be more precise, it is necessary to specify what kind of objects may appear in the entries of the matrix. For example, the general linear group over **R** (the set of real numbers) is the group of $n \times n$ invertible matrices of real numbers, and is denoted by GLn(**R**) or GL(n, **R**).

More generally, the general linear group of degree n over any field F (such as the complex numbers), or a ring R (such as the ring of integers), is the set of $n \times n$ invertible matrices with entries from F (or R), again with matrix multiplication as the group operation.[1] Typical notation is GLn(F) or GL(n, F), or simply GL(n) if the field is understood.

More generally still, the general linear group of a vector space GL(V) is the abstract automorphism group, not necessarily written as matrices.

The **special linear group**, written SL(n, F) or SLn(F), is the subgroup of GL(n, F) consisting of matrices with a determinant of 1.

The group GL(n, F) and its subgroups are often called **linear groups** or **matrix groups** (the abstract group GL(V) is a linear group but not a matrix group). These groups are important in the theory of group representations, and also arise in the study of spatial symmetries and symmetries of vector spaces in general, as well as the study of polynomials. The modular group may be realised as a quotient of the special linear group SL(2, **Z**).

If $n \geq 2$, then the group GL(n, F) is not abelian.

9.1   General linear group of a vector space

If V is a vector space over the field F, the general linear group of V, written GL(V) or Aut(V), is the group of all automorphisms of V, i.e. the set of all bijective linear transformations $V \rightarrow V$, together with functional composition as group operation. If V has finite dimension n, then GL(V) and GL(n, F) are isomorphic. The isomorphism is not canonical; it depends on a choice of basis in V. Given a basis (e_1, ..., en) of V and an automorphism T in GL(V), we have

$$Te_k = \sum_{j=1}^{n} a_{jk} e_j$$

for some constants *ajk* in *F*; the matrix corresponding to *T* is then just the matrix with entries given by the *ajk*.

In a similar way, for a commutative ring *R* the group GL(*n*, *R*) may be interpreted as the group of automorphisms of a *free R*-module *M* of rank *n*. One can also define GL(*M*) for any *R*-module, but in general this is not isomorphic to GL(*n*, *R*) (for any *n*).

9.2 In terms of determinants

Over a field *F*, a matrix is invertible if and only if its determinant is nonzero. Therefore an alternative definition of GL(*n*, *F*) is as the group of matrices with nonzero determinant.

Over a commutative ring *R*, one must be slightly more careful: a matrix over *R* is invertible if and only if its determinant is a unit in *R*, that is, if its determinant is invertible in *R*. Therefore GL(*n*, *R*) may be defined as the group of matrices whose determinants are units.

Over a non-commutative ring *R*, determinants are not at all well behaved. In this case, GL(*n*, *R*) may be defined as the unit group of the matrix ring M(*n*, *R*).

9.3 As a Lie group

9.3.1 Real case

The general linear group GL(*n*,**R**) over the field of real numbers is a real Lie group of dimension n^2. To see this, note that the set of all *n*×*n* real matrices, M*n*(**R**), forms a real vector space of dimension n^2. The subset GL(*n*,**R**) consists of those matrices whose determinant is non-zero. The determinant is a polynomial map, and hence GL(*n*,**R**) is an open affine subvariety of M*n*(**R**) (a non-empty open subset of M*n*(**R**) in the Zariski topology), and therefore[2] a smooth manifold of the same dimension.

The Lie algebra of GL(*n*,**R**), denoted $\mathfrak{gl}_n$, consists of all *n*×*n* real matrices with the commutator serving as the Lie bracket.

As a manifold, GL(*n*,**R**) is not connected but rather has two connected components: the matrices with positive determinant and the ones with negative determinant. The identity component, denoted by GL$^+$(*n*, **R**), consists of the real *n*×*n* matrices with positive determinant. This is also a Lie group of dimension n^2; it has the same Lie algebra as GL(*n*,**R**).

The group GL(*n*,**R**) is also noncompact. "The"[3] maximal compact subgroup of GL(*n*, **R**) is the orthogonal group O(*n*), while "the" maximal compact subgroup of GL$^+$(*n*, **R**) is the special orthogonal group SO(*n*). As for SO(*n*), the group GL$^+$(*n*, **R**) is not simply connected (except when *n* = 1), but rather has a fundamental group isomorphic to **Z** for *n* = 2 or $\mathbf{Z}_2$ for *n* > 2.

9.3.2 Complex case

The general linear GL(*n*,**C**) over the field of complex numbers is a *complex* Lie group of complex dimension n^2. As a real Lie group it has dimension $2n^2$. The set of all real matrices forms a real Lie subgroup. These correspond to the inclusions

$$GL(n,\mathbf{R}) < GL(n,\mathbf{C}) < GL(2n,\mathbf{R}),$$

which have real dimensions n^2, $2n^2$, and $4n^2 = (2n)^2$. Complex *n*-dimensional matrices can be characterized as real 2*n*-dimensional matrices that preserve a linear complex structure — concretely, that commute with a matrix *J* such that $J^2 = -I$, where *J* corresponds to multiplying by the imaginary unit *i*.

The Lie algebra corresponding to GL(*n*,**C**) consists of all *n*×*n* complex matrices with the commutator serving as the Lie bracket.

Unlike the real case, GL(n,**C**) is connected. This follows, in part, since the multiplicative group of complex numbers **C*** is connected. The group manifold GL(n,**C**) is not compact; rather its maximal compact subgroup is the unitary group U(n). As for U(n), the group manifold GL(n,**C**) is not simply connected but has a fundamental group isomorphic to **Z**.

9.4   Over finite fields

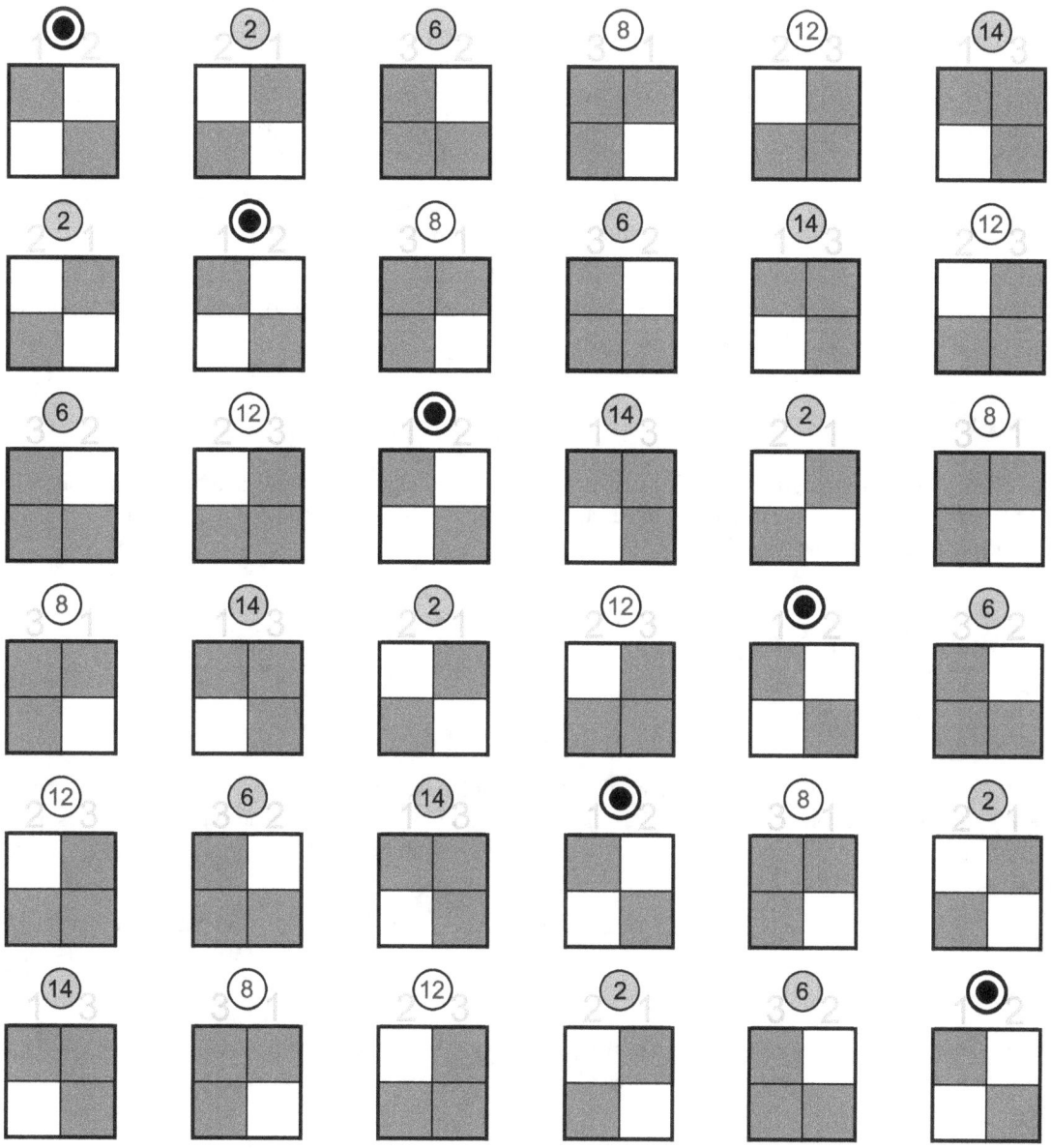

Cayley table of GL(2,2), which is isomorphic to S_3.

If F is a finite field with q elements, then we sometimes write GL(n,q) instead of GL(n,F). When p is prime, GL(n,p) is the outer automorphism group of the group **Z**n
p, and also the automorphism group, because **Z**n
p is Abelian, so the inner automorphism group is trivial.

The order of GL(n, q) is:

$$(q^n - 1)(q^n - q)(q^n - q^2) \cdots (q^n - q^{n-1})$$

This can be shown by counting the possible columns of the matrix: the first column can be anything but the zero vector; the second column can be anything but the multiples of the first column; and in general, the kth column can be any vector not in the linear span of the first $k - 1$ columns. In q-analog notation, this is $[n]_q!(q - 1)^n q^{\binom{n}{2}}$.

For example, GL(3,2) has order $(8 - 1)(8 - 2)(8 - 4) = 168$. It is the automorphism group of the Fano plane and of the group $\mathbf{Z}3$
2, and is also known as PSL(2,7).

More generally, one can count points of Grassmannian over F: in other words the number of subspaces of a given dimension k. This requires only finding the order of the stabilizer subgroup of one such subspace and dividing into the formula just given, by the orbit-stabilizer theorem.

These formulas are connected to the Schubert decomposition of the Grassmannian, and are q-analogs of the Betti numbers of complex Grassmannians. This was one of the clues leading to the Weil conjectures.

Note that in the limit $q \mapsto 1$ the order of GL(n,q) goes to 0! — but under the correct procedure (dividing by $(q-1)^n$) we see that it is the order of the symmetric group (See Lorscheid's article) — in the philosophy of the field with one element, one thus interprets the symmetric group as the general linear group over the field with one element: $S_n \cong$ GL(n,1).

9.4.1 History

The general linear group over a prime field, GL(v,p), was constructed and its order computed by Évariste Galois in 1832, in his last letter (to Chevalier) and second (of three) attached manuscripts, which he used in the context of studying the Galois group of the general equation of order p^v.[4]

9.5 Special linear group

Main article: Special linear group

The special linear group, SL(n,F), is the group of all matrices with determinant 1. They are special in that they lie on a subvariety — they satisfy a polynomial equation (as the determinant is a polynomial in the entries). Matrices of this type form a group as the determinant of the product of two matrices is the product of the determinants of each matrix. SL(n, F) is a normal subgroup of GL(n,F).

If we write $F^\times$ for the multiplicative group of F (excluding 0), then the determinant is a group homomorphism

det: GL(n,F) → $F^\times$.

that is surjective and its kernel is the special linear group. Therefore, by the first isomorphism theorem, GL(n,F)/SL(n,F) is isomorphic to $F^\times$. In fact, GL(n,F) can be written as a semidirect product:

GL(n,F) = SL(n,F) ⋊ $F^\times$

When F is $\mathbf{R}$ or $\mathbf{C}$, SL(n,F) is a Lie subgroup of GL(n,F) of dimension $n^2 - 1$. The Lie algebra of SL(n,F) consists of all $n \times n$ matrices over F with vanishing trace. The Lie bracket is given by the commutator.

The special linear group SL(n,$\mathbf{R}$) can be characterized as the group of *volume and orientation preserving* linear transformations of $\mathbf{R}^n$.

The group SL(n,$\mathbf{C}$) is simply connected while SL(n,$\mathbf{R}$) is not. SL(n,$\mathbf{R}$) has the same fundamental group as GL$^+$(n, $\mathbf{R}$), that is, $\mathbf{Z}$ for $n = 2$ and $\mathbf{Z}_2$ for $n > 2$.

9.6   Other subgroups

9.6.1   Diagonal subgroups

The set of all invertible diagonal matrices forms a subgroup of GL(n, F) isomorphic to $(F^\times)^n$. In fields like **R** and **C**, these correspond to rescaling the space; the so-called dilations and contractions.

A **scalar matrix** is a diagonal matrix which is a constant times the identity matrix. The set of all nonzero scalar matrices forms a subgroup of GL(n, F) isomorphic to $F^\times$. This group is the center of GL(n, F). In particular, it is a normal, abelian subgroup.

The center of SL(n, F) is simply the set of all scalar matrices with unit determinant, and is isomorphic to the group of nth roots of unity in the field F.

9.6.2   Classical groups

The so-called classical groups are subgroups of GL(V) which preserve some sort of bilinear form on a vector space V. These include the

- **orthogonal group**, O(V), which preserves a non-degenerate quadratic form on V,

- **symplectic group**, Sp(V), which preserves a symplectic form on V (a non-degenerate alternating form),

- **unitary group**, U(V), which, when $F = $ **C**, preserves a non-degenerate hermitian form on V.

These groups provide important examples of Lie groups.

9.7   Related groups and monoids

9.7.1   Projective linear group

Main article: Projective linear group

The projective linear group PGL(n, F) and the projective special linear group PSL(n,F) are the quotients of GL(n,F) and SL(n,F) by their centers (which consist of the multiples of the identity matrix therein); they are the induced action on the associated projective space.

9.7.2   Affine group

Main article: Affine group

The affine group Aff(n,F) is an extension of GL(n,F) by the group of translations in F^n. It can be written as a semidirect product:

$$\text{Aff}(n, F) = \text{GL}(n, F) \ltimes F^n$$

where GL(n, F) acts on F^n in the natural manner. The affine group can be viewed as the group of all affine transformations of the affine space underlying the vector space F^n.

One has analogous constructions for other subgroups of the general linear group: for instance, the special affine group is the subgroup defined by the semidirect product, SL$(n, F) \ltimes F^n$, and the Poincaré group is the affine group associated to the Lorentz group, O$(1,3,F) \ltimes F^n$.

9.7.3 General semilinear group

Main article: General semilinear group

The general semilinear group $\Gamma L(n, F)$ is the group of all invertible semilinear transformations, and contains GL. A semilinear transformation is a transformation which is linear "up to a twist", meaning "up to a field automorphism under scalar multiplication". It can be written as a semidirect product:

$$\Gamma L(n, F) = \text{Gal}(F) \ltimes GL(n, F)$$

where $\text{Gal}(F)$ is the Galois group of F (over its prime field), which acts on $GL(n, F)$ by the Galois action on the entries.

The main interest of $\Gamma L(n, F)$ is that the associated projective semilinear group $P\Gamma L(n, F)$ (which contains $PGL(n, F)$) is the collineation group of projective space, for $n > 2$, and thus semilinear maps are of interest in projective geometry.

9.7.4 Full linear monoid

If one removes the restriction of the determinant being non-zero, the resulting algebraic structure is a monoid, usually called the **full linear monoid**,[5][6][7] but occasionally also *full linear semigroup*,[8] *general linear monoid*[9][10] etc. It is actually a regular semigroup.[6]

9.8 Infinite general linear group

The **infinite general linear group** or **stable general linear group** is the direct limit of the inclusions $GL(n, F) \to GL(n+1, F)$ as the upper left block matrix. It is denoted by either $GL(F)$ or $GL(\infty, F)$, and can also be interpreted as invertible infinite matrices which differ from the identity matrix in only finitely many places.[11]

It is used in algebraic K-theory to define K_1, and over the reals has a well-understood topology, thanks to Bott periodicity.

It should not be confused with the space of (bounded) invertible operators on a Hilbert space, which is a larger group, and topologically much simpler, namely contractible — see Kuiper's theorem.

9.9 See also

- List of finite simple groups

- $SL_2(\mathbf{R})$

- Representation theory of $SL_2(\mathbf{R})$

9.10 Notes

[1] Here rings are assumed to be associative and unital.

[2] Since the Zariski topology is coarser than the metric topology; equivalently, polynomial maps are continuous.

[3] A maximal compact subgroup is not unique, but is essentially unique, hence one often refers to "the" maximal compact subgroup.

[4] Galois, Évariste (1846). "Lettre de Galois à M. Auguste Chevalier". *Journal de Mathématiques Pures et Appliquées* **XI**: 408–415. Retrieved 2009-02-04, GL(v,p) discussed on p. 410.

[5] Jan Okniński (1998). *Semigroups of Matrices*. World Scientific. Chapter 2: Full linear monoid. ISBN 978-981-02-3445-4.

[6] Meakin (2007). "Groups and Semigroups: Connections and contrast". In C. M. Campbell. *Groups St Andrews 2005*. Cambridge University Press. p. 471. ISBN 978-0-521-69470-4.

[7] John Rhodes; Benjamin Steinberg (2009). *The q-theory of Finite Semigroups*. Springer Science & Business Media. p. 306. ISBN 978-0-387-09781-7.

[8] Eric Jespers; Jan Okniski (2007). *Noetherian Semigroup Algebras*. Springer Science & Business Media. 2.3: Full linear semigroup. ISBN 978-1-4020-5810-3.

[9] Meinolf Geck (2013). *An Introduction to Algebraic Geometry and Algebraic Groups*. Oxford University Press. p. 132. ISBN 978-0-19-967616-3.

[10] Mahir Bilen Can; Zhenheng Li; Benjamin Steinberg; Qiang Wang (2014). *Algebraic Monoids, Group Embeddings, and Algebraic Combinatorics*. Springer. p. 142. ISBN 978-1-4939-0938-4.

[11] Milnor, John Willard (1971). *Introduction to algebraic K-theory*. Annals of Mathematics Studies **72**. Princeton, NJ: Princeton University Press. p. 25. MR 0349811. Zbl 0237.18005.

9.11   External links

- Hazewinkel, Michiel, ed. (2001), "General linear group", *Encyclopedia of Mathematics*, Springer, ISBN 978-1-55608-010-4

- "GL(2,*p*) and GL(3,3) Acting on Points" by Ed Pegg, Jr., Wolfram Demonstrations Project, 2007.

Chapter 10

Algebraic group

In algebraic geometry, an **algebraic group** (or **group variety**) is a group that is an algebraic variety, such that the multiplication and inversion operations are given by regular functions on the variety.

In terms of category theory, an algebraic group is a group object in the category of algebraic varieties.

10.1   Classes

Several important classes of groups are algebraic groups, including:

- Finite groups

- GL(n, **C**), the general linear group of invertible matrices over **C**

- Jet group

- Elliptic curves.

Two important classes of algebraic groups arise, that for the most part are studied separately: *abelian varieties* (the 'projective' theory) and *linear algebraic groups* (the 'affine' theory). There are certainly examples that are neither one nor the other — these occur for example in the modern theory of integrals of the second and third kinds such as the Weierstrass zeta function, or the theory of generalized Jacobians. But according to Chevalley's structure theorem any algebraic group is an extension of an abelian variety by a linear algebraic group. This is a result of Claude Chevalley: if K is a perfect field, and G an algebraic group over K, there exists a unique normal closed subgroup H in G, such that H is a linear group and G/H an abelian variety.

According to another basic theorem, any group in the category of affine varieties has a faithful finite-dimensional linear representation: we can consider it to be a matrix group over K, defined by polynomials over K and with matrix multiplication as the group operation. For that reason a concept of *affine algebraic group* is redundant over a field — we may as well use a very concrete definition. Note that this means that algebraic group is narrower than Lie group, when working over the field of real numbers: there are examples such as the universal cover of the 2×2 special linear group that are Lie groups, but have no faithful linear representation. A more obvious difference between the two concepts arises because the identity component of an affine algebraic group G is necessarily of finite index in G.

When one wants to work over a base ring R (commutative), there is the group scheme concept: that is, a group object in the category of schemes over R. *Affine group scheme* is the concept dual to a type of Hopf algebra. There is quite a refined theory of group schemes, that enters for example in the contemporary theory of abelian varieties.

10.2   Algebraic subgroup

An **algebraic subgroup** of an algebraic group is a Zariski closed subgroup. Generally these are taken to be connected (or irreducible as a variety) as well.

Another way of expressing the condition is as a subgroup which is also a subvariety.

This may also be generalized by allowing schemes in place of varieties. The main effect of this in practice, apart from allowing subgroups in which the connected component is of finite index > 1, is to admit non-reduced schemes, in characteristic p.

10.3   Coxeter groups

Main article: Coxeter group
Further information: Field with one element

There are a number of analogous results between algebraic groups and Coxeter groups – for instance, the number of elements of the symmetric group is $n!$, and the number of elements of the general linear group over a finite field is the q-factorial $[n]_q!$; thus the symmetric group behaves as though it were a linear group over "the field with one element". This is formalized by the field with one element, which considers Coxeter groups to be simple algebraic groups over the field with one element.

10.4   See also

- Algebraic topology (object)

- Borel subgroup

- Tame group

- Morley rank

- Cherlin–Zilber conjecture

- Adelic algebraic group

- Glossary of algebraic groups

10.5   Notes

10.6   References

- Chevalley, Claude, ed. (1958), *Séminaire C. Chevalley, 1956-–1958. Classification des groupes de Lie algébriques*, 2 vols, Paris: Secrétariat Mathématique, MR 0106966, Reprinted as volume 3 of Chevalley's collected works.

- Humphreys, James E. (1972), *Linear Algebraic Groups*, Graduate Texts in Mathematics **21**, Berlin, New York: Springer-Verlag, ISBN 978-0-387-90108-4, MR 0396773

- Lang, Serge (1983), *Abelian varieties*, Berlin, New York: Springer-Verlag, ISBN 978-0-387-90875-5

- Milne, J. S., *Affine Group Schemes; Lie Algebras; Lie Groups; Reductive Groups; Arithmetic Subgroups*

- Mumford, David (1970), *Abelian varieties*, Oxford University Press, ISBN 978-0-19-560528-0, OCLC 138290

- Springer, Tonny A. (1998), *Linear algebraic groups*, Progress in Mathematics **9** (2nd ed.), Boston, MA: Birkhäuser Boston, ISBN 978-0-8176-4021-7, MR 1642713

- Waterhouse, William C. (1979), *Introduction to affine group schemes*, Graduate Texts in Mathematics **66**, Berlin, New York: Springer-Verlag, ISBN 978-0-387-90421-4

- Weil, André (1971), *Courbes algébriques et variétés abéliennes*, Paris: Hermann, OCLC 322901

Chapter 11

Discrete group

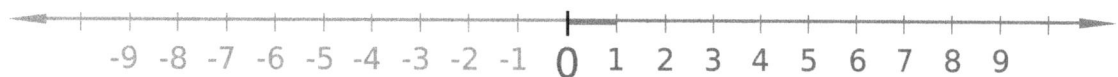

The integers with their usual topology are a discrete subgroup of the real numbers.

In mathematics, a **discrete group** is a group G equipped with the discrete topology. With this topology, G becomes a topological group. A **discrete subgroup** of a topological group G is a subgroup H whose relative topology is the discrete one. For example, the integers, **Z**, form a discrete subgroup of the reals, **R** (with the standard metric topology), but the rational numbers, **Q**, do not.

Any group can be given the discrete topology. Since every map from a discrete space is continuous, the topological homomorphisms between discrete groups are exactly the group homomorphisms between the underlying groups. Hence, there is an isomorphism between the category of groups and the category of discrete groups. Discrete groups can therefore be identified with their underlying (non-topological) groups.

There are some occasions when a topological group or Lie group is usefully endowed with the discrete topology, 'against nature'. This happens for example in the theory of the Bohr compactification, and in group cohomology theory of Lie groups.

A discrete isometry group is an isometry group such that for every point of the metric space the set of images of the point under the isometries is a discrete set. A discrete symmetry group is a symmetry group that is a discrete isometry group.

11.1   Properties

Since topological groups are homogeneous, one need only look at a single point to determine if the topological group is discrete. In particular, a topological group is discrete if and only if the singleton containing the identity is an open set.

A discrete group is the same thing as a zero-dimensional Lie group (uncountable discrete groups are not second-countable so authors who require Lie groups to satisfy this axiom do not regard these groups as Lie groups). The identity component of a discrete group is just the trivial subgroup while the group of components is isomorphic to the group itself.

Since the only Hausdorff topology on a finite set is the discrete one, a finite Hausdorff topological group must necessarily be discrete. It follows that every finite subgroup of a Hausdorff group is discrete.

A discrete subgroup H of G is **cocompact** if there is a compact subset K of G such that $HK = G$.

Discrete normal subgroups play an important role in the theory of covering groups and locally isomorphic groups. A discrete normal subgroup of a connected group G necessarily lies in the center of G and is therefore abelian.

Other properties:

- every discrete group is totally disconnected

- every subgroup of a discrete group is discrete.

- every quotient of a discrete group is discrete.

- the product of a finite number of discrete groups is discrete.

- a discrete group is compact if and only if it is finite.

- every discrete group is locally compact.

- every discrete subgroup of a Hausdorff group is closed.

- every discrete subgroup of a compact Hausdorff group is finite.

11.2   Examples

- Frieze groups and wallpaper groups are discrete subgroups of the isometry group of the Euclidean plane. Wallpaper groups are cocompact, but Frieze groups are not.

- A crystallographic group usually means a cocompact, discrete subgroup of the isometries of some Euclidean space. Sometimes, however, a crystallographic group can be a cocompact discrete subgroup of a nilpotent or solvable Lie group.

- Every triangle group T is a discrete subgroup of the isometry group of the sphere (when T is finite), the Euclidean plane (when T has a $\mathbf{Z} + \mathbf{Z}$ subgroup of finite index), or the hyperbolic plane.

- Fuchsian groups are, by definition, discrete subgroups of the isometry group of the hyperbolic plane.

  - A Fuchsian group that preserves orientation and acts on the upper half-plane model of the hyperbolic plane is a discrete subgroup of the Lie group PSL(2,$\mathbf{R}$), the group of orientation preserving isometries of the upper half-plane model of the hyperbolic plane.

  - A Fuchsian group is sometimes considered as a special case of a Kleinian group, by embedding the hyperbolic plane isometrically into three-dimensional hyperbolic space and extending the group action on the plane to the whole space.

  - The modular group PSL(2,$\mathbf{Z}$) is thought of as a discrete subgroup of PSL(2,$\mathbf{R}$). The modular group is a lattice in PSL(2,$\mathbf{R}$), but it is not cocompact.

- Kleinian groups are, by definition, discrete subgroups of the isometry group of hyperbolic 3-space. These include quasi-Fuchsian groups.

  - A Kleinian group that preserves orientation and acts on the upper half space model of hyperbolic 3-space is a discrete subgroup of the Lie group PSL(2,$\mathbf{C}$), the group of orientation preserving isometries of the upper half-space model of hyperbolic 3-space.

- A lattice in a Lie group is a discrete subgroup such that the Haar measure of the quotient space is finite.

11.3   See also

- crystallographic point group

- congruence subgroup

- arithmetic group

- geometric group theory

- computational group theory

- freely discontinuous

- free regular set

11.4 References

- Hazewinkel, Michiel, ed. (2001), "Discrete group of transformations", *Encyclopedia of Mathematics*, Springer, ISBN 978-1-55608-010-4

- Hazewinkel, Michiel, ed. (2001), "Discrete subgroup", *Encyclopedia of Mathematics*, Springer, ISBN 978-1-55608-010-4

Chapter 12

Finite group

In abstract algebra, a **finite group** is a mathematical group with a finite number of elements. A group is a set of elements together with an operation which associates, to each ordered pair of elements, an element of the set.[1] With a finite group, the set is finite.

12.1 History

During the twentieth century, mathematicians investigated some aspects of the theory of finite groups in great depth, especially the local theory of finite groups and the theory of solvable and nilpotent groups. As a consequence, the complete classification of finite simple groups was achieved, meaning that all those simple groups from which all finite groups can be built are now known.

During the second half of the twentieth century, mathematicians such as Chevalley and Steinberg also increased our understanding of finite analogs of classical groups, and other related groups. One such family of groups is the family of general linear groups over finite fields. Finite groups often occur when considering symmetry of mathematical or physical objects, when those objects admit just a finite number of structure-preserving transformations. The theory of Lie groups, which may be viewed as dealing with "continuous symmetry", is strongly influenced by the associated Weyl groups. These are finite groups generated by reflections which act on a finite-dimensional Euclidean space. The properties of finite groups can thus play a role in subjects such as theoretical physics and chemistry.

12.2 Examples

12.2.1 Permutation groups

Main article: Permutation group

The **symmetric group** Sn on a finite set of n symbols is the group whose elements are all the permutations of the n symbols, and whose group operation is the composition of such permutations, which are treated as bijective functions from the set of symbols to itself.[2] Since there are $n!$ (n factorial) possible permutations of a set of n symbols, it follows that the order (the number of elements) of the symmetric group Sn is $n!$.

12.2.2 Cyclic groups

Main article: Cyclic group

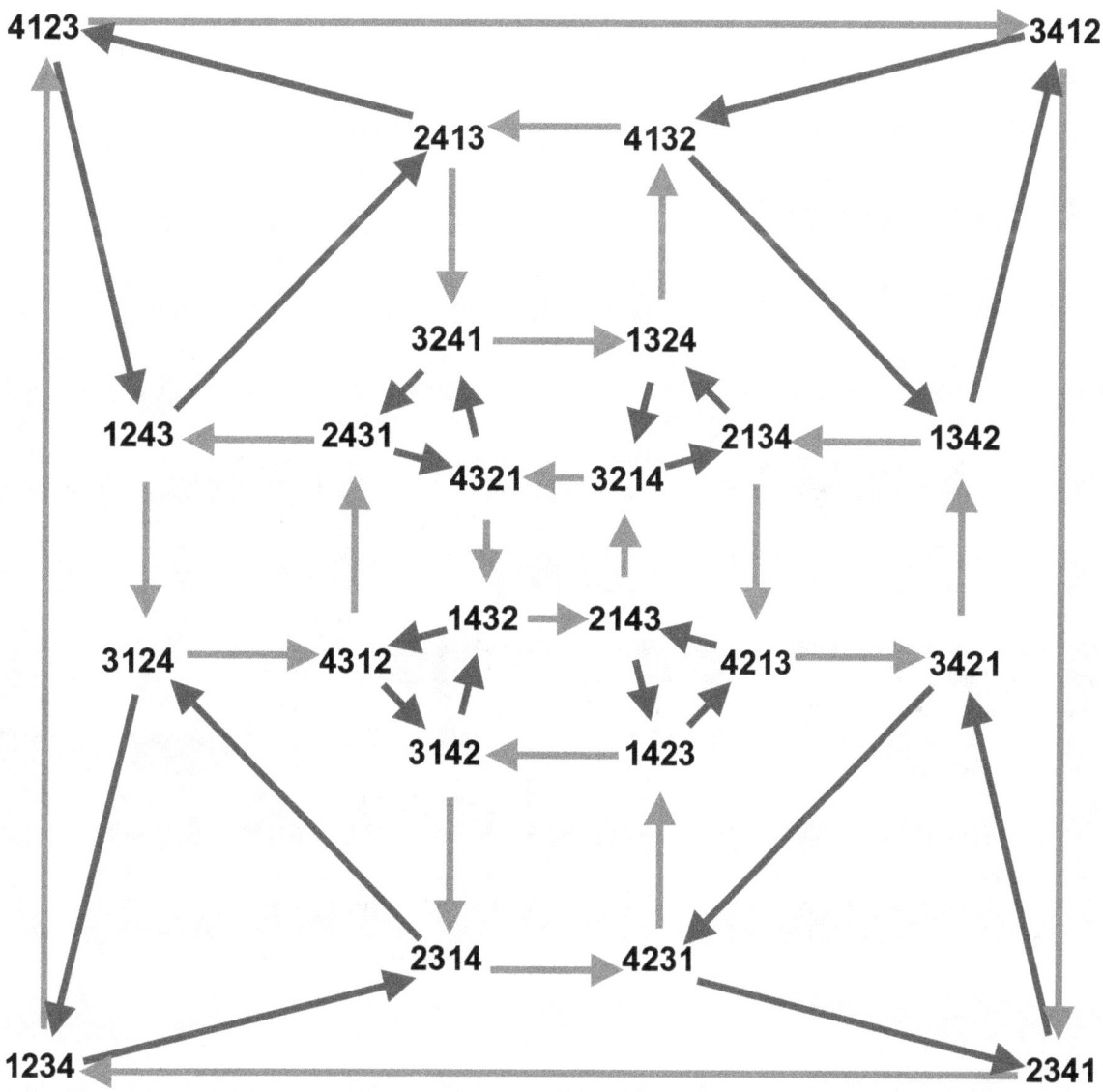

A Cayley graph of the symmetric group S$_4$

A cyclic group Zn is a group all of whose elements are powers of a particular element a where $a^n = a^0 = $ e, the identity. A typical realization of this group is as the complex nth roots of unity. Sending a to a primitive root of unity gives an isomorphism between the two. This can be done with any finite cyclic group.

12.2.3   Finite abelian groups

Main article: Finite abelian group

An **abelian group**, also called a **commutative group**, is a group in which the result of applying the group operation to two group elements does not depend on their order (the axiom of commutativity). They are named after Niels Henrik Abel.[3]

An arbitrary finite abelian group is isomorphic to a direct sum of finite cyclic groups of prime power order, and these orders are uniquely determined, forming a complete system of invariants. The automorphism group of a finite abelian group can be described directly in terms of these invariants. The theory had been first developed in the 1879 paper of Georg Frobenius and Ludwig Stickelberger and later was both simplified and generalized to finitely generated modules over a principal ideal domain, forming an important chapter of linear algebra.

12.2.4 Groups of Lie type

Main article: Group of Lie type

A **group of Lie type** is a group closely related to the group $G(k)$ of rational points of a reductive linear algebraic group G with values in the field k. Finite groups of Lie type give the bulk of nonabelian finite simple groups. Special cases include the classical groups, the Chevalley groups, the Steinberg groups, and the Suzuki–Ree groups.

Finite groups of Lie type were among the first groups to be considered in mathematics, after cyclic, symmetric and alternating groups, with the projective special linear groups over prime finite fields, $PSL(2, p)$ being constructed by Évariste Galois in the 1830s. The systematic exploration of finite groups of Lie type started with Camille Jordan's theorem that the projective special linear group $PSL(2, q)$ is simple for $q \neq 2, 3$. This theorem generalizes to projective groups of higher dimensions and gives an important infinite family $PSL(n, q)$ of finite simple groups. Other classical groups were studied by Leonard Dickson in the beginning of 20th century. In the 1950s Claude Chevalley realized that after an appropriate reformulation, many theorems about semisimple Lie groups admit analogues for algebraic groups over an arbitrary field k, leading to construction of what are now called *Chevalley groups*. Moreover, as in the case of compact simple Lie groups, the corresponding groups turned out to be almost simple as abstract groups (*Tits simplicity theorem*). Although it was known since 19th century that other finite simple groups exist (for example, Mathieu groups), gradually a belief formed that nearly all finite simple groups can be accounted for by appropriate extensions of Chevalley's construction, together with cyclic and alternating groups. Moreover, the exceptions, the sporadic groups, share many properties with the finite groups of Lie type, and in particular, can be constructed and characterized based on their *geometry* in the sense of Tits.

The belief has now become a theorem – the classification of finite simple groups. Inspection of the list of finite simple groups shows that groups of Lie type over a finite field include all the finite simple groups other than the cyclic groups, the alternating groups, the Tits group, and the 26 sporadic simple groups.

12.3 Main theorems

12.3.1 Lagrange's theorem

Main article: Lagrange's theorem (group theory)

For any finite group G, the order (number of elements) of every subgroup H of G divides the order of G. The theorem is named after Joseph-Louis Lagrange.

12.3.2 Sylow theorems

Main article: Sylow theorems

This provides a partial converse to Lagrange's theorem giving information about how many subgroups of a given order are contained in G.

12.3.3 Cayley's theorem

Main article: Cayley's theorem

Cayley's theorem, named in honour of Arthur Cayley, states that every group G is isomorphic to a subgroup of the symmetric group acting on G.[4] This can be understood as an example of the group action of G on the elements of G.[5]

12.3.4 Burnside theorem

Main article: Burnside theorem

Burnside's theorem in group theory states that if G is a finite group of order

$$p^a q^b$$

where p and q are prime numbers, and a and b are non-negative integers, then G is solvable. Hence each non-Abelian finite simple group has order divisible by at least three distinct primes.

12.3.5 Feit-Thompson theorem

The **Feit–Thompson theorem**, or **odd order theorem**, states that every finite group of odd order is solvable. It was proved by Walter Feit and John Griggs Thompson (1962, 1963)

12.3.6 Classification of finite simple groups

Main article: Classification of finite simple groups

The **classification of the finite simple groups** is a theorem stating that every finite simple group belongs to one of four categories described below. These groups can be seen as the basic building blocks of all finite groups, in a way reminiscent of the way the prime numbers are the basic building blocks of the natural numbers. The Jordan–Hölder theorem is a more precise way of stating this fact about finite groups. However, a significant difference with respect to the case of integer factorization is that such "building blocks" do not necessarily determine uniquely a group, since there might be many non-isomorphic groups with the same composition series or, put in another way, the extension problem does not have a unique solution.

The proof of the theorem consists of tens of thousands of pages in several hundred journal articles written by about 100 authors, published mostly between 1955 and 2004. Gorenstein (d.1992), Lyons, and Solomon are gradually publishing a simplified and revised version of the proof.

12.4 Number of groups of a given order

Given a positive integer n, it is not at all a routine matter to determine how many isomorphism types of groups of order n there are. Every group of prime order is cyclic, because Lagrange's theorem implies that the cyclic subgroup generated by any of its non-identity elements is the whole group. If n is the square of a prime, then there are exactly two possible isomorphism types of group of order n, both of which are abelian. If n is a higher power of a prime, then results of Graham Higman and Charles Sims give asymptotically correct estimates for the number of isomorphism types of groups of order n, and the number grows very rapidly as the power increases.

Depending on the prime factorization of n, some restrictions may be placed on the structure of groups of order n, as a consequence, for example, of results such as the Sylow theorems. For example, every group of order pq is cyclic when $q < p$ are primes with $p-1$ not divisible by q. For a necessary and sufficient condition, see cyclic number.

If n is squarefree, then any group of order n is solvable. Burnside's theorem, proved using group characters, states that every group of order n is solvable when n is divisible by fewer than three distinct primes, i.e. if $n = p^a q^b$, where p and q are prime numbers, and a and b are non-negative integers. By the Feit–Thompson theorem, which has a long and complicated proof, every group of order n is solvable when n is odd.

For every positive integer n, most groups of order n are solvable. To see this for any particular order is usually not difficult (for example, there is, up to isomorphism, one non-solvable group and 12 solvable groups of order 60) but the proof of this for all orders uses the classification of finite simple groups. For any positive integer n there are at most two simple groups of order n, and there are infinitely many positive integers n for which there are two non-isomorphic simple groups of order n.

12.4.1 Table of distinct groups of order n

Main article: oeis:A000001

12.5 See also

- Association scheme
- List of finite simple groups
- Cauchy's theorem (group theory)
- Abelian group
- Non-abelian group
- P-group
- List of small groups
- Representation theory of finite groups
- Modular representation theory
- Monstrous moonshine
- Profinite group
- Finite ring

12.6 Notes

[1] Clark, Allan (1984). *Elements of abstract algebra*. Dover. ISBN 0486647250.

[2] Jacobson 2009, p. 31

[3] Jacobson 2009, p. 41

[4] Jacobson 2009, p. 38

[5] Jacobson 2009, p. 72, ex. 1

[6] Humphreys, John F. (1996). *A Course in Group Theory*. Oxford University Press. pp. 238–242. ISBN 0198534590. Zbl 0843.20001.

12.7   Further reading

- Jacobson, Nathan (2009). *Basic Algebra I* (2nd ed.). Dover Publications. ISBN 978-0-486-47189-1.

12.8   External links

- Number of groups of order n (sequence A000001 in OEIS)

- Number of Abelian groups of order n (sequence A000688 in OEIS)

- Number of non-Abelian groups of order n (sequence A060689 in OEIS)

- A classifier for groups of small order

Chapter 13

Group action

This article is about the mathematical concept. For the sociology term, see group action (sociology).

In mathematics, a symmetry group describes all symmetries of objects. This is formalized by the notion of a **group action**: every element of the group "acts" like a bijective map (or "symmetry") on some set. In this case, the group is also called a **permutation group** (especially if the set is finite or not a vector space) or **transformation group** (especially if the set is a vector space and the group acts like linear transformations of the set). A **permutation representation** of a group G is a representation of G as a group of permutations of the set (usually if the set is finite), and may be described as a group representation of G by permutation matrices. It is the same as a group action of G on an *ordered* basis of a vector space.

A group action is an extension to the notion of a symmetry group in which every element of the group "acts" like a bijective transformation (or "symmetry") of some set, without being identified with that transformation. This allows for a more comprehensive description of the symmetries of an object, such as a polyhedron, by allowing the same group to act on several different sets of features, such as the set of vertices, the set of edges and the set of faces of the polyhedron.

If G is a group and X is a set, then a group action may be defined as a group homomorphism h from G to the symmetric group on X. The action assigns a permutation of X to each element of the group in such a way that the permutation of X assigned to

- the identity element of G is the identity transformation of X;

- a product gk of two elements of G is the composition of the permutations assigned to g and k.

The abstraction provided by group actions is a powerful one, because it allows geometrical ideas to be applied to more abstract objects. Many objects in mathematics have natural group actions defined on them. In particular, groups can act on other groups, or even on themselves. Despite this generality, the theory of group actions contains wide-reaching theorems, such as the orbit stabilizer theorem, which can be used to prove deep results in several fields.

13.1 Definition

If G is a group and X is a set, then a (*left*) *group action* φ of G on X is a function

$$\varphi : G \times X \to X : (g, x) \mapsto \varphi(g, x)$$

that satisfies the following two axioms (where we denote $\varphi(g, x)$ as $g.x$):[1]

Compatibility $(gh).x = g.(h.x)$ for all g, h in G and all x in X. (Here, gh denotes the result of applying the group operation of G to the elements g and h.)

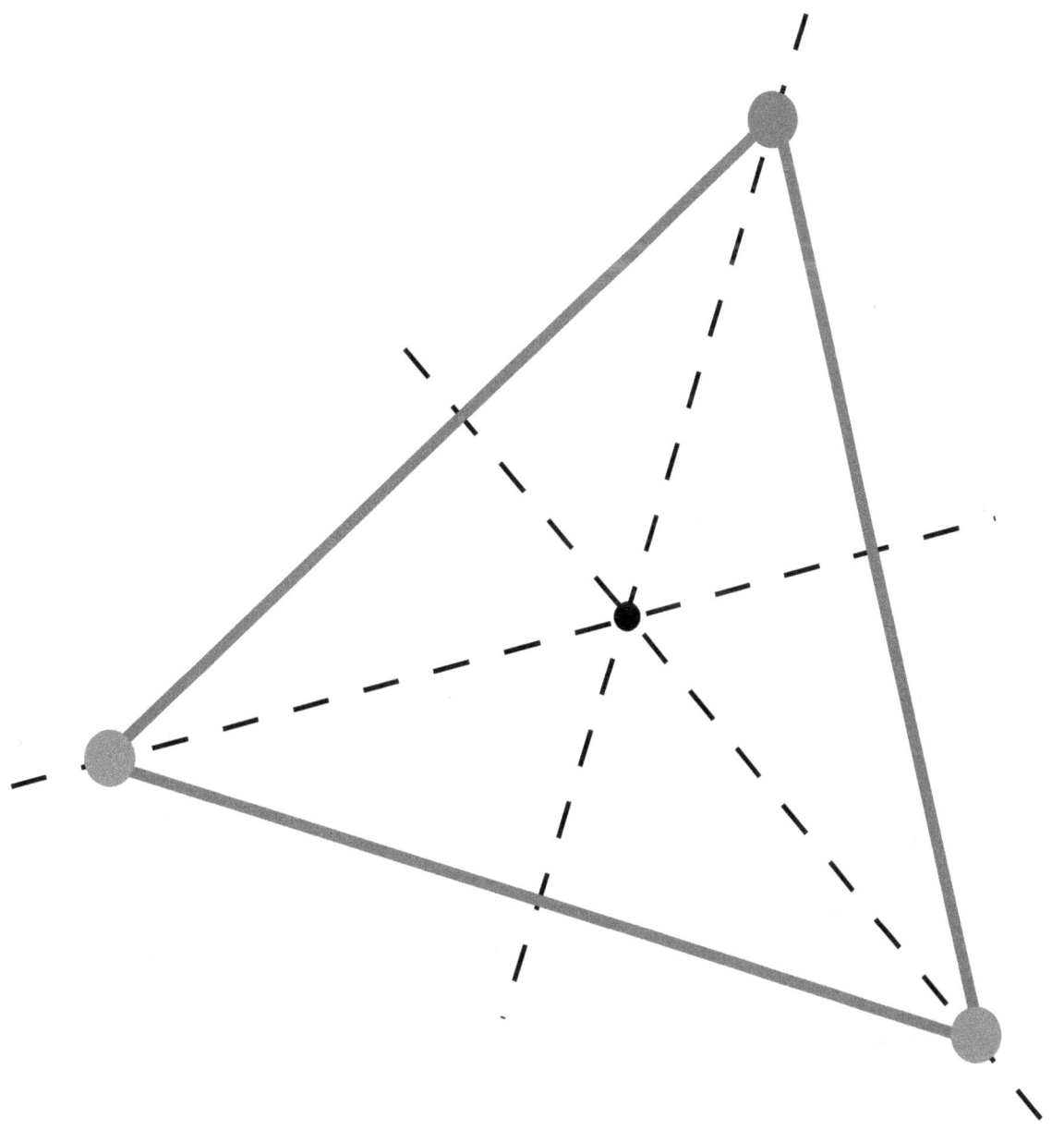

Given an equilateral triangle, the counterclockwise rotation by 120° around the center of the triangle maps every vertex of the triangle to another one. The cyclic group C_3 *consisting of the rotations by 0°, 120° and 240° acts on the set of the three vertices.*

Identity $e.x = x$ for all x in X. (Here, e denotes the neutral element of the group G.)

The set X is called a (*left*) *G-set*. The group G is said to act on X (on the left).

From these two axioms, it follows that for every g in G, the function which maps x in X to $g.x$ is a bijective map from X to X (its inverse being the function which maps x to $g^{-1}.x$). Therefore, one may alternatively define a group action of G on X as a group homomorphism from G into the symmetric group $\mathrm{Sym}(X)$ of all bijections from X to X.[2]

In complete analogy, one can define a *right group action* of G on X as an operation $X \times G \to X$ mapping (x, g) to $x.g$ and satisfying the two axioms:

Compatibility $x.(gh) = (x.g).h$ for all g, h in G and all x in X;

Identity $x.e = x$ for all x in X.

The difference between left and right actions is in the order in which a product like gh acts on x. For a left action h acts first and is followed by g, while for a right action g acts first and is followed by h. Because of the formula $(gh)^{-1} = h^{-1}g^{-1}$, one can construct a left action from a right action by composing with the inverse operation of the group. Also, a right action of a group G on X is the same thing as a left action of its opposite group G^{op} on X. It is thus sufficient to only consider left actions without any loss of generality.

13.2 Examples

- The *trivial* action of any group G on any set X is defined by $g.x = x$ for all g in G and all x in X; that is, every group element induces the identity permutation on X.[3]

- In every group G, left multiplication is an action of G on G: $g.x = gx$ for all g, x in G.

- In every group G, conjugation is an action of G on G: $g.x = gxg^{-1}$. An exponential notation is commonly used for the right-action variant: $x^g = g^{-1}xg$; it satisfies $(x^g)^h = x^{gh}$.

- The symmetric group Sn and its subgroups act on the set $\{ 1, ..., n \}$ by permuting its elements

- The symmetry group of a polyhedron acts on the set of vertices of that polyhedron. It also acts on the set of faces or the set of edges of the polyhedron.

- The symmetry group of any geometrical object acts on the set of points of that object.

- The automorphism group of a vector space (or graph, or group, or ring...) acts on the vector space (or set of vertices of the graph, or group, or ring...).

- The general linear group $GL(n, K)$ and its subgroups, particularly its Lie subgroups (including the special linear group $SL(n, K)$, orthogonal group $O(n, K)$, special orthogonal group $SO(n, K)$, and symplectic group $Sp(n, K)$) are Lie groups that act on the vector space K^n. The group operations are given by multiplying the matrices from the groups with the vectors from K^n.

- The affine group acts transitively and faithfully on the points of an affine space. This is in fact one way to define an affine space, i.e. saying that $\varphi : V \times A \to A : (v, t) \mapsto v + t$ is a transitive and faithful action is equivalent to the formal definition of an affine space: (left) identity and associativity are equivalent to the definition of a (left) action and uniqueness is equivalent to the property of the action of being transitive and faithful.

- The projective linear group $PGL(n+1, K)$ and its subgroups, particularly its Lie subgroups, which are Lie groups that act on the projective space $\mathbf{P}^n(K)$. This is a quotient of the action of the general linear group on projective space. Particularly notable is $PGL(2, K)$, the symmetries of the projective line, which is sharply 3-transitive, preserving the cross ratio; the Möbius group $PGL(2, \mathbf{C})$ is of particular interest.

- The isometries of the plane act on the set of 2D images and patterns, such as wallpaper patterns. The definition can be made more precise by specifying what is meant by image or pattern; e.g., a function of position with values in a set of colors. Isometries are in fact one example of affine group (action).

- The sets acted on by a group G comprise the category of G-sets in which the objects are G-sets and the morphisms are G-set homomorphisms: functions $f : X \to Y$ such that $(g.f)(x) = f(g.x)$ for every g in G.

- The Galois group of a field extension L/K acts on the field L but has only a trivial action on elements of the subfield K. Subgroups of $Gal(L/K)$ correspond to subfields of L that contain K, i.e. intermediate field extensions between L and K.

- The additive group of the real numbers $(\mathbf{R}, +)$ acts on the phase space of "well-behaved" systems in classical mechanics (and in more general dynamical systems) by time translation: if t is in $\mathbf{R}$ and x is in the phase space, then x describes a state of the system, and $t+x$ is defined to be the state of the system t seconds later if t is positive or $-t$ seconds ago if t is negative.

- The additive group of the real numbers ($\mathbf{R}$, +) acts on the set of real functions of a real variable in various ways, with $(t.f)(x)$ equal to e.g. $f(x + t)$, $f(x) + t$, $f(xe^t)$, $f(x)e^t$, $f(x + t)e^t$, or $f(xe^t) + t$, but not $f(xe^t + t)$.

- Given a group action of G on X, we can define an induced action of G on the power set of X, by setting $g.U = \{g.u : u \in U\}$ for every subset U of X and every g in G. This is useful, for instance, in studying the action of the large Mathieu group on a 24-set and in studying symmetry in certain models of finite geometries.

- The quaternions with norm 1 (the versors), as a multiplicative group, act on $\mathbf{R}^3$: for any such quaternion $z = \cos \frac{1}{2}\alpha + \sin \frac{1}{2}\alpha \hat{v}$, the mapping $f(\mathbf{x}) = z\mathbf{x}z^*$ is a counterclockwise rotation through an angle α about an axis $\mathbf{v}$; z is the same rotation; see quaternions and spatial rotation.

13.3   Types of actions

The action of G on X is called

- *Transitive* if X is non-empty and if for any x, y in X there exists a g in G such that $g.x = y$.

- *Faithful* (or *effective*) if for any two distinct g, h in G there exists an x in X such that $g.x \neq h.x$; or equivalently, if for any $g \neq e$ in G there exists an x in X such that $g.x \neq x$. Intuitively, in a faithful group action, different elements of G induce different permutations of X.

- *Free* (or *semiregular* or *fixed point free*) if, given g, h in G, the existence of an x in X with $g.x = h.x$ implies $g = h$. Equivalently: if g is a group element and there exists an x in X with $g.x = x$ (that is, if g has at least one fixed point), then g is the identity.

- *Regular* (or *simply transitive* or *sharply transitive*) if it is both transitive and free; this is equivalent to saying that for any two x, y in X there exists precisely one g in G such that $g.x = y$. In this case, X is known as a principal homogeneous space for G or as a G-torsor.

- *n-transitive* if X has at least n elements and for any pairwise distinct $x_1, ..., x_n$ and pairwise distinct $y_1, ..., y_n$ there is a g in G such that $g \cdot x_k = y_k$ for $1 \leq k \leq n$. A 2-transitive action is also called *doubly transitive*, a 3-transitive action is also called *triply transitive*, and so on. Such actions define 2-transitive groups, 3-transitive groups, and multiply transitive groups.

  - *Sharply n-transitive* if there is exactly one such g.

- *Primitive* if it is transitive and preserves no non-trivial partition of X. See primitive permutation group for details.

- *Locally free* if G is a topological group, and there is a neighbourhood U of e in G such that the restriction of the action to U is free; that is, if $g.x = x$ for some x and some g in U then $g = e$.

- *Irreducible* if X is a non-zero module over a ring R, the action of G is R-linear, and there is no nonzero proper invariant submodule.

Every free action on a non-empty set is faithful. A group G acts faithfully on X if and only if the corresponding homomorphism $G \to \text{Sym}(X)$ has a trivial kernel. Thus, for a faithful action, G embeds into to a permutation group on X; specifically, G is isomorphic to its image in $\text{Sym}(X)$.

The action of any group G on itself by left multiplication is regular, and thus faithful as well. Every group can, therefore, be embedded in the symmetric group on its own elements, $\text{Sym}(G)$. This result is known as Cayley's theorem.

If G does not act faithfully on X, one can easily modify the group to obtain a faithful action. If we define $N = \{g$ in $G : g.x = x$ for all x in $X\}$, then N is a normal subgroup of G; indeed, it is the kernel of the homomorphism $G \to \text{Sym}(X)$. The factor group G/N acts faithfully on X by setting $(gN).x = g.x$. The original action of G on X is faithful if and only if $N = \{e\}$.

In the compound of five tetrahedra, the symmetry group is the (rotational) icosahedral group I of order 60, while the stabilizer of a single chosen tetrahedron is the (rotational) tetrahedral group T of order 12, and the orbit space I/T (of order 60/12 = 5) is naturally identified with the 5 tetrahedra – the coset gT corresponds to which tetrahedron g sends the chosen tetrahedron to.

13.4 Orbits and stabilizers

Consider a group G acting on a set X. The *orbit* of a point x in X is the set of elements of X to which x can be moved by the elements of G. The orbit of x is denoted by $G.x$:

$$G.x = \{g.x \mid g \in G\}.$$

The defining properties of a group guarantee that the set of orbits of (points x in) X under the action of G form a partition of X. The associated equivalence relation is defined by saying $x \sim y$ if and only if there exists a g in G with $g.x = y$. The orbits are then the equivalence classes under this relation; two elements x and y are equivalent if and only if their orbits are the same; i.e., $G.x = G.y$.

The group action is transitive if and only if it has only one orbit, i.e. if there exists x in X with $G.x = X$. This is the case if and only if $G.x = X$ for *all* x in X.

The set of all orbits of X under the action of G is written as X/G (or, less frequently: $G\backslash X$), and is called the *quotient* of the action. In geometric situations it may be called the *orbit space*, while in algebraic situations it may be called the space of *coinvariants*, and written XG, by contrast with the invariants (fixed points), denoted X^G: the coinvariants are a *quotient* while the invariants are a *subset*. The coinvariant terminology and notation are used particularly in group cohomology and group homology, which use the same superscript/subscript convention.

13.4.1   Invariant subsets

If Y is a subset of X, we write GY for the set $\{\, g.y : y \in Y$ and $g \in G \}$. We call the subset Y *invariant under G* if $G.Y = Y$ (which is equivalent to $G.Y \subseteq Y$). In that case, G also operates on Y by restricting the action to Y. The subset Y is called *fixed under G* if $g.y = y$ for all g in G and all y in Y. Every subset that is fixed under G is also invariant under G, but not vice versa.

Every orbit is an invariant subset of X on which G acts transitively. The action of G on X is *transitive* if and only if all elements are equivalent, meaning that there is only one orbit.

A *G-invariant* element of X is $x \in X$ such that $g.x = x$ for all $g \in G$. The set of all such x is denoted X^G and called the *G-invariants* of X. When X is a G-module, X^G is the zeroth group cohomology group of G with coefficients in X, and the higher cohomology groups are the derived functors of the functor of G-invariants.

13.4.2   Fixed points and stabilizer subgroups

Given g in G and x in X with $g.x = x$, we say x is a fixed point of g and g fixes x.

For every x in X, we define the *stabilizer subgroup* of x (also called the *isotropy group*) as the set of all elements in G that fix x:

$$G_x = \{ g \in G \mid g.x = x \}.$$

This is a subgroup of G, though typically not a normal one. The action of G on X is free if and only if all stabilizers are trivial. The kernel N of the homomorphism $G \to \mathrm{Sym}(X)$ is given by the intersection of the stabilizers Gx for all x in X. If N is trivial, the action is said to be faithful (or effective).

Let x and y be two elements in X, and let g be a group element such that $y = g.x$. Then the two stabilizer groups Gx and Gy are related by $Gy = g\, Gx\, g^{-1}$. Proof: by definition, $h \in Gy$ if and only if $h.(g.x) = g.x$. Applying g^{-1} to both sides of this equality yields $(g^{-1}hg).x = (g^{-1}g).x = x$; that is, $g^{-1}hg \in Gx$.

The above says that the stabilizers of elements in the same orbit are conjugate to each other. Thus, to each orbit, one can associate a conjugacy class of a subgroup of G (i.e., the set of all conjugates of the subgroup). Let (H) denote the conjugacy class of H. Then one says that the orbit O has type (H) if the stabilizer G_x of some/any x in O belongs to (H). A maximal orbit type is often called a principal orbit type.

13.4.3   Orbit-stabilizer theorem and Burnside's lemma

Orbits and stabilizers are closely related. For a fixed x in X, consider the map from G to X given by $g \mapsto g.x$ for all $g \in G$. The image of this map is the orbit of x and the coimage is the set of all left cosets of Gx. The standard quotient theorem of set theory then gives a natural bijection between G/Gx and $G.x$. Specifically, the bijection is given by $hGx \mapsto h.x$. This result is known as the *orbit-stabilizer theorem*. From a more categorical perspective, the orbit-stabilizer theorem comes from the fact that every G-set is a sum of quotients of the G-set G.

If G and X are finite then the orbit-stabilizer theorem, together with Lagrange's theorem, gives

$$|G.x| = [G : G_x] = |G|/|G_x|.$$

This result is especially useful since it can be employed for counting arguments.

A result closely related to the orbit-stabilizer theorem is Burnside's lemma:

$$|X/G| = \frac{1}{|G|} \sum_{g \in G} |X^g|$$

where X^g is the set of points fixed by g. This result is mainly of use when G and X are finite, when it can be interpreted as follows: the number of orbits is equal to the average number of points fixed per group element.

Fixing a group G, the set of formal differences of finite G-sets forms a ring called the Burnside ring of G, where addition corresponds to disjoint union, and multiplication to Cartesian product.

13.5 Group actions and groupoids

The notion of group action can be put in a broader context by using the *action groupoid* $G' = G \ltimes X$ associated to the group action, thus allowing techniques from groupoid theory such as presentations and fibrations. Further the stabilisers of the action are the vertex groups, and the orbits of the action are the components, of the action groupoid. For more details, see the book *Topology and groupoids* referenced below.

This action groupoid comes with a morphism $p\colon G' \to G$ which is a *covering morphism of groupoids*. This allows a relation between such morphisms and covering maps in topology.

13.6 Morphisms and isomorphisms between *G*-sets

If X and Y are two G-sets, we define a *morphism* from X to Y to be a function $f : X \to Y$ such that $f(g.x) = g.f(x)$ for all g in G and all x in X. Morphisms of G-sets are also called *equivariant maps* or *G-maps*.

The composition of two morphisms is again a morphism.

If a morphism f is bijective, then its inverse is also a morphism, and we call f an *isomorphism* and the two G-sets X and Y are called *isomorphic*; for all practical purposes, they are indistinguishable in this case.

Some example isomorphisms:

- Every regular G action is isomorphic to the action of G on G given by left multiplication.

- Every free G action is isomorphic to $G \times S$, where S is some set and G acts on $G \times S$ by left multiplication on the first coordinate. (S can be taken to be the set of orbits X/G.)

- Every transitive G action is isomorphic to left multiplication by G on the set of left cosets of some subgroup H of G. (H can be taken to be the stabilizer group of any element of the original G-set.the original action.)

With this notion of morphism, the collection of all G-sets forms a category; this category is a Grothendieck topos (in fact, assuming a classical metalogic, this topos will even be Boolean).

13.7 Continuous group actions

Main article: continuous group action

One often considers *continuous group actions*: the group G is a topological group, X is a topological space, and the map $G \times X \to X$ is continuous with respect to the product topology of $G \times X$. The space X is also called a *G-space* in this case. This is indeed a generalization, since every group can be considered a topological group by using the discrete topology. All the concepts introduced above still work in this context, however we define morphisms between G-spaces to be *continuous* maps compatible with the action of G. The quotient X/G inherits the quotient topology from X, and is called the *quotient space* of the action. The above statements about isomorphisms for regular, free and transitive actions are no longer valid for continuous group actions.

If G is a discrete group acting on a topological space X, the action is properly discontinuous if for any point x in X there is an open neighborhood U of x in X, such that the set of all g in G for which $g(U) \cap U \neq \emptyset$ consists of the identity only. If X is a regular covering space of another topological space Y, then the action of the deck transformation group on X is properly discontinuous as well as being free. Every free, properly discontinuous action of a group G on a path-connected topological space X arises in this manner: the quotient map $X \mapsto X/G$ is a regular covering map, and the deck transformation group is the given action of G on X. Furthermore, if X is simply connected, the fundamental group of X/G will be isomorphic to G.

These results have been generalised in the book *Topology and Groupoids* referenced below to obtain the fundamental groupoid of the orbit space of a discontinuous action of a discrete group on a Hausdorff space, as, under reasonable local conditions, the orbit groupoid of the fundamental groupoid of the space. This allows calculations such as the fundamental group of the symmetric square of a space X, namely the orbit space of the product of X with itself under the twist action of the cyclic group of order 2 sending (x, y) to (y, x).

An action of a group G on a locally compact space X is *cocompact* if there exists a compact subset A of X such that $GA = X$. For a properly discontinuous action, cocompactness is equivalent to compactness of the quotient space X/G.

The action of G on X is said to be *proper* if the mapping $G \times X \to X \times X$ that sends $(g, x) \mapsto (g.x, x)$ is a proper map.

13.7.1 Strongly continuous group action and smooth points

A group action of a topological group G on a topological space X is said to be *strongly continuous* if for all x in X, the map $g \mapsto g.x$ is continuous with respect to the respective topologies. Such an action induces an action on the space of continuous functions on X by defining $(g.f)(x) = f(g^{-1}.x)$ for every g in G, f a continuous function on X, and x in X. Note that, while every continuous group action is strongly continuous, the converse is not in general true.[4]

The subspace of *smooth points* for the action is the subspace of X of points x such that $g \mapsto g.x$ is smooth; i.e., it is continuous and all derivatives are continuous.

13.8 Variants and generalizations

One can also consider actions of monoids on sets, by using the same two axioms as above. This does not define bijective maps and equivalence relations however. See semigroup action.

Instead of actions on sets, one can define actions of groups and monoids on objects of an arbitrary category: start with an object X of some category, and then define an action on X as a monoid homomorphism into the monoid of endomorphisms of X. If X has an underlying set, then all definitions and facts stated above can be carried over. For example, if we take the category of vector spaces, we obtain group representations in this fashion.

One can view a group G as a category with a single object in which every morphism is invertible. A group action is then nothing but a functor from G to the category of sets, and a group representation is a functor from G to the category of vector spaces. A morphism between G-sets is then a natural transformation between the group action functors. In analogy, an action of a groupoid is a functor from the groupoid to the category of sets or to some other category.

In addition to continuous actions of topological groups on topological spaces, one also often considers smooth actions of Lie groups on smooth manifolds, regular actions of algebraic groups on algebraic varieties, and actions of group schemes on schemes. All of these are examples of group objects acting on objects of their respective category.

13.9 See also

- Gain graph
- Group with operators
- Monoid action
- Lie group action

13.10 Notes

[1] Eie & Chang (2010). *A Course on Abstract Algebra*. p. 144.

[2] This is done e.g. by Smith (2008). *Introduction to abstract algebra*. p. 253.

[3] Eie & Chang (2010). *A Course on Abstract Algebra*. p. 145.

[4] Yuan, Qiaochu (27 February 2013). "wiki's definition of "strongly continuous group action" wrong?". Mathematics Stack Exchange. Retrieved 1 April 2013.

13.11 References

- Aschbacher, Michael (2000). *Finite Group Theory*. Cambridge University Press. ISBN 978-0-521-78675-1. MR 1777008.

- Brown, Ronald (2006). *Topology and groupoids*, Booksurge PLC, ISBN 1-4196-2722-8.

- Categories and groupoids, P.J. Higgins, downloadable reprint of van Nostrand Notes in Mathematics, 1971, which deal with applications of groupoids in group theory and topology.

- Dummit, David; Richard Foote (2004). *Abstract Algebra* ((3rd ed.) ed.). Wiley. ISBN 0-471-43334-9.

- Eie, Minking; Chang, Shou-Te (2010). *A Course on Abstract Algebra*. World Scientific. ISBN 978-981-4271-88-2.

- Rotman, Joseph (1995). *An Introduction to the Theory of Groups*. Graduate Texts in Mathematics **148** ((4th ed.) ed.). Springer-Verlag. ISBN 0-387-94285-8.

- Smith, Jonathan D.H. (2008). *Introduction to abstract algebra*. Textbooks in mathematics. CRC Press. ISBN 978-1-4200-6371-4.

13.12 External links

- Hazewinkel, Michiel, ed. (2001), "Action of a group on a manifold", *Encyclopedia of Mathematics*, Springer, ISBN 978-1-55608-010-4

- Weisstein, Eric W., "Group Action", *MathWorld*.

Chapter 14

Classical group

For the book by Weyl, see The Classical Groups.

In mathematics, the **classical groups** are defined as the special linear groups over the reals **R**, the complex numbers **C** and the quaternions **H** together with special[1] automorphism groups of symmetric or skew-symmetric bilinear forms and Hermitian or skew-Hermitian sesquilinear forms defined on real, complex and quaternionic finite-dimensional vector spaces.[2] Of these, the **complex classical Lie groups** are four infinite families of Lie groups that together with the exceptional groups exhaust the classification of simple Lie groups. The **compact classical groups** are compact real forms of the complex classical groups. The finite analogues of the classical groups are the **classical groups of Lie type**. The term "classical group" was coined by Hermann Weyl, it being the title of his 1939 monograph *The Classical Groups*.[3]

The classical groups form the deepest and most useful part of the subject of linear Lie groups.[4] Most types of classical groups find application in classical and modern physics. A few examples are the following. The rotation group SO(3) is a symmetry of Euclidean space and all fundamental laws of physics, the Lorentz group O(3,1) is a symmetry group of spacetime of special relativity. The special unitary group SU(3) is the symmetry group of quantum chromodynamics and the symplectic group Sp(m) finds application in hamiltonian mechanics and quantum mechanical versions of it.

14.1   The classical groups

The **classical groups** are exactly the general linear groups over **R**, **C** and **H** together with the automorphism groups of non-degenerate forms discussed below.[5] These groups are usually additionally restricted to the subgroups whose elements have determinant 1. The classical groups, with the determinant 1 condition, are listed in the table below. In the sequel, the determinant 1 condition is *not* used consistently in the interest of greater generality.

The **complex classical groups** are SL(n, **C**), SO(n, **C**) and Sp(m, **C**). A group is complex according to whether its Lie algebra is complex. The **real classical groups** refers to all of the classical groups since any Lie algebra is a real algebra. The **compact classical groups** are the compact real forms of the complex classical groups. These are, in turn, SU(n), SO(n) and Sp(m). One characterization of the compact real form is in terms of the Lie algebra **g**. If **g** = **u** + i**u**, the complexification of **u**, then if the connected group K generated by exp(X): $X \in$ **u** is a compact, K is a compact real form.[6]

The classical groups can uniformly be characterized in a different way using real forms. The classical groups (here with the determinant 1 condition, but this is not necessary) are the following:

The complex linear algebraic groups SL(n, **C**), SO(n, **C**), and Sp(n, **C**) together with their real forms.[7]

For instance, SO*($2n$) is a real form of SO($2n$, **C**), SU(p, q) is a real form of Sl(n, **C**), and Sl(n, **H**) is a real form of SO($2n$, **C**). Without the determinant 1 condition, replace the special linear groups with the corresponding general linear groups in the characterization. The algebraic groups in question are Lie groups, but the "algebraic" qualifier is needed to get the right notion of "real form".

14.2 Bilinear and sesquilinear forms

Main articles: Bilinear form and Sesquilinear form

The classical groups are defined in terms of forms defined on $\mathbf{R}^n$, $\mathbf{C}^n$, and $\mathbf{H}^n$, where $\mathbf{R}$ and $\mathbf{C}$ are the fields of the real and complex numbers. The quaternions, $\mathbf{H}$, do not constitute a field because multiplication does not commute; they form a division ring or a **skew field** or **non-commutative field**. However, it is still possible to define matrix quaternionic groups. For this reason, a vector space V is allowed to be defined over $\mathbf{R}$, $\mathbf{C}$, as well as $\mathbf{H}$ below. In the case of $\mathbf{H}$, V is a *right* vector space to make possible the representation of the group action as matrix multiplication from the *left*, just as for $\mathbf{R}$ and $\mathbf{C}$.[8]

A form φ: $V \times V \to F$ on some finite-dimensional right vector space over $F = \mathbf{R}$, $\mathbf{C}$, or $\mathbf{H}$ is bilinear if

$$\varphi(x\alpha, y\beta) = \alpha\varphi(x, y)\beta, \quad \forall x, y \in V, \forall \alpha, \beta \in F.$$

It is called **sesquilinear** if

$$\varphi(x\alpha, y\beta) = \bar{\alpha}\varphi(x, y)\beta, \quad \forall x, y \in V, \forall \alpha, \beta \in F.$$

These conventions are chosen because they work in all cases considered. An automorphism of φ is a map A in the set of linear operators on V such that

The set of all automorphisms of φ form a group, it is called the automorphism group of φ, denoted Aut(φ). This leads to a preliminary definition of a classical group:

> *A classical group is a group that preserves a bilinear or sesquilinear form on finite-dimensional vector spaces over* $\mathbf{R}$, $\mathbf{C}$ *or* $\mathbf{H}$.

This definition has shortcomings because there is some unnecessary redundancy. In the case of $F = \mathbf{R}$, bilinear is equivalent to sesquilinear. In the case of $F = \mathbf{H}$, there are no non-zero bilinear forms.[9]

14.2.1 Symmetric, skew-symmetric, Hermitian, and skew-Hermitian forms

A form is **symmetric** if

$$\varphi(x, y) = \varphi(y, x).$$

It is **skew-symmetric** if

$$\varphi(x, y) = -\varphi(y, x).$$

It is **Hermitian** if

$$\varphi(x,y) = \overline{\varphi(y,x)}$$

Finally, it is **skew-Hermitian** if

$$\varphi(x,y) = -\overline{\varphi(y,x)}.$$

A bilinear form φ is uniquely a sum of a symmetric form and a skew-symmetric form. A transformation preserving φ preserves both parts separately. The groups preserving symmetric and skew-symmetric forms can thus be studied separately. The same applies, mutatis mutandis, to Hermitian and skew-Hermitian forms. For this reason, for the purposes of classification, only purely symmetric, skew-symmetric, Hermitian, or skew-Hermitian forms are considered. The **normal forms** of the forms correspond to specific suitable choices of bases. These are bases giving the following normal forms in coordinates:

basis: (pseudo-)orthonormal in form symmetric Bilinear	$\varphi(x,y) = \pm\xi_1\eta_1 \pm \xi_2\eta_2 \pm \cdots \pm \xi_n\eta_n,\ (\mathbf{R})$
basis: orthonormal in form symmetric Bilinear	$\varphi(x,y) = \xi_1\eta_1 + \xi_2\eta_2 + \cdots + \xi_n\eta_n,\ (\mathbf{C})$
basis: symplectic in skew-symmetric Bilinear	$\varphi(x,y) = \xi_1\eta_{m+1} + \xi_2\eta_{m+2} + \cdots + \xi_m\eta_{2m=n}$
	$\quad -\xi_{m+1}\eta_1 - \xi_{m+2}\eta_2 - \cdots - \xi_{2m=n}\eta_m,\ (\mathbf{R},\mathbf{C})$
Hermitian: Sesquilinear	$\varphi(x,y) = \pm\bar{\xi}_1\eta_1 \pm \bar{\xi}_2\eta_2 \pm \cdots \pm \bar{\xi}_n\eta_n,\ (\mathbf{C},\mathbf{H})$
skew-Hermitian: Sesquilinear	$\varphi(x,y) = \bar{\xi}_1\mathbf{j}\eta_1 + \bar{\xi}_2\mathbf{j}\eta_2 + \cdots + \bar{\xi}_n\mathbf{j}\eta_n,\ (\mathbf{H}).$

The $\mathbf{j}$ in the skew-Hermitian form is the third basis element in the basis $(\mathbf{1}, \mathbf{i}, \mathbf{j}, \mathbf{k})$ for $\mathbf{H}$. Proof of existence of these bases and Sylvester's law of inertia, the independence of the number of plus- and minus-signs, p and q, in the symmetric and Hermitian forms, as well as the presence or absence of the fields in each expression, can be found in Rossmann (2002) or Goodman & Wallach (2009). The pair (p, q), and sometimes $p - q$, is called the **signature** of the form.

Explanation of occurrence of the fields R, C, H: There are no nontrivial bilinear forms over H. In the symmetric bilinear case, only forms over $\mathbf{R}$ have a signature. In other words, a complex bilinear form with "signature" (p, q) can, by a change of basis, be reduced to a form where all signs are "+" in the above expression, whereas this is impossible in the real case, in which $p - q$ is independent of the basis when put into this form. However, Hermitian forms have basis-independent signature in both the complex and the quaternionic case. (The real case reduces to the symmetric case.) A skew-Hermitian form on a complex vector space is rendered Hermitian by multiplication by i, so in this case, only $\mathbf{H}$ is interesting.

14.3 Automorphism groups

The first section presents the general framework. The other sections exhaust the qualitatively different cases that arise as automorphism groups of bilinear and sesquilinear forms on finite-dimensional vector spaces over $\mathbf{R}$, $\mathbf{C}$ and $\mathbf{H}$.

14.3.1 Aut(φ) – the automorphism group

Assume that φ is a non-degenerate form on a finite-dimensional vector space V over $\mathbf{R}$, $\mathbf{C}$ or $\mathbf{H}$. The automorphism group is defined, based on condition (1), as

$$\text{Aut}(\varphi) = \{A \in \text{GL}(V) : \varphi(Au, Av) = \varphi(x,y), \quad \forall x, y \in V\}.$$

Every $A \in Mn(V)$ has an adjoint A^φ with respect to φ defined by

Hermann Weyl, the author of The Classical Groups. Weyl made substantial contributions to the representation theory of the classical groups.

Using this definition in condition (**1**), the automorphism group is seen to be given by

Fix a basis for V. In terms of this basis, put

$$\varphi(x, y) = \sum \xi_i \varphi_{ij} \eta_j$$

where ξi, ηj are the components of x, y. This is appropriate for the bilinear forms. Sesquilinear forms have similar expressions and are treated separately later. In matrix notation one finds

$$\varphi(x,y) = x^{\mathsf{T}}\Phi y$$

and

from (**2**) where Φ is the matrix (φij). The non-degeneracy condition means precisely that Φ is invertible, so the adjoint always exists. Aut(φ) expressed with this becomes

$$\mathrm{Aut}(\varphi) = \{A \in \mathrm{GL}(V) : \Phi^{-1}A^{\mathsf{T}}\Phi A = 1\}.$$

The Lie algebra $\mathbf{aut}(\varphi)$ of the automorphism groups can be written down immediately. Abstractly, $X \in \mathbf{aut}(\varphi)$ if and only if

$$(e^{tX})^\varphi e^{tX} = 1$$

for all t, corresponding to the condition in (**3**) under the exponential mapping of Lie algebras, so that

$$\mathfrak{aut}(\varphi) = \{X \in M_n(V) : X^\varphi = -X\},$$

or in a basis

as is seen using the power series expansion of the exponential mapping and the linearity of the involved operations. Conversely, suppose that $X \in \mathbf{aut}(\varphi)$. Then, using the above result, $\varphi(Xx, y) = \varphi(x, X^\varphi y) = -\varphi(x, Xy)$. Thus the Lie algebra can be characterized without reference to a basis, or the adjoint, as

$$\mathfrak{aut}(\varphi) = \{X \in M_n(V) : \varphi(Xx,y) = -\varphi(x,Xy), \quad \forall x,y \in V\}.$$

The normal form for φ will be given for each classical group below. From that normal form, the matrix Φ can be read off directly. Consequently, expressions for the adjoint and the Lie algebras can be obtained using formulas (**4**) and (**5**). This is demonstrated below in most of the non-trivial cases.

14.3.2 Bilinear case

When the form is symmetric, Aut(φ) is called O(φ). When it is skew-symmetric then Aut(φ) is called Sp(φ). This applies to the real and the complex cases. The quaternionic case is empty since no nonzero bilinear forms exists on quaternionic vector spaces.[12]

Real case

The real case breaks up into two cases, the symmetric and the antisymmetric forms that should be treated separately.

O(p, q) and O(n) – the orthogonal groups   Main articles: Orthogonal group and Indefinite orthogonal group

If φ is symmetric and the vector space is real, a basis may be chosen so that

$$\varphi(x, y) = \pm \xi_1 \eta_1 \pm \xi_1 \eta_1 \cdots \pm \xi_n \eta_n.$$

The number of plus and minus-signs are independent of the particular basis.[13] In the case $V = \mathbf{R}^n$ one writes O(φ) = O(p, q) where p is the number of plus signs and q is the number of minus-signs, $p + q = n$. If $q = 0$ the notation is O(n). The matrix Φ is in this case

$$\Phi = \begin{pmatrix} I_p & 0 \\ 0 & -I_q \end{pmatrix} \equiv I_{p,q}$$

after reordering the basis if necessary. The adjoint operation (4) then becomes

$$A^\varphi = \begin{pmatrix} I_p & 0 \\ 0 & -I_q \end{pmatrix} \begin{pmatrix} A_{11} & \cdots \\ \cdots & A_{nn} \end{pmatrix}^{\mathrm{T}} \begin{pmatrix} I_p & 0 \\ 0 & -I_q \end{pmatrix},$$

which reduces to the usual transpose when p or q is 0. The Lie algebra is found using equation (5) and a suitable ansatz (this is detailed for the case of Sp(m, $\mathbf{R}$) below),

$$\mathfrak{o}(p, q) = \left\{ \begin{pmatrix} X_{p \times p} & Y_{p \times q} \\ Y^{\mathrm{T}} & W_{q \times q} \end{pmatrix} \middle| X^{\mathrm{T}} = -X, \quad W^{\mathrm{T}} = -W \right\},$$

and the group according to (3) is given by

$$\mathrm{O}(p, q) = \{ g \in \mathrm{GL}(n, \mathbb{R}) | I_{p,q}^{-1} g^{\mathrm{T}} I_{p,q} g = I \}.$$

The groups O(p, q) and O(q, p) are isomorphic through the map

$$\mathrm{O}(p, q) \to \mathrm{O}(q, p), \quad g \to \sigma g \sigma^{-1}, \quad \sigma = \begin{bmatrix} 0 & 0 & \cdots & 1 \\ \vdots & \vdots & \ddots & \vdots \\ 0 & 1 & \cdots & 0 \\ 1 & 0 & \cdots & 0 \end{bmatrix}.$$

For example, the Lie algebra of the Lorentz group could be written as

$$\mathfrak{o}(3, 1) = \mathrm{span}\left\{ \begin{pmatrix} 0 & 1 & 0 & 0 \\ -1 & 0 & 0 & 0 \\ 0 & 0 & 0 & 0 \\ 0 & 0 & 0 & 0 \end{pmatrix}, \begin{pmatrix} 0 & 0 & -1 & 0 \\ 0 & 0 & 0 & 0 \\ 1 & 0 & 0 & 0 \\ 0 & 0 & 0 & 0 \end{pmatrix}, \begin{pmatrix} 0 & 0 & 0 & 0 \\ 0 & 0 & 1 & 0 \\ 0 & -1 & 0 & 0 \\ 0 & 0 & 0 & 0 \end{pmatrix}, \begin{pmatrix} 0 & 0 & 0 & 1 \\ 0 & 0 & 0 & 0 \\ 0 & 0 & 0 & 0 \\ 1 & 0 & 0 & 0 \end{pmatrix}, \begin{pmatrix} 0 & 0 & 0 & 0 \\ 0 & 0 & 0 & 1 \\ 0 & 0 & 0 & 0 \\ 0 & 1 & 0 & 0 \end{pmatrix}, \begin{pmatrix} 0 & 0 & 0 & 0 \\ 0 & 0 & 0 & 0 \\ 0 & 0 & 0 & 1 \\ 0 & 0 & 1 & 0 \end{pmatrix} \right\}.$$

Naturally, it is possible to rearrange so that the q-block is the upper left (or any other block). Here the "time component" end up as the fourth coordinate in a physical interpretation, and not the first as may be more common.

Sp(m, $\mathbf{R}$) – the real symplectic group   Main article: Symplectic group

If φ is skew-symmetric and the vector space is real, there is a basis giving

$$\varphi(x, y) = \xi_1 \eta_{m+1} + \xi_2 \eta_{m+2} \cdots + \xi_m \eta_{2m=n} - \xi_{m+1} \eta_1 - \xi_{m+2} \eta_2 \cdots - \xi_{2m=n} \eta_m,$$

where $n = 2m$. For $\mathrm{Aut}(\varphi)$ one writes $\mathrm{Sp}(\varphi) = \mathrm{Sp}(V)$ In case $V = \mathbf{R}^n = \mathbf{R}^{2m}$ one writes $\mathrm{Sp}(m, \mathbf{R})$ or $\mathrm{Sp}(2m, \mathbf{R})$. From the normal form one reads off

$$\Phi = \begin{pmatrix} 0_m & I_m \\ -I_m & 0_m \end{pmatrix} = J_m.$$

By making the ansatz

$$V = \begin{pmatrix} X & Y \\ Z & W \end{pmatrix},$$

where X, Y, Z, W are m-dimensional matrices and considering (**5**),

$$\begin{pmatrix} 0_m & -I_m \\ I_m & 0_m \end{pmatrix} \begin{pmatrix} X & Y \\ Z & W \end{pmatrix}^{\mathrm{T}} \begin{pmatrix} 0_m & I_m \\ -I_m & 0_m \end{pmatrix} = - \begin{pmatrix} X & Y \\ Z & W \end{pmatrix}$$

one finds the Lie algebra of $\mathrm{Sp}(m, \mathbf{R})$,

$$\mathfrak{sp}(m, \mathbb{R}) = \{ X \in M_n(\mathbb{R}) : J_m X + X^{\mathrm{T}} J_m = 0 \} = \left\{ \begin{pmatrix} X & Y \\ Z & -X^{\mathrm{T}} \end{pmatrix} \middle| Y^{\mathrm{T}} = Y, Z^{\mathrm{T}} = Z \right\},$$

and the group is given by

$$\mathrm{Sp}(m, \mathbb{R}) = \{ g \in M_n(\mathbb{R}) | g^{\mathrm{T}} J_m g = J_m \}.$$

Complex case

Like in the real case, there are two cases, the symmetric and the antisymmetric case that each yield a family of classical groups.

O(n, C) – the complex orthogonal group   Main article: Complex orthogonal group

If case φ is symmetric and the vector space is complex, a basis

$$\varphi(x, y) = \xi_1 \eta_1 + \xi_1 \eta_1 \cdots + \xi_n \eta_n$$

with only plus-signs can be used. The automorphism group is in the case of $V = \mathbf{C}^n$ called O(n, **C**). The lie algebra is simply a special case of that for $\mathbf{o}(p, q)$,

$$\mathfrak{o}(n, \mathbb{C}) = \mathfrak{so}(n, \mathbb{C}) = \{ X | X^{\mathrm{T}} = -X \},$$

and the group is given by

$$\mathrm{O}(n, \mathbb{C}) = \{ g | g^{\mathrm{T}} g = I_n \}.$$

In terms of classification of simple Lie algebras, the $\mathbf{so}(n)$ are split into two classes, those with n odd with root system Bn and n even with root system Dn.

Sp(*m*, C) – the complex symplectic group   Main article: Symplectic group

For φ skew-symmetric and the vector space complex, the same formula,

$$\varphi(x,y) = \xi_1\eta_{m+1} + \xi_2\eta_{m+2}\cdots + \xi_m\eta_{2m=n} - \xi_{m+1}\eta_1 - \xi_{m+2}\eta_2\cdots - \xi_{2m=n}\eta_m,$$

applies as in the real case. For Aut(φ) one writes Sp(φ) = Sp(V) In case $V = \mathbb{C}^n = \mathbb{C}^{2m}$ one writes Sp(m, $\mathbb{C}$) or Sp($2m$, $\mathbb{C}$). The Lie algebra parallels that of **sp**(*m*, $\mathbb{R}$),

$$\mathfrak{sp}(m,\mathbb{C}) = \{X \in M_n(\mathbb{C}) : J_m X + X^T J_m = 0\} = \left\{ \begin{pmatrix} X & Y \\ Z & X^T \end{pmatrix} \middle| Y^T = Y, Z^T = Z \right\},$$

and the group is given by

$$\mathrm{Sp}(m,\mathbb{C}) = \{g \in M_n(\mathbb{C}) | g^T J_m g = J_m\}.$$

14.3.3 Sesquilinear case

In the sequilinear case, one makes a slightly different ansatz for the form in terms of a basis,

$$\varphi(x,y) = \sum \bar{\xi}_i \varphi_{ij} \eta_j.$$

The other expressions that get modified are

$$\varphi(x,y) = x^*\Phi y, \qquad A^\varphi = \Phi^{-1}A^*\Phi, \text{[14]}$$
$$\mathrm{Aut}(\varphi) = \{A \in \mathrm{GL}(V) : \Phi^{-1}A^*\Phi A = 1\},$$

The real case, of course, provides nothing new. The complex and the quaternionic case will be considered below.

Complex case

From a qualitative point of view, consideration of skew-Hermitean forms (up to isomorphism) provide no new groups; multiplication by *i* renders a skew-Hermitean form Hermitean, and vice versa. Thus only the Hermitian case needs to be considered.

U(*p*, *q*) and U(*n*) – the unitary groups   Main article: Unitary group

A non-degenerate hermitian form has the normal form

$$\varphi(x,y) = \pm\bar{\xi}_1\eta_1 \pm \bar{\xi}_2\eta_2 \cdots \pm \bar{\xi}_n\eta_n.$$

As in the bilinear case, the signature (p, q) is independent of the basis. The automorphism group is denoted U(V), or, in the case of $V = \mathbf{C}^n$, U(p, q). If $q = 0$ the notation is U(n). In this case, Φ takes the form

$$\Phi = \begin{pmatrix} 1_p & 0 \\ 0 & -1_q \end{pmatrix} = I_{p,q},$$

and the Lie algebra is given by

$$\mathfrak{u}(p,q) = \left\{ \begin{pmatrix} X_{p\times p} & Z_{p\times q} \\ \overline{Z}^{\mathrm{T}} & Y_{q\times q} \end{pmatrix} \middle| \overline{X}^{\mathrm{T}} = -X, \quad \overline{Y}^{\mathrm{T}} = -Y \right\}.$$

The group is given by

$$\mathrm{U}(p,q) = \{g | I_{p,q}^{-1} g^* I_{p,q} g = I\}.$$

Quaternionic case

The space $\mathbf{H}^n$ is considered as a *right* vector space over $\mathbf{H}$. This way, $A(vh) = (Av)h$ for a quaternion h, a quaternion column vector v and quaternion matrix A. If $\mathbf{H}^n$ was a *left* vector space over $\mathbf{H}$, then matrix multiplication from the *right* on row vectors would be required to maintain linearity. This does not correspond to the usual linear operation of a group on a vector space when a basis is given, which is matrix multiplication from the *left* on column vectors. Thus V is henceforth a right vector space over $\mathbf{H}$. Even so, care must be taken due to the non-commutative nature of $\mathbf{H}$. The (mostly obvious) details are skipped because complex representations will be used.

When dealing with quaternionic groups it is convenient to represent quaternions using complex 2×2-matrices,

With this representation, quaternionic multiplication becomes matrix multiplication and quaternionic conjugation becomes taking the Hermitian adjoint. Moreover, if a quaternion according to the complex encoding $q = x + \mathbf{j}y$ is given as a column vector $(x, y)^{\mathrm{T}}$, then multiplication from the left by a matrix representation of a quaternion produces a new column vector representing the correct quaternion. This representation differs slightly from a more common representation found in the quaternion article. The more common convention would force multiplication from the right on a row matrix to achieve the same thing.

Incidentally, the representation above makes it clear that the group of unit quaternions ($\alpha\bar{\alpha} + \beta\bar{\beta} = 1 = \det Q$) is isomorphic to SU(2).

Quaternionic $n\times n$-matrices matrices can, by obvious extension, be represented by $2n\times 2n$ block-matrices of complex numbers.[16] If one agrees to represent a quaternionic $n\times 1$ column vector by a $2n\times 1$ column vector with complex numbers according to the encoding of above, with the upper n numbers being the αi and the lower n the βi, then a quaternionic $n\times n$-matrix becomes a complex $2n\times 2n$-matrix exactly of the form given above, but now with α and β $n\times n$-matrices. More formally

A matrix $T \in \mathrm{GL}(2n, \mathbf{C})$ has the form displayed in (**8**) if and only if $JnT = TJn$. With these identifications,

$$\mathbb{H}^n \approx \mathbb{C}^{2n}, M_n(\mathbb{H}) \approx \left\{ T \in M_{2n}(\mathbb{C}) \middle| J_n T = \overline{T} J_n, \quad J_n = \begin{pmatrix} 0 & I_n \\ -I_n & 0 \end{pmatrix} \right\}.$$

The space $Mn(\mathbf{H}) \subset M_{2n}(\mathbf{C})$ is a real algebra, but it is not a complex subspace of $M_2n(\mathbf{C})$. Multiplication (from the left) by $\mathbf{i}$ in $Mn(\mathbf{H})$ using entry-wise quaternionic multiplication and then mapping to the image in $M_2n(\mathbf{C})$ yields a different result than multiplying entry-wise by i directly in $M_2n(\mathbf{C})$. The quaternionic multiplication rules give $\mathbf{i}(X + \mathbf{j}Y) = (\mathbf{i}X) + \mathbf{j}(-\mathbf{i}Y)$ where the new X and Y are inside the parentheses.

The action of the quaternionic matrices on quaternionic vectors is now represented by complex quantities, but otherwise it is the same as for "ordinary" matrices and vectors. The quaternionic groups are thus embedded in $M_2n(C)$ where n is the dimension of the quaternionic matrices.

The determinant of a quaternionic matrix is defined in this representation as being the ordinary complex determinant of its representative matrix. The non-commutative nature of quaternionic multiplication would, in the quaternionic representation of matrices, be ambiguous. The way $Mn(\mathbf{H})$ is embedded in $M_2n(\mathbf{C})$ is not unique, but all such embeddings are related through $g \mapsto AgA^{-1}, g \in GL(2n, \mathbf{C})$ for $A \in O(2n, \mathbf{C})$, leaving the determinant unaffected.[17] The name of SL(n, $\mathbf{H}$) in this complex guise is SU*($2n$).

As opposed to in the case of $\mathbf{C}$, both the Hermitian and the skew-Hermitean case bring in something new when $\mathbf{H}$ is considered, so these cases are considered separately.

GL(n, H) and SL(n, H)   Under the identification above,

$$GL(n, \mathbb{H}) = \{g \in GL(2n, \mathbb{C}) | Jg = \bar{g}J, \det \ g \neq 0\} \equiv U^*(2n).$$

Its Lie algebra $\mathbf{gl}(n, \mathbf{H})$ is the set of all matrices in the image of the mapping $Mn(\mathbf{H}) \leftrightarrow M_2n(\mathbf{C})$ of above,

$$\mathfrak{gl}(n, \mathbb{H}) = \left\{ \begin{pmatrix} X & -\overline{Y} \\ Y & \overline{X} \end{pmatrix} \middle| X, Y \in \mathfrak{gl}(n, \mathbb{C}) \right\} \equiv \mathfrak{u}^*(2n).$$

The quaternionic special linear group is given by

$$SL(n, \mathbb{H}) = \{g \in GL(n, \mathbb{H}) | \det g = 1\} \equiv SU^*(2n),$$

where the determinant is taken on the matrices in $\mathbf{C}^{2n}$. The Lie algebra is

$$\mathfrak{sl}(n, \mathbb{H}) = \left\{ \begin{pmatrix} X & -\overline{Y} \\ Y & \overline{X} \end{pmatrix} \middle| \operatorname{Tr} X = 0 \right\} \equiv \mathfrak{su}^*(2n).$$

Sp(p, q) – the quaternionic unitary group   As above in the complex case, the normal form is

$$\varphi(x, y) = \pm \bar{\xi}_1 \eta_1 \pm \bar{\xi}_2 \eta_2 \cdots \pm \bar{\xi}_n \eta_n$$

and the number of plus-signs is independent of basis. When $V = \mathbf{H}^n$ with this form, Sp(φ) = Sp(p, q). The reason for the notation is that the group can be represented, using the above prescription, as a subgroup of Sp(n, $\mathbf{C}$) preserving a complex-hermitian form of signature $(2p, 2q)$[18] If p or $q = 0$ the group is denoted U(n, $\mathbf{H}$). It is sometimes called the **hyperunitary group**.

In quaternionic notation,

$$\Phi = \begin{pmatrix} I_p & 0 \\ 0 & -I_q \end{pmatrix} = I_{p,q}$$

meaning that *quaternionic* matrices of the form

will satisfy

$$\Phi^{-1}\mathcal{Q}^*\Phi = -\mathcal{Q},$$

see the section about $\mathbf{u}(p, q)$. Caution needs to be exercised when dealing with quaternionic matrix multiplication, but here only I and $-I$ are involved and these commute with every quaternion matrix. Now apply prescription (**8**) to each block,

$$\mathcal{X} = \begin{pmatrix} X_{1(p\times p)} & -\overline{X}_2 \\ X_2 & \overline{X}_1 \end{pmatrix}, \mathcal{Y} = \begin{pmatrix} Y_{1(q\times q)} & -\overline{Y}_2 \\ Y_2 & \overline{Y}_1 \end{pmatrix}, \mathcal{Z} = \begin{pmatrix} Z_{1(p\times q)} & -\overline{Z}_2 \\ Z_2 & \overline{Z}_1 \end{pmatrix},$$

and the relations in (**9**) will be satisfied if

$$X_1^* = -X, Y_1^* = -Y.$$

The Lie algebra becomes

$$\mathfrak{sp}(p, q) = \left\{ \left(\begin{bmatrix} X_{1(p\times p)} & -\overline{X}_2 \\ X_2 & \overline{X}_1 \\ Z_{1(p\times q)} & -\overline{Z}_2 \\ Z_2 & \overline{Z}_1 \end{bmatrix}^* \quad \begin{bmatrix} Z_{1(p\times q)} & -\overline{Z}_2 \\ Z_2 & \overline{Z}_1 \\ Y_{1(q\times q)} & -\overline{Y}_2 \\ Y_2 & \overline{Y}_1 \end{bmatrix} \right) \middle| X_1^* = -X, Y_1^* = -Y \right\}.$$

The group is given by

$$\mathrm{Sp}(p, q) = \{g \in \mathrm{GL}(n, \mathbb{H}) | I_{p,q}^{-1} g^* I_{p,q} g = I_{p+q}\} = \{g \in \mathrm{GL}(2n, \mathbb{C}) | K_{p,q}^{-1} g^* K_{p,q} g = I_{2(p+q)}, \qquad K = \mathrm{diag}(I_{p,q}, I_{p,q})\}.$$

Returning to the normal form of $\varphi(w, z)$ for $\mathrm{Sp}(p, q)$, make the substitutions $w \to u + jv$ and $z \to x + jy$ with $u, v, x, y \in \mathbf{C}^n$. Then

$$\varphi(w, z) = \begin{bmatrix} u^* & v^* \end{bmatrix} K_{p,q} \begin{bmatrix} x \\ y \end{bmatrix} + j \begin{bmatrix} u & -v \end{bmatrix} K_{p,q} \begin{bmatrix} y \\ x \end{bmatrix} = \varphi_1(w, z) + \mathbf{j}\varphi_2(w, z), \qquad K_{p,q} = \mathrm{diag}(I_{p,q}, I_{p,q})$$

viewed as a $\mathbf{H}$-valued form on $\mathbf{C}^{2n}$.[19] Thus the elements of $\mathrm{Sp}(p, q)$, viewed as linear transformations of $\mathbf{C}^{2n}$, preserve both a Hermitian form of signature $(2p, 2q)$ and a non-degenerate skew-symmetric form. Both forms take purely complex values and due to the prefactor of $\mathbf{j}$ of the second form, they are separately conserved. This means that

$$\mathrm{Sp}(p, q) = \mathrm{U}(\mathbb{C}^{2n}, \varphi_1) \cap \mathrm{Sp}(\mathbb{C}^{2n}, \varphi_2)$$

and this explains both the name of the group and the notation.

O*(2n)= O(n, H)- quaternionic orthogonal group   The normal form for a skew-hermitian form is given by

$$\varphi(x, y) = \bar{\xi}_1 \mathbf{j}\eta_1 + \bar{\xi}_2 \mathbf{j}\eta_2 \cdots + \bar{\xi}_n \mathbf{j}\eta_n,$$

where $\mathbf{j}$ is the third basis quaternion in the ordered listing $(\mathbf{1}, \mathbf{i}, \mathbf{j}, \mathbf{k})$. In this case, $\text{Aut}(\varphi) = O^*(2n)$ may be realized, using the complex matrix encoding of above, as a subgroup of $O(2n, \mathbf{C})$ which preserves a non-degenerate complex skew-hermitian form of signature (n, n).[20] From the normal form one sees that in quaternionic notation

$$\Phi = \begin{pmatrix} \mathbf{j} & 0 & \cdots & 0 \\ 0 & \mathbf{j} & \cdots & \vdots \\ \vdots & & \ddots & \\ 0 & \cdots & 0 & \mathbf{j} \end{pmatrix} \equiv \mathbf{j}_n$$

and from (6) follows that

for $V \in \mathbf{o}(2n)$. Now put

$$V = X + \mathbf{j}Y \leftrightarrow \begin{pmatrix} X & -\overline{Y} \\ Y & \overline{X} \end{pmatrix}$$

according to prescription (8). The same prescription yields for Φ,

$$\Phi \leftrightarrow \begin{pmatrix} 0 & -I_n \\ I_n & 0 \end{pmatrix} \equiv J_n.$$

Now the last condition in (9) in complex notation reads

$$\begin{pmatrix} X & -\overline{Y} \\ Y & \overline{X} \end{pmatrix}^* = \begin{pmatrix} 0 & -I_n \\ I_n & 0 \end{pmatrix} \begin{pmatrix} X & -\overline{Y} \\ Y & \overline{X} \end{pmatrix} \begin{pmatrix} 0 & -I_n \\ I_n & 0 \end{pmatrix} \Leftrightarrow X^{\mathsf{T}} = -X, \quad \overline{Y}^{\mathsf{T}} = Y.$$

The Lie algebra becomes

$$\mathbf{o}^*(2n) = \left\{ \begin{pmatrix} X & -\overline{Y} \\ Y & \overline{X} \end{pmatrix} \middle| X^{\mathsf{T}} = -X, \quad \overline{Y}^{\mathsf{T}} = Y \right\},$$

and the group is given by

$$O^*(2n) = \{g \in GL(n, \mathbb{H}) | \mathbf{j}_n^{-1} g^* \mathbf{j}_n g = I_n\} = \{g \in GL(2n, \mathbb{C}) | J_n^{-1} g^* J_n g = I_{2n}\}.$$

The group $SO^*(2n)$ can be characterized as

$$O^*(2n) = \{g \in O(2n, \mathbb{C}) | \theta(\overline{g}) = g\}, \text{[21]}$$

where the map $\theta \colon GL(2n, \mathbf{C}) \to GL(2n, \mathbf{C})$ is defined by $g \mapsto -J_{2n} g J_{2n}$. Also, the form determining the group can be viewed as a $\mathbf{H}$-valued form on $\mathbf{C}^{2n}$.[22] Make the substitutions $x \to w_1 + iw_2$ and $y \to z_1 + iz_2$ in the expression for the form. Then

$$\varphi(x, y) = \overline{w}_2 I_n z_1 - \overline{w}_1 I_n z_2 + \mathbf{j}(w_1 I_n z_1 + w_2 I_n z_2) = \overline{\varphi_1(w, z)} + \mathbf{j}\varphi_2(w, z).$$

The form φ_1 is Hermitian (while the first form on the left hand side is skew-Hermitian) of signature (n, n). The signature is made evident by a change of basis from $(\mathbf{e}, \mathbf{f})$ to $((\mathbf{e} + i\mathbf{f})/\sqrt{2}, (\mathbf{e} - i\mathbf{f})/\sqrt{2})$ where $\mathbf{e}$, $\mathbf{f}$ are the first and last n basis vectors respectively. The second form, φ_2 is symmetric positive definite. Thus, due to the factor $\mathbf{j}$, $\mathrm{O}^*(2n)$ preserves both separately and it may be concluded that

$$\mathrm{O}^*(2n) = \mathrm{O}(2n, \mathbb{C}) \cap \mathrm{U}(\mathbb{C}^{2n}, \varphi_1),$$

and the notation "O" is explained.

14.4 Classical groups over general fields or algebras

Classical groups, more broadly considered in algebra, provide particularly interesting matrix groups. When the field F of coefficients of the matrix group is either real number or complex numbers, these groups are just the classical Lie groups. When the ground field is a finite field, then the classical groups are groups of Lie type. These groups play an important role in the classification of finite simple groups. Also, one may consider classical groups over a unital associative algebra R over F; where $R = \mathbf{H}$ (an algebra over reals) represents an important case. For the sake of generality the article will refer to groups over R, where R may be the ground field F itself.

Considering their abstract group theory, many linear groups have a "**special**" subgroup, usually consisting of the elements of determinant 1 over the ground field, and most of them have associated "**projective**" quotients, which are the quotients by the center of the group. For orthogonal groups in characteristic 2 "S" has a different meaning.

The word "**general**" in front of a group name usually means that the group is allowed to multiply some sort of form by a constant, rather than leaving it fixed. The subscript n usually indicates the dimension of the module on which the group is acting; it is a vector space if $R = F$. Caveat: this notation clashes somewhat with the n of Dynkin diagrams, which is the rank.

14.4.1 General and special linear groups

The general linear group $\mathrm{GL}n(R)$ is the group of all R-linear automorphisms of R^n. There is a subgroup: the special linear group $\mathrm{SL}n(R)$, and their quotients: the projective general linear group $\mathrm{PGL}n(R) = \mathrm{GL}n(R)/\mathrm{Z}(\mathrm{GL}n(R))$ and the projective special linear group $\mathrm{PSL}n(R) = \mathrm{SL}n(R)/\mathrm{Z}(\mathrm{SL}n(R))$. The projective special linear group $\mathrm{PSL}n(F)$ over a field F is simple for $n \geq 2$, except for the two cases when $n = 2$ and the field has order 2 or 3.

14.4.2 Unitary groups

The unitary group $\mathrm{U}n(R)$ is a group preserving a sesquilinear form on a module. There is a subgroup, the special unitary group $\mathrm{SU}n(R)$ and their quotients the projective unitary group $\mathrm{PU}n(R) = \mathrm{U}n(R)/\mathrm{Z}(\mathrm{U}n(R))$ and the projective special unitary group $\mathrm{PSU}n(R) = \mathrm{SU}n(R)/\mathrm{Z}(\mathrm{SU}n(R))$

14.4.3 Symplectic groups

The symplectic group $\mathrm{Sp}_{2}n(R)$ preserves a skew symmetric form on a module. It has a quotient, the projective symplectic group $\mathrm{PSp}_{2}n(R)$. The general symplectic group $\mathrm{GSp}_{2}n(R)$ consists of the automorphisms of a module multiplying a skew symmetric form by some invertible scalar. The projective symplectic group $\mathrm{PSp}_{2}n(\mathbf{F}q)$ over a finite field is simple for $n \geq 1$, except for the two cases when $n = 1$ and the field has order 2 or 3.

14.4.4 Orthogonal groups

The orthogonal group $\mathrm{O}n(R)$ preserves a non-degenerate quadratic form on a module. There is a subgroup, the special orthogonal group $\mathrm{SO}n(R)$ and quotients, the projective orthogonal group $\mathrm{PO}n(R)$, and the projective special orthogonal group PSO$n(R)$. In characteristic 2 the determinant is always 1, so the special orthogonal group is often defined as the subgroup of elements of Dickson invariant 1.

There is a nameless group often denoted by $\Omega n(R)$ consisting of the elements of the orthogonal group of elements of spinor norm 1, with corresponding subgroup and quotient groups S$\Omega n(R)$, P$\Omega n(R)$, PS$\Omega n(R)$. (For positive definite quadratic forms over the reals, the group Ω happens to be the same as the orthogonal group, but in general it is smaller.) There is also a double cover of $\Omega n(R)$, called the pin group Pin$n(R)$, and it has a subgroup called the spin group Spin$n(R)$. The general orthogonal group GO$n(R)$ consists of the automorphisms of a module multiplying a quadratic form by some invertible scalar.

14.4.5   Notational conventions

For more details on this topic, see Group of Lie type § Notation issues.

14.5   Contrast with exceptional Lie groups

Contrasting with the classical Lie groups are the exceptional Lie groups, G_2, F_4, E_6, E_7, E_8, which share their abstract properties, but not their familiarity.[23] These were only discovered around 1890 in the classification of the simple Lie algebras over the complex numbers by Wilhelm Killing and Élie Cartan.

14.6   Notes

[1] Here, *special* means the subgroup of the full automorphism group whose elements have determinant 1.

[2] Rossmann 2002 p. 94.

[3] Weyl 1939

[4] Rossmann 2002 p. 91.

[5] Rossmann 2002 p, 94

[6] Rossmann 2002 p. 103.

[7] Goodman & Wallach 2009 See end of chapter 1.

[8] Rossmann 2002p. 93.

[9] Rossmann 2002 p. 105

[10] Rossmann 2002 p. 91

[11] Rossmann 2002 p. 92

[12] Rossmann 2002 p. 105

[13] Rossmann 2002 p. 107.

[14] Rossmann 2002 p. 93

[15] Rossmann 2002 p. 95.

[16] Rossmann 2002 p. 94.

[17] Goodman & Wallach 2009 Exercise 14, Section 1.1.

[18] Rossmann 2002 p. 94.

[19] Goodman & Wallach 2009Exercise 11, Chapter 1.

[20] Rossmann 2002 p. 94.

[21] Goodman & Wallach 2009 p.11.

[22] Goodman & Wallach 2009 Exercise 12 Chapter 1.

[23] Wybourne, B. G. (1974). *Classical Groups for Physicists*, Wiley-Interscience. ISBN 0471965057 .

14.7 References

- E. Artin, *Geometric algebra*, Interscience (1957)

- Dieudonné, Jean (1955), *La géométrie des groupes classiques*, Ergebnisse der Mathematik und ihrer Grenzgebiete (N.F.), Heft 5, Berlin, New York: Springer-Verlag, ISBN 978-0-387-05391-2, MR 0072144

- Goodman, Roe; Wallach, Nolan R. (2009), *Symmetry, Representations,and Invariants*, Graduate texts in mathematics **255**, Springer-Verlag, ISBN 978-0-387-79851-6

- Knapp, A. W. (2002). *Lie groups beyond an introduction.* Progress in Mathematics **120** (2nd ed.). Boston·Basel·Berlin: Birkhäuser. ISBN 0-8176-4259-5.

- V. L. Popov (2001), "Classical group", in Hazewinkel, Michiel, *Encyclopedia of Mathematics*, Springer, ISBN 978-1-55608-010-4

- Rossmann, Wulf (2002), *Lie Groups - An Introduction Through Linear Groups*, Oxford Graduate Texts in Mathematics, Oxford Science Publications, ISBN 0 19 859683 9

Chapter 15

Subgroup

This article is about the mathematical concept. For the galaxy-related concept, see galaxy group.

In mathematics, given a group G under a binary operation $*$, a subset H of G is called a **subgroup** of G if H also forms a group under the operation $*$. More precisely, H is a subgroup of G if the restriction of $*$ to $H \times H$ is a group operation on H. This is usually denoted $H \leq G$, read as "H is a subgroup of G".

A **proper subgroup** of a group G is a subgroup H which is a proper subset of G (i.e. $H \neq G$). This is usually represented notationally by $H < G$, read as "H is a proper subgroup of G".

The **trivial subgroup** of any group is the subgroup $\{e\}$ consisting of just the identity element.

If H is a subgroup of G, then G is sometimes called an **overgroup** of H.

The same definitions apply more generally when G is an arbitrary semigroup, but this article will only deal with subgroups of groups. The group G is sometimes denoted by the ordered pair $(G, *)$, usually to emphasize the operation $*$ when G carries multiple algebraic or other structures.

This article will write ab for $a * b$, as is usual.

15.1   Basic properties of subgroups

- A subset H of the group G is a subgroup of G if and only if it is nonempty and closed under products and inverses. (The closure conditions mean the following: whenever a and b are in H, then ab and a^{-1} are also in H. These two conditions can be combined into one equivalent condition: whenever a and b are in H, then ab^{-1} is also in H.) In the case that H is finite, then H is a subgroup if and only if H is closed under products. (In this case, every element a of H generates a finite cyclic subgroup of H, and the inverse of a is then $a^{-1} = a^{n-1}$, where n is the order of a.)

- The above condition can be stated in terms of a homomorphism; that is, H is a subgroup of a group G if and only if H is a subset of G and there is an inclusion homomorphism (i.e., i$(a) = a$ for every a) from H to G.

- The identity of a subgroup is the identity of the group: if G is a group with identity eG, and H is a subgroup of G with identity eH, then $eH = eG$.

- The inverse of an element in a subgroup is the inverse of the element in the group: if H is a subgroup of a group G, and a and b are elements of H such that $ab = ba = eH$, then $ab = ba = eG$.

- The intersection of subgroups A and B is again a subgroup.[1] The union of subgroups A and B is a subgroup if and only if either A or B contains the other, since for example 2 and 3 are in the union of 2Z and 3Z but their sum 5 is not. Another example is the union of the x-axis and the y-axis in the plane (with the addition operation); each of these objects is a subgroup but their union is not. This also serves as an example of two subgroups, whose intersection is precisely the identity.

- If S is a subset of G, then there exists a minimum subgroup containing S, which can be found by taking the intersection of all of subgroups containing S; it is denoted by $<S>$ and is said to be the subgroup generated by S. An element of G is in $<S>$ if and only if it is a finite product of elements of S and their inverses.

- Every element a of a group G generates the cyclic subgroup $<a>$. If $<a>$ is isomorphic to $\mathbf{Z}/n\mathbf{Z}$ for some positive integer n, then n is the smallest positive integer for which $a^n = e$, and n is called the *order* of a. If $<a>$ is isomorphic to $\mathbf{Z}$, then a is said to have *infinite order*.

- The subgroups of any given group form a complete lattice under inclusion, called the lattice of subgroups. (While the infimum here is the usual set-theoretic intersection, the supremum of a set of subgroups is the subgroup *generated by* the set-theoretic union of the subgroups, not the set-theoretic union itself.) If e is the identity of G, then the trivial group $\{e\}$ is the minimum subgroup of G, while the maximum subgroup is the group G itself.

15.2   Cosets and Lagrange's theorem

Given a subgroup H and some a in G, we define the **left coset** $aH = \{ah : h \text{ in } H\}$. Because a is invertible, the map $\varphi : H \rightarrow aH$ given by $\varphi(h) = ah$ is a bijection. Furthermore, every element of G is contained in precisely one left coset of H; the left cosets are the equivalence classes corresponding to the equivalence relation $a_1 \sim a_2$ if and only if $a_1^{-1}a_2$ is in H. The number of left cosets of H is called the index of H in G and is denoted by $[G : H]$.

Lagrange's theorem states that for a finite group G and a subgroup H,

$$[G : H] = \frac{|G|}{|H|}$$

where $|G|$ and $|H|$ denote the orders of G and H, respectively. In particular, the order of every subgroup of G (and the order of every element of G) must be a divisor of $|G|$.

Right cosets are defined analogously: $Ha = \{ha : h \text{ in } H\}$. They are also the equivalence classes for a suitable equivalence relation and their number is equal to $[G : H]$.

If $aH = Ha$ for every a in G, then H is said to be a normal subgroup. Every subgroup of index 2 is normal: the left cosets, and also the right cosets, are simply the subgroup and its complement. More generally, if p is the lowest prime dividing the order of a finite group G, then any subgroup of index p (if such exists) is normal.

15.3   Example: Subgroups of $\mathbf{Z}_8$

Let G be the cyclic group $\mathbf{Z}_8$ whose elements are

$$G = \{0, 2, 4, 6, 1, 3, 5, 7\}$$

and whose group operation is addition modulo eight. Its Cayley table is

This group has two nontrivial subgroups: $J=\{0,4\}$ and $H=\{0,2,4,6\}$, where J is also a subgroup of H. The Cayley table for H is the top-left quadrant of the Cayley table for G. The group G is cyclic, and so are its subgroups. In general, subgroups of cyclic groups are also cyclic.

15.4   Example: Subgroups of S_4 (the symmetric group on 4 elements)

Every group has as many small subgroups as neutral elements on the main diagonal:

The trivial group and two-element groups Z_2. These small subgroups are not counted in the following list.

G

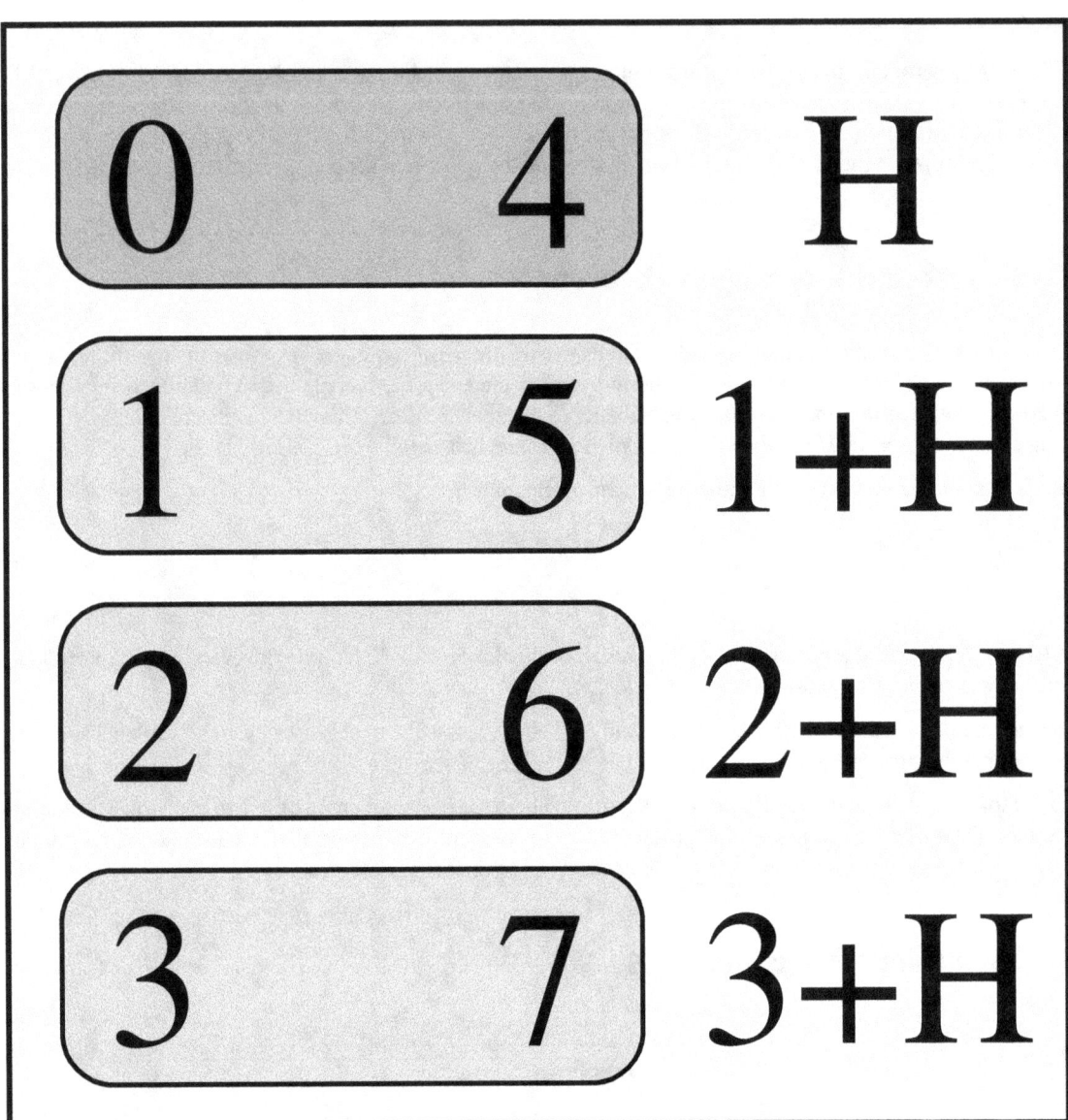

G is the group $\mathbb{Z}/8\mathbb{Z}$, the integers mod 8 under addition. The subgroup H contains only 0 and 4, and is isomorphic to $\mathbb{Z}/2\mathbb{Z}$. There are four left cosets of H: H itself, 1+H, 2+H, and 3+H (written using additive notation since this is an additive group). Together they partition the entire group G into equal-size, non-overlapping sets. The index [G : H] is 4.

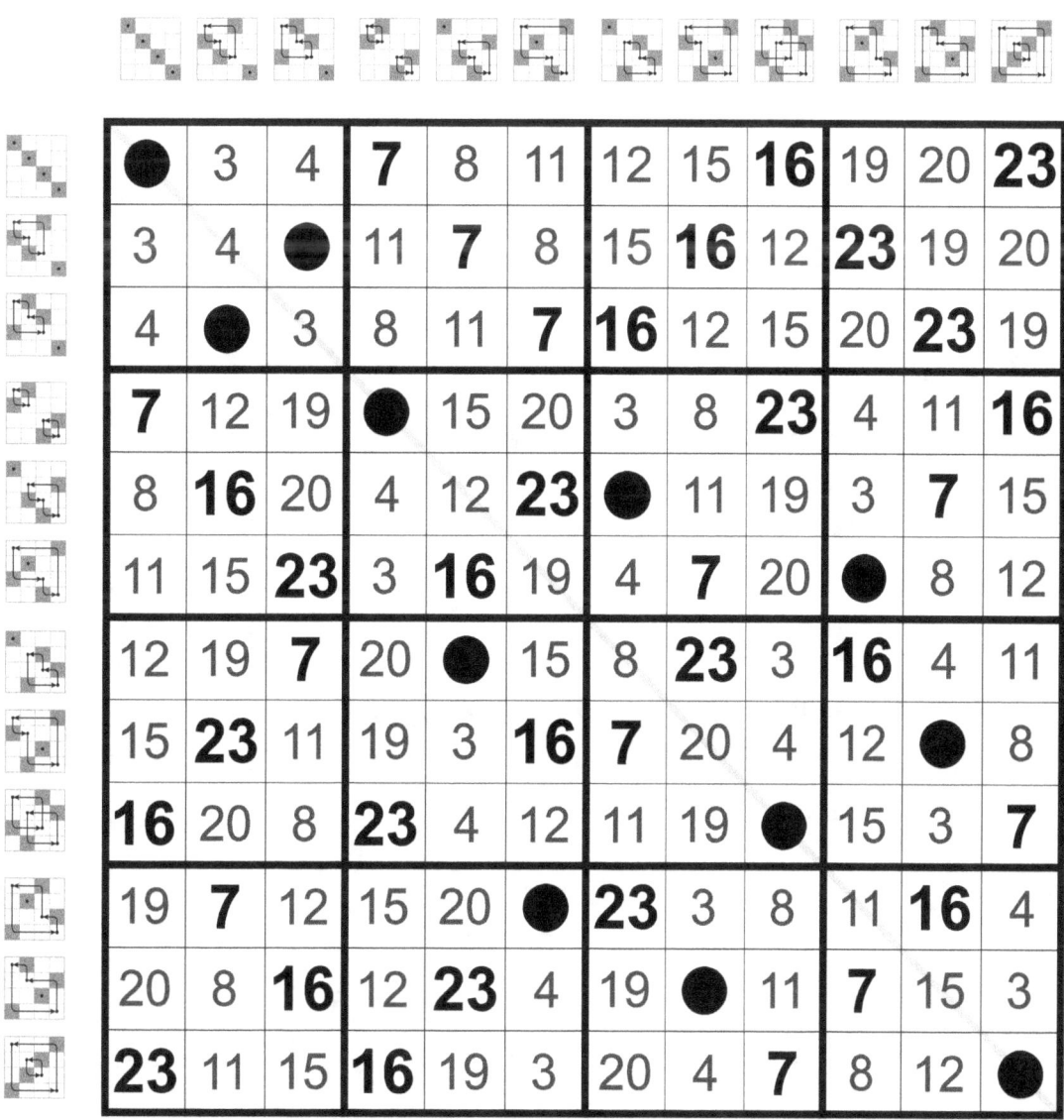

The alternating group A$_4$ showing only the even permutations
Subgroups:

15.4.1   12 elements

15.4.2   8 elements

15.4.3   6 elements

15.4.4   4 elements

15.4.5   3 elements

15.5   See also

- Cartan subgroup

- Fitting subgroup
- Stable subgroup
- Fixed-point subgroup

15.6   Notes

[1] Jacobson (2009), p. 41

15.7   References

- Jacobson, Nathan (2009), *Basic algebra* **1** (2nd ed.), Dover, ISBN 978-0-486-47189-1.

Chapter 16

Topological group

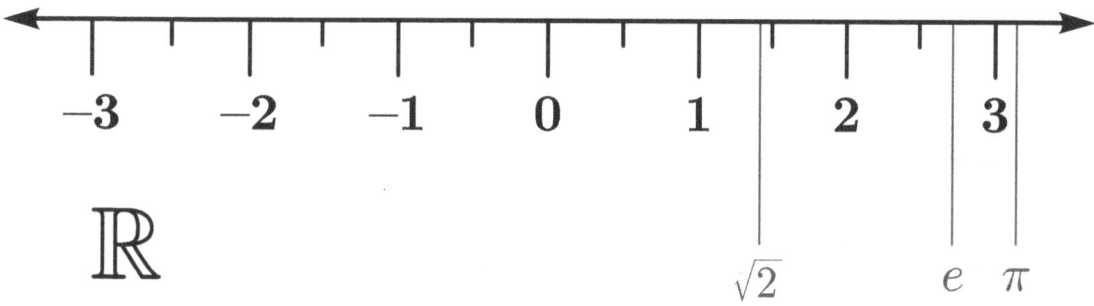

The real numbers form a topological group under addition

In mathematics, a **topological group** is a group G together with a topology on G such that the group's binary operation and the group's inverse function are continuous functions with respect to the topology.[1] A topological group is a mathematical object with both an algebraic structure and a topological structure. Thus, one may perform algebraic operations, because of the group structure, and one may talk about continuous functions, because of the topology.

Topological groups, along with continuous group actions, are used to study continuous symmetries, which have many applications, for example in physics.

16.1 Formal definition

A **topological group** G is a topological space and group such that the group operations of product:

$$G \times G \to G : (x, y) \mapsto xy$$

and taking inverses:

$$G \to G : x \mapsto x^{-1}$$

are continuous functions. Here, $G \times G$ is viewed as a topological space by using the product topology.

Although not part of this definition, many authors[2] require that the topology on G be Hausdorff; this corresponds to the identity map $* \to G$ being a closed inclusion (hence also a cofibration). The reasons, and some equivalent conditions, are discussed below. In the end, this is not a serious restriction—any topological group can be made Hausdorff in a canonical fashion.[3]

In the language of category theory, topological groups can be defined concisely as group objects in the category of topological spaces, in the same way that ordinary groups are group objects in the category of sets. Note that the axioms are given in terms of the maps (binary product, unary inverse, and nullary identity), hence are categorical definitions. Adding the further requirement of Hausdorff (and cofibration) corresponds to refining to a model category.

16.1.1 Homomorphisms

A homomorphism between two topological groups G and H is just a continuous group homomorphism $G \to H$. An isomorphism of topological groups is a group isomorphism which is also a homeomorphism of the underlying topological spaces. This is stronger than simply requiring a continuous group isomorphism—the inverse must also be continuous. There are examples of topological groups which are isomorphic as ordinary groups but not as topological groups. Indeed, any nondiscrete topological group is also a topological group when considered with the discrete topology. The underlying groups are the same, but as topological groups there is not an isomorphism.

Topological groups, together with their homomorphisms, form a category.

16.2 Examples

Every group can be trivially made into a topological group by considering it with the discrete topology; such groups are called discrete groups. In this sense, the theory of topological groups subsumes that of ordinary groups.

The real numbers $\mathbf{R}$, together with addition as operation and its usual topology, form a topological group. More generally, Euclidean n-space $\mathbf{R}^n$ with addition and standard topology is a topological group. More generally yet, the additive groups of all topological vector spaces, such as Banach spaces or Hilbert spaces, are topological groups.

The above examples are all abelian. Examples of non-abelian topological groups are given by the classical groups. For instance, the general linear group $GL(n,\mathbf{R})$ of all invertible n-by-n matrices with real entries can be viewed as a topological group with the topology defined by viewing $GL(n,\mathbf{R})$ as a subset of Euclidean space $\mathbf{R}^{n \times n}$.

An example of a topological group which is not a Lie group is given by the rational numbers $\mathbf{Q}$ with the topology inherited from $\mathbf{R}$. This is a countable space and it does not have the discrete topology. For a nonabelian example, consider the subgroup of rotations of $\mathbf{R}^3$ generated by two rotations by irrational multiples of 2π about different axes.

In every Banach algebra with multiplicative identity, the set of invertible elements forms a topological group under multiplication.

16.3 Properties

The algebraic and topological structures of a topological group interact in non-trivial ways. For example, in any topological group the identity component (i.e. the connected component containing the identity element) is a closed normal subgroup. This is because if C is the identity component, $a*C$ is the component of G (the group) containing a. In fact, the collection of all left cosets (or right cosets) of C in G is equal to the collection of all components of G. Therefore, the quotient topology induced by the quotient map from G to G/C is totally disconnected.[4]

The inversion operation on a topological group G is a homeomorphism from G to itself. Likewise, if a is any element of G, then left or right multiplication by a yields a homeomorphism $G \to G$.

Every topological group can be viewed as a uniform space in two ways; the *left uniformity* turns all left multiplications into uniformly continuous maps while the *right uniformity* turns all right multiplications into uniformly continuous maps.

If G is not abelian, then these two need not coincide. The uniform structures allow one to talk about notions such as completeness, uniform continuity and uniform convergence on topological groups.

As a uniform space, every topological group is completely regular. It follows that if a topological group is T_0 (Kolmogorov) then it is already T_2 (Hausdorff), even $T_3\frac{1}{2}$ (Tychonoff).

Every subgroup of a topological group is itself a topological group when given the subspace topology. If H is a subgroup of G, the set of left or right cosets G/H is a topological space when given the quotient topology (the finest topology on G/H which makes the natural projection $q : G \to G/H$ continuous). One can show that the quotient map $q : G \to G/H$ is always open.

Every open subgroup H is also closed, since the complement of H is the open set given by the union of open sets gH for g in $G \setminus H$.

If H is a normal subgroup of G, then the factor group, G/H becomes a topological group when given the quotient topology. However, if H is not closed in the topology of G, then G/H will not be T_0 even if G is. It is therefore natural to restrict oneself to the category of T_0 topological groups, and restrict the definition of *normal* to *normal and closed*.

The isomorphism theorems known from ordinary group theory are not always true in the topological setting. This is because a bijective homomorphism need not be an isomorphism of topological groups. The theorems are valid if one places certain restrictions on the maps involved. For example, the first isomorphism theorem states that if $f : G \to H$ is a homomorphism then $G/\ker(f)$ is isomorphic to $\mathrm{im}(f)$ if and only if the map f is open onto its image.

If H is a subgroup of G then the closure of H is also a subgroup. Likewise, if H is a normal subgroup, the closure of H is normal.

A topological group G is Hausdorff if and only if the trivial one-element subgroup is closed in G. If G is not Hausdorff then one can obtain a Hausdorff group by passing to the quotient space G/K where K is the closure of the identity. This is equivalent to taking the Kolmogorov quotient of G.

The fundamental group of a topological group is always abelian. This is a special case of the fact that the fundamental group of an H-space is abelian, since topological groups are H-spaces.

16.4 Relationship to other areas of mathematics

A compact group is a topological group whose topology is compact. Compact groups are a natural generalisation of finite groups with the discrete topology and have properties that carry over in significant fashion. Compact groups have a well-understood theory, in relation to group actions and representation theory.

Of particular importance in harmonic analysis are the locally compact groups, because they admit a natural notion of measure and integral, given by the Haar measure. The theory of group representations is almost identical for finite groups and for compact topological groups. In general, σ-compact Baire topological groups are locally compact.

In topology, the homeomorphism group of a topological space is the group consisting of all homeomorphisms from the space to itself with function composition as the group operation. The homeomorphism group can be given a topology, such as the compact-open topology (in the case of regular, locally compact spaces), making it into a topological group.

16.5 Generalizations

Various generalizations of topological groups can be obtained by weakening the continuity conditions:[5]

- A *semitopological group* is a group G with a topology such that for each c in G the two functions $G \to G$ defined by $x \mapsto xc$ and $x \mapsto cx$ are continuous.

- A *quasitopological group* is a semitopological group in which the function mapping elements to their inverses is also continuous.

- A *paratopological group* is a group with a topology such that the group operation is continuous.

16.6   See also

- Lie group

- Algebraic group

- Profinite group

- Topological ring

- Locally compact group

16.7   Notes

[1] van der Waerden, Bartel Leendert, et al (2003). "Topological algebra". *Algebra*. Vol. 2. Springer. p. 256. ISBN 978-0-387-40625-1.

[2] Armstrong, p. 73; Bredon, p. 51; Willard, p. 91.

[3] D. Ramakrishnan and R. Valenza (1999). "Fourier Analysis on Number Fields". Springer-Verlag, Graduate Texts in Mathematics. Pp. 6–7.

[4] O.V. Mel'nikov (2001), "Topological group", in Hazewinkel, Michiel, *Encyclopedia of Mathematics*, Springer, ISBN 978-1-55608-010-4

[5] Arhangel'skii & Tkachenko, p12

16.8   References

- Arhangel'skii, Alexander; Tkachenko, Mikhail (2008). *Topological Groups and Related Structures*. Atlantis Press. ISBN 90-78677-06-6.

- Armstrong, M. A. (1997). *Basic Topology* (1st ed.). Springer Verlag. ISBN 0-387-90839-0.

- Bourbaki, Nicolas (1966). *General Topology*, Part 1, chapter 3: "Topological Groups", Hermann, Paris.

- Bredon, Glen E. (1997). *Topology and Geometry*. Graduate Texts in Mathematics (1 ed.). Springer. ISBN 0-387-97926-3.

- Husain, Taqdir (1981). *Introduction to Topological Groups*. Philadelphia: R.E. Krieger Pub. Co. ISBN 0-89874-193-9.

- Pontryagin, Lev S. (1986). *Topological Groups*. trans. from Russian by Arlen Brown and P.S.V. Naidu (3rd ed.). New York: Gordon and Breach Science Publishers. ISBN 2-88124-133-6.

- Porteous, I.R. (1969). *Topological Geometry*. Van Nostrand Reinhold. pp. 336–352. ISBN 0-442-06606-6. Zbl 0186.06304.

- Willard, Stephen (2004). *General Topology*. Dover Publications. ISBN 0-486-43479-6.

Chapter 17

Abelian group

For the group described by the archaic use of this term, see Symplectic group.

In abstract algebra, an **abelian group**, also called a **commutative group**, is a group in which the result of applying the group operation to two group elements does not depend on the order in which they are written (the axiom of commutativity). Abelian groups generalize the arithmetic of addition of integers. They are named after Niels Henrik Abel.[1]

The concept of an abelian group is one of the first concepts encountered in undergraduate abstract algebra, with many other basic objects, such as a module and a vector space, being its refinements. The theory of abelian groups is generally simpler than that of their non-abelian counterparts, and finite abelian groups are very well understood. On the other hand, the theory of infinite abelian groups is an area of current research.

17.1   Definition

An abelian group is a set, A, together with an operation • that combines any two elements a and b to form another element denoted $a • b$. The symbol • is a general placeholder for a concretely given operation. To qualify as an abelian group, the set and operation, $(A, •)$, must satisfy five requirements known as the *abelian group axioms*:

Closure   For all a, b in A, the result of the operation $a • b$ is also in A.

Associativity   For all a, b and c in A, the equation $(a • b) • c = a • (b • c)$ holds.

Identity element   There exists an element e in A, such that for all elements a in A, the equation $e • a = a • e = a$ holds.

Inverse element   For each a in A, there exists an element b in A such that $a • b = b • a = e$, where e is the identity element.

Commutativity   For all a, b in A, $a • b = b • a$.

More compactly, an abelian group is a commutative group. A group in which the group operation is not commutative is called a "non-abelian group" or "non-commutative group".

17.2   Facts

17.2.1   Notation

See also: Additive group and Multiplicative group

There are two main notational conventions for abelian groups – additive and multiplicative.

Generally, the multiplicative notation is the usual notation for groups, while the additive notation is the usual notation for modules and rings. The additive notation may also be used to emphasize that a particular group is abelian, whenever both abelian and non-abelian groups are considered, some notable exceptions being near-rings and partially ordered groups, where an operation is written additively even when non-abelian.

17.2.2 Multiplication table

To verify that a finite group is abelian, a table (matrix) – known as a Cayley table – can be constructed in a similar fashion to a multiplication table. If the group is $G = \{g_1 = e, g_2, ..., gn\}$ under the operation ·, the (i, j)th entry of this table contains the product $gi \cdot gj$. The group is abelian if and only if this table is symmetric about the main diagonal.

This is true since if the group is abelian, then $gi \cdot gj = gj \cdot gi$. This implies that the (i, j)th entry of the table equals the (j, i)th entry, thus the table is symmetric about the main diagonal.

17.3 Examples

- For the integers and the operation addition "+", denoted $(\mathbf{Z}, +)$, the operation + combines any two integers to form a third integer, addition is associative, zero is the additive identity, every integer n has an additive inverse, $-n$, and the addition operation is commutative since $m + n = n + m$ for any two integers m and n.

- Every cyclic group G is abelian, because if x, y are in G, then $xy = a^m a^n = a^{m+n} = a^{n+m} = a^n a^m = yx$. Thus the integers, $\mathbf{Z}$, form an abelian group under addition, as do the integers modulo n, $\mathbf{Z}/n\mathbf{Z}$.

- Every ring is an abelian group with respect to its addition operation. In a commutative ring the invertible elements, or units, form an abelian multiplicative group. In particular, the real numbers are an abelian group under addition, and the nonzero real numbers are an abelian group under multiplication.

- Every subgroup of an abelian group is normal, so each subgroup gives rise to a quotient group. Subgroups, quotients, and direct sums of abelian groups are again abelian.

- The concepts of abelian group and $\mathbf{Z}$-module agree. More specifically, every $\mathbf{Z}$-module is an abelian group with its operation of addition, and every abelian group is a module over the ring of integers $\mathbf{Z}$ in a unique way.

In general, matrices, even invertible matrices, do not form an abelian group under multiplication because matrix multiplication is generally not commutative. However, some groups of matrices are abelian groups under matrix multiplication – one example is the group of 2×2 rotation matrices.

17.4 Historical remarks

Abelian groups were named after Norwegian mathematician Niels Henrik Abel by Camille Jordan because Abel found that the commutativity of the group of a polynomial implies that the roots of the polynomial can be calculated by using radicals. See Section 6.5 of Cox (2004) for more information on the historical background.

17.5 Properties

If n is a natural number and x is an element of an abelian group G written additively, then nx can be defined as $x + x + ... + x$ (n summands) and $(-n)x = -(nx)$. In this way, G becomes a module over the ring $\mathbf{Z}$ of integers. In fact, the modules over $\mathbf{Z}$ can be identified with the abelian groups.

Theorems about abelian groups (i.e. modules over the principal ideal domain **Z**) can often be generalized to theorems about modules over an arbitrary principal ideal domain. A typical example is the classification of finitely generated abelian groups which is a specialization of the structure theorem for finitely generated modules over a principal ideal domain. In the case of finitely generated abelian groups, this theorem guarantees that an abelian group splits as a direct sum of a torsion group and a free abelian group. The former may be written as a direct sum of finitely many groups of the form $\mathbf{Z}/p^k\mathbf{Z}$ for p prime, and the latter is a direct sum of finitely many copies of **Z**.

If $f, g : G \to H$ are two group homomorphisms between abelian groups, then their sum $f + g$, defined by $(f + g)(x) = f(x) + g(x)$, is again a homomorphism. (This is not true if H is a non-abelian group.) The set $\text{Hom}(G, H)$ of all group homomorphisms from G to H thus turns into an abelian group in its own right.

Somewhat akin to the dimension of vector spaces, every abelian group has a *rank*. It is defined as the cardinality of the largest set of linearly independent elements of the group. The integers and the rational numbers have rank one, as well as every subgroup of the rationals.

17.6   Finite abelian groups

Cyclic groups of integers modulo n, $\mathbf{Z}/n\mathbf{Z}$, were among the first examples of groups. It turns out that an arbitrary finite abelian group is isomorphic to a direct sum of finite cyclic groups of prime power order, and these orders are uniquely determined, forming a complete system of invariants. The automorphism group of a finite abelian group can be described directly in terms of these invariants. The theory had been first developed in the 1879 paper of Georg Frobenius and Ludwig Stickelberger and later was both simplified and generalized to finitely generated modules over a principal ideal domain, forming an important chapter of linear algebra.

17.6.1   Classification

The **fundamental theorem of finite abelian groups** states that every finite abelian group G can be expressed as the direct sum of cyclic subgroups of prime-power order. This is a special case of the fundamental theorem of finitely generated abelian groups when G has zero rank.

The cyclic group $\mathbf{Z}mn$ of order mn is isomorphic to the direct sum of $\mathbf{Z}m$ and $\mathbf{Z}n$ if and only if m and n are coprime. It follows that any finite abelian group G is isomorphic to a direct sum of the form

in either of the following canonical ways:

- the numbers $k_1, ..., ku$ are powers of primes

- k_1 divides k_2, which divides k_3, and so on up to ku.

For example, $\mathbf{Z}_{15}$ can be expressed as the direct sum of two cyclic subgroups of order 3 and 5: $\mathbf{Z}_{15} \cong \{0, 5, 10\} \oplus \{0, 3, 6, 9, 12\}$. The same can be said for any abelian group of order 15, leading to the remarkable conclusion that all abelian groups of order 15 are isomorphic.

For another example, every abelian group of order 8 is isomorphic to either $\mathbf{Z}_8$ (the integers 0 to 7 under addition modulo 8), $\mathbf{Z}_4 \oplus \mathbf{Z}_2$ (the odd integers 1 to 15 under multiplication modulo 16), or $\mathbf{Z}_2 \oplus \mathbf{Z}_2 \oplus \mathbf{Z}_2$.

See also list of small groups for finite abelian groups of order 16 or less.

17.6.2 Automorphisms

One can apply the fundamental theorem to count (and sometimes determine) the automorphisms of a given finite abelian group G. To do this, one uses the fact that if G splits as a direct sum $H \oplus K$ of subgroups of coprime order, then $\text{Aut}(H \oplus K) \cong \text{Aut}(H) \oplus \text{Aut}(K)$.

Given this, the fundamental theorem shows that to compute the automorphism group of G it suffices to compute the automorphism groups of the Sylow p-subgroups separately (that is, all direct sums of cyclic subgroups, each with order a power of p). Fix a prime p and suppose the exponents ei of the cyclic factors of the Sylow p-subgroup are arranged in increasing order:

$$e_1 \leq e_2 \leq \cdots \leq e_n$$

for some $n > 0$. One needs to find the automorphisms of

$$\mathbf{Z}_{p^{e_1}} \oplus \cdots \oplus \mathbf{Z}_{p^{e_n}}.$$

One special case is when $n = 1$, so that there is only one cyclic prime-power factor in the Sylow p-subgroup P. In this case the theory of automorphisms of a finite cyclic group can be used. Another special case is when n is arbitrary but $ei = 1$ for $1 \leq i \leq n$. Here, one is considering P to be of the form

$$\mathbf{Z}_p \oplus \cdots \oplus \mathbf{Z}_p,$$

so elements of this subgroup can be viewed as comprising a vector space of dimension n over the finite field of p elements $\mathbf{F}p$. The automorphisms of this subgroup are therefore given by the invertible linear transformations, so

$$\text{Aut}(P) \cong \text{GL}(n, \mathbf{F}_p),$$

where GL is the appropriate general linear group. This is easily shown to have order

$$|\text{Aut}(P)| = (p^n - 1) \cdots (p^n - p^{n-1}).$$

In the most general case, where the ei and n are arbitrary, the automorphism group is more difficult to determine. It is known, however, that if one defines

$$d_k = \max\{r | e_r = e_k\}$$

and

$$c_k = \min\{r | e_r = e_k\}$$

then one has in particular $dk \geq k$, $ck \leq k$, and

$$|\text{Aut}(P)| = \prod_{k=1}^{n} (p^{d_k} - p^{k-1}) \prod_{j=1}^{n} (p^{e_j})^{n-d_j} \prod_{i=1}^{n} (p^{e_i-1})^{n-c_i+1}.$$

One can check that this yields the orders in the previous examples as special cases (see [Hillar,Rhea]).

17.7   Infinite abelian groups

The simplest infinite abelian group is the infinite cyclic group $\mathbf{Z}$. Any finitely generated abelian group A is isomorphic to the direct sum of r copies of $\mathbf{Z}$ and a finite abelian group, which in turn is decomposable into a direct sum of finitely many cyclic groups of primary orders. Even though the decomposition is not unique, the number r, called the **rank** of A, and the prime powers giving the orders of finite cyclic summands are uniquely determined.

By contrast, classification of general infinitely generated abelian groups is far from complete. Divisible groups, i.e. abelian groups A in which the equation $nx = a$ admits a solution $x \in A$ for any natural number n and element a of A, constitute one important class of infinite abelian groups that can be completely characterized. Every divisible group is isomorphic to a direct sum, with summands isomorphic to $\mathbf{Q}$ and Prüfer groups $\mathbf{Q}p/\mathbf{Z}p$ for various prime numbers p, and the cardinality of the set of summands of each type is uniquely determined.[2] Moreover, if a divisible group A is a subgroup of an abelian group G then A admits a direct complement: a subgroup C of G such that $G = A \oplus C$. Thus divisible groups are injective modules in the category of abelian groups, and conversely, every injective abelian group is divisible (Baer's criterion). An abelian group without non-zero divisible subgroups is called **reduced**.

Two important special classes of infinite abelian groups with diametrically opposite properties are *torsion groups* and *torsion-free groups*, exemplified by the groups $\mathbf{Q}/\mathbf{Z}$ (periodic) and $\mathbf{Q}$ (torsion-free).

17.7.1   Torsion groups

An abelian group is called **periodic** or **torsion** if every element has finite order. A direct sum of finite cyclic groups is periodic. Although the converse statement is not true in general, some special cases are known. The first and second Prüfer theorems state that if A is a periodic group and either it has **bounded exponent**, i.e. $nA = 0$ for some natural number n, or if A is countable and the p-heights of the elements of A are finite for each p, then A is isomorphic to a direct sum of finite cyclic groups.[3] The cardinality of the set of direct summands isomorphic to $\mathbf{Z}/p^m\mathbf{Z}$ in such a decomposition is an invariant of A. These theorems were later subsumed in the **Kulikov criterion**. In a different direction, Helmut Ulm found an extension of the second Prüfer theorem to countable abelian p-groups with elements of infinite height: those groups are completely classified by means of their Ulm invariants.

17.7.2   Torsion-free and mixed groups

An abelian group is called **torsion-free** if every non-zero element has infinite order. Several classes of torsion-free abelian groups have been studied extensively:

- Free abelian groups, i.e. arbitrary direct sums of $\mathbf{Z}$

- Cotorsion and algebraically compact torsion-free groups such as the p-adic integers

- Slender groups

An abelian group that is neither periodic nor torsion-free is called **mixed**. If A is an abelian group and $T(A)$ is its torsion subgroup then the factor group $A/T(A)$ is torsion-free. However, in general the torsion subgroup is not a direct summand of A, so A is *not* isomorphic to $T(A) \oplus A/T(A)$. Thus the theory of mixed groups involves more than simply combining the results about periodic and torsion-free groups.

17.7.3   Invariants and classification

One of the most basic invariants of an infinite abelian group A is its rank: the cardinality of the maximal linearly independent subset of A. Abelian groups of rank 0 are precisely the periodic groups, while torsion-free abelian groups of rank 1 are necessarily subgroups of $\mathbf{Q}$ and can be completely described. More generally, a torsion-free abelian group of finite rank r is a subgroup of $\mathbf{Q}^r$. On the other hand, the group of p-adic integers $\mathbf{Z}p$ is a torsion-free abelian group of infinite

Z-rank and the groups $\mathbf{Z}p^n$ with different n are non-isomorphic, so this invariant does not even fully capture properties of some familiar groups.

The classification theorems for finitely generated, divisible, countable periodic, and rank 1 torsion-free abelian groups explained above were all obtained before 1950 and form a foundation of the classification of more general infinite abelian groups. Important technical tools used in classification of infinite abelian groups are pure and basic subgroups. Introduction of various invariants of torsion-free abelian groups has been one avenue of further progress. See the books by Irving Kaplansky, László Fuchs, Phillip Griffith, and David Arnold, as well as the proceedings of the conferences on Abelian Group Theory published in Lecture Notes in Mathematics for more recent results.

17.7.4 Additive groups of rings

The additive group of a ring is an abelian group, but not all abelian groups are additive groups of rings (with nontrivial multiplication). Some important topics in this area of study are:

- Tensor product

- Corner's results on countable torsion-free groups

- Shelah's work to remove cardinality restrictions.

17.8 Relation to other mathematical topics

Many large abelian groups possess a natural topology, which turns them into topological groups.

The collection of all abelian groups, together with the homomorphisms between them, forms the category **Ab**, the prototype of an abelian category.

Nearly all well-known algebraic structures other than Boolean algebras are undecidable. Hence it is surprising that Tarski's student Szmielew (1955) proved that the first order theory of abelian groups, unlike its nonabelian counterpart, is decidable. This decidability, plus the fundamental theorem of finite abelian groups described above, highlight some of the successes in abelian group theory, but there are still many areas of current research:

- Amongst torsion-free abelian groups of finite rank, only the finitely generated case and the rank 1 case are well understood;

- There are many unsolved problems in the theory of infinite-rank torsion-free abelian groups;

- While countable torsion abelian groups are well understood through simple presentations and Ulm invariants, the case of countable mixed groups is much less mature.

- Many mild extensions of the first order theory of abelian groups are known to be undecidable.

- Finite abelian groups remain a topic of research in computational group theory.

Moreover, abelian groups of infinite order lead, quite surprisingly, to deep questions about the set theory commonly assumed to underlie all of mathematics. Take the Whitehead problem: are all Whitehead groups of infinite order also free abelian groups? In the 1970s, Saharon Shelah proved that the Whitehead problem is:

- Undecidable in ZFC (Zermelo–Fraenkel axioms), the conventional axiomatic set theory from which nearly all of present day mathematics can be derived. The Whitehead problem is also the first question in ordinary mathematics proved undecidable in ZFC;

- Undecidable even if ZFC is augmented by taking the generalized continuum hypothesis as an axiom;

- Positively answered if ZFC is augmented with the axiom of constructibility (see statements true in L).

17.9   A note on the typography

Among mathematical adjectives derived from the proper name of a mathematician, the word "abelian" is rare in that it is often spelled with a lowercase **a**, rather than an uppercase **A**, indicating how ubiquitous the concept is in modern mathematics.[4]

17.10   See also

- Abelianization

- Class field theory

- Commutator subgroup

- Dihedral group of order 6, the smallest non-Abelian group

- Elementary abelian group

- Pontryagin duality

- Pure injective module

- Pure projective module

17.11   Notes

[1] Jacobson (2009), p. 41

[2] For example, $\mathbf{Q}/\mathbf{Z} \cong \sum_p \mathbf{Q}p/\mathbf{Z}p$.

[3] Countability assumption in the second Prüfer theorem cannot be removed: the torsion subgroup of the direct product of the cyclic groups $\mathbf{Z}/p^m\mathbf{Z}$ for all natural m is not a direct sum of cyclic groups.

[4] Abel Prize Awarded: The Mathematicians' Nobel

17.12   References

- Cox, David (2004). *Galois Theory*. Wiley-Interscience. MR 2119052.

- Fuchs, László (1970). *Infinite Abelian Groups*. Pure and Applied Mathematics **36–I**. Academic Press. MR 0255673.

- Fuchs, László (1973). *Infinite Abelian Groups*. Pure and Applied Mathematics. 36-II. Academic Press. MR 0349869.

- Griffith, Phillip A. (1970). *Infinite Abelian group theory*. Chicago Lectures in Mathematics. University of Chicago Press. ISBN 0-226-30870-7.

- Herstein, I. N. (1975). *Topics in Algebra* (2nd ed.). John Wiley & Sons. ISBN 0-471-02371-X.

- Hillar, Christopher; Rhea, Darren (2007). "Automorphisms of finite abelian groups". *American Mathematical Monthly* **114** (10): 917–923. arXiv:math/0605185.

- Jacobson, Nathan (2009). *Basic Algebra I* (2nd ed.). Dover Publications. ISBN 978-0-486-47189-1.

- Szmielew, Wanda (1955). "Elementary properties of abelian groups". *Fundamenta Mathematicae* **41**: 203–271.

17.13 External links

- Hazewinkel, Michiel, ed. (2001), "Abelian group", *Encyclopedia of Mathematics*, Springer, ISBN 978-1-55608-010-4

Chapter 18

Orthogonal group

"Rotation group" redirects here. For other uses, see Rotation group (disambiguation).

In mathematics, the **orthogonal group** of dimension n, denoted O(n), is the group of distance-preserving transformations of a Euclidean space of dimension n that preserve a fixed point, where the group operation is given by composing transformations. Equivalently, it is the group of $n \times n$ orthogonal matrices, where the group operation is given by matrix multiplication, and an orthogonal matrix is a real matrix whose inverse equals its transpose.

The determinant of an orthogonal matrix being either 1 or −1, an important subgroup of O(n) is the **special orthogonal group**, denoted SO(n), of the orthogonal matrices of determinant 1. This group is also called the **rotation group**, because, in dimensions 2 and 3, its elements are the usual rotations around a point (in dimension 2) or a line (in dimension 3). In low dimension, these groups have been widely studied, see SO(2), SO(3) and SO(4).

The term "orthogonal group" may also refer to a generalization of the above case: the group of invertible linear operators that preserve a non-degenerate symmetric bilinear form or quadratic form[1] on a vector space over a field. In particular, when the bilinear form is the scalar product on the vector space F^n of dimension n over a field F, with quadratic form the sum of squares, then the corresponding orthogonal group, denoted O(n, F), is the set of $n \times n$ orthogonal matrices with entries from F, with the group operation of matrix multiplication. This is a subgroup of the general linear group GL(n, F) given by

$$\mathrm{O}(n, F) = \{Q \in \mathrm{GL}(n, F) \mid Q^\mathsf{T} Q = Q Q^\mathsf{T} = I\}$$

where Q^T is the transpose of Q and I is the identity matrix.

This article mainly discusses the orthogonal groups of quadratic forms that may be expressed over some bases as the dot product; over the reals, they are the positive definite quadratic forms. Over the reals, for any non-degenerate quadratic form, there is a basis, on which the matrix of the form is a diagonal matrix such that the diagonal entries are either 1 or −1. Thus the orthogonal group depends only on the numbers of 1 and of −1, and is denoted O(p, q), where p is the number of ones and q the number of negative ones. For details, see indefinite orthogonal group.

The derived subgroup $\Omega(n, F)$ of O(n, F) is an often studied object because, when F is a finite field, $\Omega(n, F)$ is often a central extension of a finite simple group.

Both O(n, F) and SO(n, F) are algebraic groups, because the condition that a matrix be orthogonal, i.e. have its own transpose as inverse, can be expressed as a set of polynomial equations in the entries of the matrix. The Cartan–Dieudonné theorem describes the structure of the orthogonal group for a non-singular form.

18.1 Name

The determinant of any orthogonal matrix is either 1 or −1. The orthogonal *n*-by-*n* matrices with determinant 1 form a normal subgroup of O(*n*, *F*) known as the **special orthogonal group** SO(*n*, *F*), consisting of all proper rotations. (More precisely, SO(*n*, *F*) is the kernel of the Dickson invariant, discussed below.). By analogy with GL–SL (general linear group, special linear group), the orthogonal group is sometimes called the *general* **orthogonal group** and denoted GO, though this term is also sometimes used for *indefinite* orthogonal groups O(*p*, *q*). The term **rotation group** can be used to describe either the special or general orthogonal group.

18.2 Even and odd dimension

The structure of the orthogonal group differs in certain aspects between even and odd dimensions – for example, over ordered fields (such as **R**) the −*I* element is orientation-preserving in even dimensions, but orientation-reversing in odd dimensions. When this distinction wishes to be emphasized, the groups are generally denoted O(2*k*) and O(2*k* + 1), reserving *n* for the dimension of the space (*n* = 2*k* or *n* = 2*k* + 1). The letters *p* or *r* are also used, indicating the rank of the corresponding Lie algebra; in odd dimension the corresponding Lie algebra is $\mathfrak{so}(2r + 1)$, while in even dimension the Lie algebra is $\mathfrak{so}(2r)$.

18.2.1 Difference between O(*n*) and SO(*n*) in even dimensions

In two dimensions O(2) is all such rotations about the origin and all reflections along a line through the origin. while SO(2) is the group of all rotations about the origin.

These groups are closely related: not only is SO(2) a subgroup of O(2) because any two reflections gives a rotation.

More generally an even number of reflections gives a rotation, and in *n*-dimensions all rotations can be generated this way.

To get a "reflection through the origin" you can reflect along each of the axes. Two axes in two dimensions, so two reflections which is a rotation, and the same is true in any dimension. The 'reflection through the origin' is not a reflection in the usual sense in even dimensions, it's a rotation. It's usually a particularly interesting rotation: in 2D it's the only rotation which when done twice gives the identity, so is its own inverse - this is true in higher dimensions though other rotations have the same property. In 4D it's isoclinic, and if that classification were generalised it would be isoclinic in higher dimensions too.

18.3 Over the real number field

Over the field **R** of real numbers, the orthogonal group O(*n*, **R**) and the special orthogonal group SO(*n*, **R**) are often simply denoted by O(*n*) and SO(*n*) if no confusion is possible. They form real compact Lie groups of dimension *n*(*n* − 1)/2. O(*n*, **R**) has two connected components, with SO(*n*, **R**) being the identity component, i.e., the connected component containing the identity matrix.

18.3.1 Geometric interpretation

The real orthogonal and real special orthogonal groups have the following geometric interpretations:

O(*n*, **R**) is a subgroup of the Euclidean group *E*(*n*), the group of isometries of **R**^{n}; it contains those that leave the origin fixed – O(*n*, **R**) = *E*(*n*) ∩ GL(*n*, **R**). It is the symmetry group of the sphere (*n* = 3) or (*n* − 1)-sphere and all objects with spherical symmetry, if the origin is chosen at the center.

SO(*n*, **R**) is a subgroup of *E*$^+$(*n*), which consists of *direct* isometries, i.e., isometries preserving orientation; it contains those that leave the origin fixed – SO(*n*, **R**) = *E*$^+$(*n*) ∩ GL(*n*, **R**) = *E*(*n*) ∩ GL$^+$(*n*, **R**). It is the rotation group of the sphere and all objects with spherical symmetry, if the origin is chosen at the center.

$\{\pm I\}$ is a normal subgroup and even a characteristic subgroup of O(n, **R**), and, if n is even, also of SO(n, **R**). If n is odd, O(n, **R**) is the internal direct product of SO(n, **R**) and $\{\pm I\}$. For every positive integer k the cyclic group Ck of k-fold rotations is a normal subgroup of O(2, **R**) and SO(2, **R**).

Relative to suitable orthogonal bases, the isometries are of the form:

$$\begin{bmatrix} R_1 & & & & & \\ & \ddots & & & 0 & \\ & & R_k & & & \\ & & & \pm 1 & & \\ & 0 & & & \ddots & \\ & & & & & \pm 1 \end{bmatrix}$$

where the matrices R_1, ..., Rk are 2-by-2 rotation matrices in orthogonal planes of rotation. As a special case, known as Euler's rotation theorem, any (non-identity) element of SO(3, **R**) is rotation about a uniquely defined axis.

The orthogonal group is generated by reflections (two reflections give a rotation), as in a Coxeter group,[note 1] and elements have length at most n (require at most n reflections to generate; this follows from the above classification, noting that a rotation is generated by 2 reflections, and is true more generally for indefinite orthogonal groups, by the Cartan–Dieudonné theorem). A longest element (element needing the most reflections) is reflection through the origin (the map $v \mapsto -v$), though so are other maximal combinations of rotations (and a reflection, in odd dimension).

The symmetry group of a circle is O(2, **R**). The orientation preserving subgroup SO(2, **R**) is isomorphic (as a *real* Lie group) to the circle group, also known as U(1). This isomorphism sends the complex number $\exp(\varphi\, i) = \cos\varphi + i \sin\varphi$ of absolute value 1 to the special orthogonal matrix

$$\begin{bmatrix} \cos(\phi) & -\sin(\phi) \\ \sin(\phi) & \cos(\phi) \end{bmatrix}.$$

The group SO(3, **R**), understood as the set of rotations of 3-dimensional space, is of major importance in the sciences and engineering, and there are numerous charts on SO(3).

18.3.2   Maximal tori and Weyl groups

A maximal torus T for SO($2n$), of rank n, is given by the block-diagonal matrices

$$\begin{bmatrix} R_1 & & 0 \\ & \ddots & \\ 0 & & R_n \end{bmatrix},$$

where the Rj are 2-by-2 rotation matrices. The image $T \times \{1\}$ of the same torus under the block-diagonal inclusion

$$\mathrm{SO}(2n) \cong \mathrm{SO}(2n) \times \{1\} < \mathrm{SO}(2n+1)$$

is a maximal torus for SO($2n$+1). The Weyl group of SO($2n$+1) is the semidirect product $\{\pm 1\}^n \rtimes S_n$ of a normal elementary abelian 2-subgroup and a symmetric group, where the nontrivial element of each $\{\pm 1\}$ factor of $\{\pm 1\}^n$ acts on the corresponding circle factor of $T \times \{1\}$ by inversion, and the symmetric group Sn acts on both $\{\pm 1\}^n$ and $T \times \{1\}$ by permuting factors. The elements of the Weyl group are represented by matrices in O($2n$) $\times \{1\}$. The Sn factor is represented by block permutation matrices with 2-by-2 blocks, and a final 1 on the diagonal. The $\{\pm 1\}^n$ component is represented by block-diagonal matrices with 2-by-2 blocks either

$$\begin{bmatrix} 1 & 0 \\ 0 & 1 \end{bmatrix} \quad \text{or} \quad \begin{bmatrix} 0 & 1 \\ 1 & 0 \end{bmatrix},$$

with the last component ± 1 chosen to make the determinant 1.

The Weyl group of SO($2n$) is the subgroup $H_{n-1} \rtimes S_n < \{\pm 1\}^n \rtimes S_n$ of that of SO($2n + 1$), where $Hn-1 < \{\pm 1\}^n$ is the kernel of the product homomorphism $\{\pm 1\}^n \to \{\pm 1\}$ given by $(\epsilon_1, \ldots, \epsilon_n) \mapsto \epsilon_1 \cdots \epsilon_n$; that is $Hn-1 < \{\pm 1\}^n$ is the subgroup with an even number of minus signs. The Weyl group of SO($2n$) is represented in SO($2n$) by the preimages under the standard injection SO($2n$) $\to$ SO($2n+1$) of the representatives for the Weyl group of SO($2n + 1$). Those matrices with an odd number of $\begin{bmatrix} 0 & 1 \\ 1 & 0 \end{bmatrix}$ blocks have no remaining final -1 coordinate to make their determinants positive, and hence cannot be represented in SO($2n$).

18.3.3 Low-dimensional topology

The low-dimensional (real) orthogonal groups are familiar spaces:

- O(1) = S^0, a two-point discrete space

- SO(1) = \{1\}

- SO(2) is S^1

- SO(3) is $\mathbf{RP}^3$

- SO(4) is double covered by SU(2) $\times$ SU(2) = $S^3 \times S^3$.

18.3.4 Homotopy groups

In terms of algebraic topology, for $n > 2$ the fundamental group of SO(n, $\mathbf{R}$) is cyclic of order 2, and the spin group Spin(n) is its universal cover. For $n = 2$ the fundamental group is infinite cyclic and the universal cover corresponds to the real line (the group Spin(2) is the unique 2-fold cover).

Generally, the homotopy groups $\pi k(O)$ of the real orthogonal group are related to homotopy groups of spheres, and thus are in general hard to compute. However, one can compute the homotopy groups of the stable orthogonal group (aka the infinite orthogonal group), defined as the direct limit of the sequence of inclusions:

$$O(0) \subset O(1) \subset O(2) \subset \cdots \subset O = \bigcup_{k=0}^{\infty} O(k)$$

Since the inclusions are all closed, hence cofibrations, this can also be interpreted as a union. On the other hand S^n is a homogeneous space for O($n + 1$), and one has the following fiber bundle:

$$O(n) \to O(n + 1) \to S^n,$$

which can be understood as "The orthogonal group O($n + 1$) acts transitively on the unit sphere S^n, and the stabilizer of a point (thought of as a unit vector) is the orthogonal group of the perpendicular complement, which is an orthogonal group one dimension lower. Thus the natural inclusion O(n) $\to$ O($n + 1$) is ($n - 1$)-connected, so the homotopy groups stabilize, and $\pi k(O(n+1)) = \pi k(O(n))$ for $n > k + 1$: thus the homotopy groups of the stable space equal the lower homotopy groups of the unstable spaces.

From Bott periodicity we obtain $\Omega^8 O \cong O$, therefore the homotopy groups of O are 8-fold periodic, meaning $\pi k + 8(O) = \pi k(O)$, and one needs only to compute the lower 8 homotopy groups:

$$\pi_0(O) = \mathbf{Z}/2$$
$$\pi_1(O) = \mathbf{Z}/2$$
$$\pi_2(O) = 0$$
$$\pi_3(O) = \mathbf{Z}$$
$$\pi_4(O) = 0$$
$$\pi_5(O) = 0$$
$$\pi_6(O) = 0$$
$$\pi_7(O) = \mathbf{Z}$$

Relation to KO-theory

Via the clutching construction, homotopy groups of the stable space O are identified with stable vector bundles on spheres (up to isomorphism), with a dimension shift of 1: $\pi k(O) = \pi k_{+1}(BO)$. Setting $KO = BO \times \mathbf{Z} = \Omega^{-1}O \times \mathbf{Z}$ (to make π_0 fit into the periodicity), one obtains:

$$\pi_0(KO) = \mathbf{Z}$$
$$\pi_1(KO) = \mathbf{Z}/2$$
$$\pi_2(KO) = \mathbf{Z}/2$$
$$\pi_3(KO) = 0$$
$$\pi_4(KO) = \mathbf{Z}$$
$$\pi_5(KO) = 0$$
$$\pi_6(KO) = 0$$
$$\pi_7(KO) = 0$$

Computation and interpretation of homotopy groups

Low-dimensional groups   The first few homotopy groups can be calculated by using the concrete descriptions of low-dimensional groups.

- $\pi_0(O) = \pi_0(O(1)) = \mathbf{Z}/2$, from orientation-preserving/reversing (this class survives to O(2) and hence stably)

- $\pi_1(O) = \pi_1(SO(3)) = \mathbf{Z}/2$, which is spin comes from $SO(3) = \mathbf{RP}^3 = S^3/(\mathbf{Z}/2)$.

- $\pi_2(O) = \pi_2(SO(3)) = 0$, which surjects onto $\pi_2(SO(4))$; this latter thus vanishes.

Lie groups   From general facts about Lie groups, $\pi_2(G)$ always vanishes, and $\pi_3(G)$ is free (free abelian).

Vector bundles   From the vector bundle point of view, $\pi_0(KO)$ is vector bundles over S^0, which is two points. Thus over each point, the bundle is trivial, and the non-triviality of the bundle is the difference between the dimensions of the vector spaces over the two points, so $\pi_0(KO) = \mathbf{Z}$ is dimension.

Loop spaces   Using concrete descriptions of the loop spaces in Bott periodicity, one can interpret higher homotopy of O as lower homotopy of simple to analyze spaces. Using π_0, O and O/U have two components, $KO = BO \times \mathbf{Z}$ and $KSp = BSp \times \mathbf{Z}$ have countably many components, and the rest are connected.

Interpretation of homotopy groups

In a nutshell:[2]

- $\pi_0(KO) = \mathbf{Z}$ is about dimension

- $\pi_1(KO) = \mathbf{Z}/2$ is about orientation

- $\pi_2(KO) = \mathbf{Z}/2$ is about spin

- $\pi_4(KO) = \mathbf{Z}$ is about topological quantum field theory.

Let R be any of the four division algebras $\mathbf{R}$, $\mathbf{C}$, $\mathbf{H}$, $\mathbf{O}$, and let LR be the tautological line bundle over the projective line $R\mathrm{P}^1$, and $[LR]$ its class in K-theory. Noting that $\mathbf{R}\mathrm{P}^1 = S^1$, $\mathbf{C}\mathrm{P}^1 = S^2$, $\mathbf{H}\mathrm{P}^1 = S^4$, $\mathbf{O}\mathrm{P}^1 = S^8$, these yield vector bundles over the corresponding spheres, and

- $\pi_1(KO)$ is generated by $[L\mathbf{R}]$

- $\pi_2(KO)$ is generated by $[L\mathbf{C}]$

- $\pi_4(KO)$ is generated by $[L\mathbf{H}]$

- $\pi_8(KO)$ is generated by $[L\mathbf{O}]$

From the point of view of symplectic geometry, $\pi_0(KO) \cong \pi_8(KO) = \mathbf{Z}$ can be interpreted as the Maslov index, thinking of it as the fundamental group $\pi_1(\mathrm{U}/\mathrm{O})$ of the stable Lagrangian Grassmannian as $\mathrm{U}/\mathrm{O} \cong \Omega^7(KO)$, so $\pi_1(\mathrm{U}/\mathrm{O}) = \pi_{1+7}(KO)$.

18.4 Over the complex number field

Over the field $\mathbf{C}$ of complex numbers, $\mathrm{O}(n, \mathbf{C})$ and $\mathrm{SO}(n, \mathbf{C})$ are complex Lie groups of dimension $n(n-1)/2$ over $\mathbf{C}$ (it means the dimension over $\mathbf{R}$ is twice that). $\mathrm{O}(n, \mathbf{C})$ has two connected components, and $\mathrm{SO}(n, \mathbf{C})$ is the connected component containing the identity matrix. For $n \geq 2$ these groups are noncompact.

Just as in the real case $\mathrm{SO}(n, \mathbf{C})$ is not simply connected. For $n > 2$ the fundamental group of $\mathrm{SO}(n, \mathbf{C})$ is cyclic of order 2 whereas the fundamental group of $\mathrm{SO}(2, \mathbf{C})$ is infinite cyclic.

18.5 Over finite fields

Orthogonal groups can also be defined over finite fields $\mathbf{F}q$, where is a power of a prime p.

Over finite fields of characteristic not equal to 2, orthogonal groups come in two types in even dimension: $\mathrm{O}^+(2n, q)$ and $\mathrm{O}^-(2n, q)$; and one type in odd dimension: $\mathrm{O}(2n + 1, q)$.[3]

If V is the vector space on which the orthogonal group G acts, it can be written as a direct orthogonal sum as follows:

$$V = L_1 \oplus L_2 \oplus \cdots \oplus L_m \oplus W,$$

where Li are hyperbolic lines and W contains no singular vectors. If W is the zero subspace, then G is of plus type. If W is one-dimensional then G has odd dimension. If W has dimension 2, G is of minus type.

In the special case where $n = 1$, $\mathrm{O}^\epsilon(2, q)$ is a dihedral group of order $2(q - \epsilon)$.

We have the following formulas for the order of $\mathrm{O}(n, q)$, when the characteristic is not two:

$$|O(2n + 1, q)| = 2q^n \prod_{i=0}^{n-1}(q^{2n} - q^{2i}).$$

If -1 is a square in $\mathbf{F}q$

$$|O(2n, q)| = 2(q^n - 1) \prod_{i=1}^{n-1}(q^{2n} - q^{2i}).$$

If -1 is a non-square in $\mathbf{F}q$

$$|O(2n, q)| = 2(q^n + (-1)^{n+1}) \prod_{i=1}^{n-1}(q^{2n} - q^{2i}).$$

18.6   The Dickson invariant

For orthogonal groups, the **Dickson invariant** is a homomorphism from the orthogonal group to the quotient group $\mathbf{Z}/2\mathbf{Z}$ (integers modulo 2), taking the value 0 in case the element is the product of an even number of reflections, and the value of 1 otherwise.[4]

Algebraically, the Dickson invariant can be defined as $D(f) = \operatorname{rank}(I - f)$ modulo 2, where I is the identity (Taylor 1992, Theorem 11.43). Over fields that are not of characteristic 2 it is equivalent to the determinant: the determinant is -1 to the power of the Dickson invariant. Over fields of characteristic 2, the determinant is always 1, so the Dickson invariant gives more information than the determinant.

The special orthogonal group is the kernel of the Dickson invariant[4] and usually has index 2 in $O(n, F)$.[5] When the characteristic of F is not 2, the Dickson Invariant is 0 whenever the determinant is 1. Thus when the characteristic is not 2, $SO(n, F)$ is commonly defined to be the elements of $O(n, F)$ with determinant 1. Each element in $O(n, F)$ has determinant ± 1. Thus in characteristic 2, the determinant is always 1.

The Dickson invariant can also be defined for Clifford groups and Pin groups in a similar way (in all dimensions).

18.7   Orthogonal groups of characteristic 2

Over fields of characteristic 2 orthogonal groups often exhibit special behaviors, some of which are listed in this section. (Formerly these groups were known as the **hypoabelian groups** but this term is no longer used.)

- Any orthogonal group over any field is generated by reflections, except for a unique example where the vector space is 4-dimensional over the field with 2 elements and the Witt index is 2.[6] Note that a reflection in characteristic two has a slightly different definition. In characteristic two, the reflection orthogonal to a vector $\mathbf{u}$ takes a vector $\mathbf{v}$ to $\mathbf{v} + B(\mathbf{v}, \mathbf{u})/Q(\mathbf{u}) \cdot \mathbf{u}$ where B is the bilinear form and Q is the quadratic form associated to the orthogonal geometry. Compare this to the Householder reflection of odd characteristic or characteristic zero, which takes $\mathbf{v}$ to $\mathbf{v} - 2 \cdot B(\mathbf{v}, \mathbf{u})/Q(\mathbf{u}) \cdot \mathbf{u}$.

- The center of the orthogonal group usually has order 1 in characteristic 2, rather than 2, since $I = -I$.

- In odd dimensions $2n + 1$ in characteristic 2, orthogonal groups over perfect fields are the same as symplectic groups in dimension $2n$. In fact the symmetric form is alternating in characteristic 2, and as the dimension is odd it must have a kernel of dimension 1, and the quotient by this kernel is a symplectic space of dimension $2n$, acted upon by the orthogonal group.

- In even dimensions in characteristic 2 the orthogonal group is a subgroup of the symplectic group, because the symmetric bilinear form of the quadratic form is also an alternating form.

18.8   The spinor norm

The **spinor norm** is a homomorphism from an orthogonal group over a field F to the quotient group F^*/F^{*2} (the multiplicative group of the field F up to square elements), that takes reflection in a vector of norm n to the image of n in F^*/F^{*2}.[7]

For the usual orthogonal group over the reals it is trivial, but it is often non-trivial over other fields, or for the orthogonal group of a quadratic form over the reals that is not positive definite.

18.9   Galois cohomology and orthogonal groups

In the theory of Galois cohomology of algebraic groups, some further points of view are introduced. They have explanatory value, in particular in relation with the theory of quadratic forms; but were for the most part *post hoc*, as far as the discovery of the phenomena is concerned. The first point is that quadratic forms over a field can be identified as a Galois H^1, or twisted forms (torsors) of an orthogonal group. As an algebraic group, an orthogonal group is in general neither connected nor simply-connected; the latter point brings in the spin phenomena, while the former is related to the discriminant.

The 'spin' name of the spinor norm can be explained by a connection to the spin group (more accurately a pin group). This may now be explained quickly by Galois cohomology (which however postdates the introduction of the term by more direct use of Clifford algebras). The spin covering of the orthogonal group provides a short exact sequence of algebraic groups.

$$1 \to \mu_2 \to \mathrm{Pin}_V \to \mathrm{O}_V \to 1$$

Here μ_2 is the algebraic group of square roots of 1; over a field of characteristic not 2 it is roughly the same as a two-element group with trivial Galois action. The connecting homomorphism from $H^0(\mathrm{OV})$, which is simply the group $\mathrm{OV}(F)$ of F-valued points, to $H^1(\mu_2)$ is essentially the spinor norm, because $H^1(\mu_2)$ is isomorphic to the multiplicative group of the field modulo squares.

There is also the connecting homomorphism from H^1 of the orthogonal group, to the H^2 of the kernel of the spin covering. The cohomology is non-abelian, so that this is as far as we can go, at least with the conventional definitions.

18.10   Lie algebra

The Lie algebra corresponding to Lie groups $\mathrm{O}(n, F)$ and $\mathrm{SO}(n, F)$ consists of the skew-symmetric $n \times n$ matrices, with the Lie bracket $[\,,\,]$ given by the commutator. One Lie algebra corresponds to both groups. It is often denoted by $\mathfrak{o}(n, F)$ or $\mathfrak{so}(n, F)$, and called the **orthogonal Lie algebra** or **special orthogonal Lie algebra**. Over real numbers, these Lie algebras for different n are the compact real forms of two of the four families of semisimple Lie algebras: in odd dimension Bk, where $n = 2k + 1$, while in even dimension Dr, where $n = 2r$.

More intrinsically, given a vector space with an inner product, the special orthogonal Lie algebra is given by the bivectors on the space, which are sums of simple bivectors (2-blades) $\mathbf{v} \wedge \mathbf{w}$. The correspondence is given by the map $\mathbf{v} \wedge \mathbf{w} \mapsto \mathbf{v}^* \otimes \mathbf{w} - \mathbf{w}^* \otimes \mathbf{v}$, where $\mathbf{v}^*$ is the covector dual to the vector $\mathbf{v}$; in coordinates these are exactly the elementary skew-symmetric matrices.

Over real numbers, this characterization is used in interpreting the curl of a vector field (naturally a 2-vector) as an infinitesimal rotation or "curl", hence the name. Generalizing the inner product with a nondegenerate form yields the indefinite orthogonal Lie algebras $\mathfrak{so}(p, q)$.

The representation theory of the orthogonal Lie algebras includes both representations corresponding to linear representations of the orthogonal groups, and representations corresponding to projective representations of the orthogonal groups (linear representations of spin groups), the so-called spin representation, which are important in physics.

18.11 Related groups

The orthogonal groups and special orthogonal groups have a number of important subgroups, supergroups, quotient groups, and covering groups. These are listed below.

The inclusions $O(n) \subset U(n) \subset Sp(n) = USp(2n)$ and $USp(n) \subset U(n) \subset O(2n)$ are part of a sequence of 8 inclusions used in a geometric proof of the Bott periodicity theorem, and the corresponding quotient spaces are symmetric spaces of independent interest – for example, $U(n)/O(n)$ is the Lagrangian Grassmannian.

18.11.1 Lie subgroups

In physics, particularly in the areas of Kaluza–Klein compactification, it is important to find out the subgroups of the orthogonal group. The main ones are:

$O(n) \supset O(n-1)$ – preserve an axis

$O(2n) \supset U(n) \supset SU(n)$ – $U(n)$ are those that preserve a compatible complex structure *or* a compatible symplectic structure – see 2-out-of-3 property; $SU(n)$ also preserves a complex orientation.

$O(2n) \supset USp(n)$

$O(7) \supset G_2$

18.11.2 Lie supergroups

The orthogonal group $O(n)$ is also an important subgroup of various Lie groups:

$U(n) \supset SU(n) \supset O(n)$

$USp(2n) \supset O(n)$

$G_2 \supset O(3)$

$F_4 \supset O(9)$

$E_6 \supset O(10)$

$E_7 \supset O(12)$

$E_8 \supset O(16)$

Conformal group

Main article: Conformal group

Being isometries, real orthogonal transforms preserve angles, and are thus conformal maps, though not all conformal linear transforms are orthogonal. In classical terms this is the difference between congruence and similarity, as exemplified by SSS (Side-Side-Side) congruence of triangles and AAA (Angle-Angle-Angle) similarity of triangles. The group of conformal linear maps of $\mathbf{R}^n$ is denoted $CO(n)$ for the **conformal orthogonal group**, and consists of the product of the orthogonal group with the group of dilations. If n is odd, these two subgroups do not intersect, and they are a direct product: $CO(2k + 1) = O(2k + 1) \times \mathbf{R}^*$, where $\mathbf{R}^* = \mathbf{R}\backslash\{0\}$ is the real multiplicative group, while if n is even, these subgroups intersect in ± 1, so this is not a direct product, but it is a direct product with the subgroup of dilation by a positive scalar: $CO(2k) = O(2k) \times \mathbf{R}^+$.

Similarly one can define $CSO(n)$; note that this is always: $CSO(n) = CO(n) \cap GL^+(n) = SO(n) \times \mathbf{R}^+$.

18.11.3 Discrete subgroups

As the orthogonal group is compact, discrete subgroups are equivalent to finite subgroups.[note 2] These subgroups are known as point group and can be realized as the symmetry groups of polytopes. A very important class of examples are the finite Coxeter groups, which include the symmetry groups of regular polytopes.

Dimension 3 is particularly studied – see point groups in three dimensions, polyhedral groups, and list of spherical symmetry groups. In 2 dimensions, the finite groups are either cyclic or dihedral – see point groups in two dimensions.

Other finite subgroups include:

- Permutation matrices (the Coxeter group An)

- Signed permutation matrices (the Coxeter group Bn); also equals the intersection of the orthogonal group with the integer matrices.[note 3]

18.11.4 Covering and quotient groups

The orthogonal group is neither simply connected nor centerless, and thus has both a covering group and a quotient group, respectively:

- Two covering Pin groups, $\text{Pin}_+(n) \to O(n)$ and $\text{Pin}_-(n) \to O(n)$,

- The quotient projective orthogonal group, $O(n) \to PO(n)$.

These are all 2-to-1 covers.

For the special orthogonal group, the corresponding groups are:

- Spin group, $\text{Spin}(n) \to SO(n)$,

- Projective special orthogonal group, $SO(n) \to PSO(n)$.

Spin is a 2-to-1 cover, while in even dimension, $PSO(2k)$ is a 2-to-1 cover, and in odd dimension $PSO(2k + 1)$ is a 1-to-1 cover, i.e., isomorphic to $SO(2k + 1)$. These groups, $\text{Spin}(n)$, $SO(n)$, and $PSO(n)$ are Lie group forms of the compact special orthogonal Lie algebra, $\mathfrak{so}(n, \mathbb{R})$ – Spin is the simply connected form, while PSO is the centerless form, and SO is in general neither.[note 4]

In dimension 3 and above these are the covers and quotients, while dimension 2 and below are somewhat degenerate; see specific articles for details.

18.12 Principal homogeneous space: Stiefel manifold

Main article: Stiefel manifold

The principal homogeneous space for the orthogonal group $O(n)$ is the Stiefel manifold $V_n(\mathbf{R}^n)$ of orthonormal bases (orthonormal n-frames).

In other words, the space of orthonormal bases is like the orthogonal group, but without a choice of base point: given an orthogonal space, there is no natural choice of orthonormal basis, but once one is given one, there is a one-to-one correspondence between bases and the orthogonal group. Concretely, a linear map is determined by where it sends a basis: just as an invertible map can take any basis to any other basis, an orthogonal map can take any *orthogonal* basis to any other *orthogonal* basis.

The other Stiefel manifolds $Vk(\mathbf{R}^n)$ for $k < n$ of *incomplete* orthonormal bases (orthonormal k-frames) are still homogeneous spaces for the orthogonal group, but not *principal* homogeneous spaces: any k-frame can be taken to any other k-frame by an orthogonal map, but this map is not uniquely determined.

18.13   See also

18.13.1   Specific transforms

- Coordinate rotations and reflections
- Reflection through the origin

18.13.2   Specific groups

- rotation group, SO(3, **R**)
- SO(8)

18.13.3   Related groups

- indefinite orthogonal group
- unitary group
- symplectic group

18.13.4   Lists of groups

- list of finite simple groups
- list of simple Lie groups

18.14   Notes

[1] The analogy is stronger: Weyl groups, a class of (representations of) Coxeter groups, can be considered as simple algebraic groups over the field with one element, and there are a number of analogies between algebraic groups and vector spaces on the one hand, and Weyl groups and sets on the other.

[2] Infinite subsets of a compact space have an accumulation point and are not discrete.

[3] $O(n) \cap GL(n, \mathbf{Z})$ equals the signed permutation matrices because an integer vector of norm 1 must have a single non-zero entry, which must be ± 1 (if it has two non-zero entries or a larger entry, the norm will be larger than 1), and in an orthogonal matrix these entries must be in different coordinates, which is exactly the signed permutation matrices.

[4] In odd dimension, $SO(2k + 1) \cong PSO(2k + 1)$ is centerless (but not simply connected), while in even dimension $SO(2k)$ is neither centerless nor simply connected.

18.15 References

[1] For base fields of characteristic not 2, it is equivalent to use symmetric bilinear forms or quadratic forms. But in characteristic 2 these notions differ.

[2] John Baez "This Week's Finds in Mathematical Physics" week 105

[3] Wilson, Robert A. (2009). *The finite simple groups*. Graduate Texts in Mathematics **251**. London: Springer. pp. 69–75. ISBN 978-1-84800-987-5. Zbl 1203.20012.

[4] Knus, Max-Albert (1991), *Quadratic and Hermitian forms over rings*, Grundlehren der Mathematischen Wissenschaften **294**, Berlin etc.: Springer-Verlag, p. 224, ISBN 3-540-52117-8, Zbl 0756.11008

[5] (Taylor 1992, page 160)

[6] (Grove 2002, Theorem 6.6 and 14.16)

[7] Cassels 1978, p. 178

- Cassels, J.W.S. (1978), *Rational Quadratic Forms*, London Mathematical Society Monographs **13**, Academic Press, ISBN 0-12-163260-1, Zbl 0395.10029

- Grove, Larry C. (2002), *Classical groups and geometric algebra*, Graduate Studies in Mathematics **39**, Providence, R.I.: American Mathematical Society, ISBN 978-0-8218-2019-3, MR 1859189

- Taylor, Donald E. (1992), *The Geometry of the Classical Groups*, Sigma Series in Pure Mathematics **9**, Berlin: Heldermann Verlag, ISBN 3-88538-009-9, MR 1189139, Zbl 0767.20001

18.16 External links

- Hazewinkel, Michiel, ed. (2001), "Orthogonal group", *Encyclopedia of Mathematics*, Springer, ISBN 978-1-55608-010-4

- John Baez "This Week's Finds in Mathematical Physics" week 105

- John Baez on Octonions

- (Italian) n-dimensional Special Orthogonal Group parametrization

Chapter 19

Normal subgroup

"Invariant subgroup" redirects here. It is not to be confused with Fully invariant subgroup.

In abstract algebra, a **normal subgroup** is a subgroup which is invariant under conjugation by members of the group of which it is a part. In other words, a subgroup H of a group G is normal in G if and only if $gH = Hg$ for all g in G, i.e., the sets of left and right cosets coincide. Normal subgroups (and *only* normal subgroups) can be used to construct quotient groups from a given group.

Évariste Galois was the first to realize the importance of the existence of normal subgroups.[1]

19.1   Definitions

A subgroup N of a group G is called a **normal subgroup** if it is invariant under conjugation; that is, for each element n in N and each g in G, the element gng^{-1} is still in N.[2] We write

$$N \triangleleft G \iff \forall n \in N, \forall g \in G, gng^{-1} \in N.$$

For any subgroup, the following conditions are equivalent to normality. Therefore any one of them may be taken as the definition:

- For all g in G, $gNg^{-1} \subseteq N$.

- For all g in G, $gNg^{-1} = N$.

- The sets of left and right cosets of N in G coincide.

- For all g in G, $gN = Ng$.

- N is a union of conjugacy classes of G.

- There is some homomorphism on G for which N is the kernel.

The last condition accounts for some of the importance of normal subgroups; they are a way to internally classify all homomorphisms defined on a group. For example, a non-identity finite group is simple if and only if it is isomorphic to all of its non-identity homomorphic images,[3] a finite group is perfect if and only if it has no normal subgroups of prime index, and a group is imperfect if and only if the derived subgroup is not supplemented by any proper normal subgroup.

19.2 Examples

- The subgroup $\{e\}$ consisting of just the identity element of G and G itself are always normal subgroups of G. The former is called the trivial subgroup, and if these are the only normal subgroups, then G is said to be simple.

- The center of a group is a normal subgroup.

- The commutator subgroup is a normal subgroup.

- More generally, any characteristic subgroup is normal, since conjugation is always an automorphism.

- All subgroups N of an abelian group G are normal, because $gN = Ng$. A group that is not abelian but for which every subgroup is normal is called a Hamiltonian group.

- The translation group in any dimension is a normal subgroup of the Euclidean group (with the orthogonal group as the quotient group); for example in 3D rotating, translating, and rotating back results in only translation; also reflecting, translating, and reflecting again results in only translation (a translation seen in a mirror looks like a translation, with a reflected translation vector). The translations by a given distance in any direction form a conjugacy class; the translation group is the union of those for all distances.

- In the Rubik's Cube group, the subgroup consisting of operations which only affect the corner pieces is normal, because no conjugate transformation can make such an operation affect an edge piece instead of a corner. By contrast, the subgroup consisting of turns of the top face only is not normal, because a conjugate transformation can move parts of the top face to the bottom and hence not all conjugates of elements of this subgroup are contained in the subgroup.

19.3 Properties

- Normality is preserved upon surjective homomorphisms, and is also preserved upon taking inverse images.

- Normality is preserved on taking direct products

- A normal subgroup of a normal subgroup of a group need not be normal in the group. That is, normality is not a transitive relation. The smallest group exhibiting this phenomenon is the dihedral group of order 8. However, a characteristic subgroup of a normal subgroup is normal. Also, a normal subgroup of a central factor is normal. In particular, a normal subgroup of a direct factor is normal. A group in which normality is transitive is called a T-group.

- Every subgroup of index 2 is normal. More generally, a subgroup H of finite index n in G contains a subgroup K normal in G and of index dividing $n!$ called the normal core. In particular, if p is the smallest prime dividing the order of G, then every subgroup of index p is normal.

19.3.1 Lattice of normal subgroups

The normal subgroups of a group G form a lattice under subset inclusion with least element $\{e\}$ and greatest element G. Given two normal subgroups N and M in G, meet is defined as

$$N \wedge M := N \cap M$$

and join is defined as

$$N \vee M := NM = \{nm \mid n \in N \text{ and }, m \in M\}.$$

The lattice is complete and modular.

19.4   Normal subgroups and homomorphisms

If N is normal subgroup, we can define a multiplication on cosets by

$$(a_1 N)(a_2 N) := (a_1 a_2)N.$$

This turns the set of cosets into a group called the quotient group G/N. There is a natural homomorphism $f \colon G \to G/N$ given by $f(a) = aN$. The image $f(N)$ consists only of the identity element of G/N, the coset $eN = N$.

In general, a group homomorphism $f \colon G \to H$ sends subgroups of G to subgroups of H. Also, the preimage of any subgroup of H is a subgroup of G. We call the preimage of the trivial group $\{e\}$ in H the **kernel** of the homomorphism and denote it by $\ker(f)$. As it turns out, the kernel is always normal and the image $f(G)$ of G is always isomorphic to $G/\ker(f)$ (the first isomorphism theorem). In fact, this correspondence is a bijection between the set of all quotient groups G/N of G and the set of all homomorphic images of G (up to isomorphism). It is also easy to see that the kernel of the quotient map, $f \colon G \to G/N$, is N itself, so we have shown that the normal subgroups are precisely the kernels of homomorphisms with domain G.

19.5   See also

19.6   References

[1] C.D. Cantrell, *Modern Mathematical Methods for Physicists and Engineers*. Cambridge University Press, 200, p 160.

[2] Dummit, David S.; Foote, Richard M. (2004), *Abstract Algebra* (3rd ed.), John Wiley & Sons, ISBN 0-471-43334-9

[3] Pál Dömösi and Chrystopher L. Nehaniv, *Algebraic Theory of Automata Networks (SIAM Monographs on Discrete Mathematics and Applications, 11)*, SIAM, 2004, p.7

19.7   Further reading

- I. N. Herstein, *Topics in algebra*. Second edition. Xerox College Publishing, Lexington, Mass.-Toronto, Ont., 1975. xi+388 pp.

19.8   External links

- Weisstein, Eric W., "normal subgroup", *MathWorld*.

- Normal subgroup in Springer's Encyclopedia of Mathematics

- Robert Ash: Group Fundamentals in *Abstract Algebra. The Basic Graduate Year*

- Timothy Gowers, Normal subgroups and quotient groups

- John Baez, What's a Normal Subgroup?

Chapter 20

Lattice (discrete subgroup)

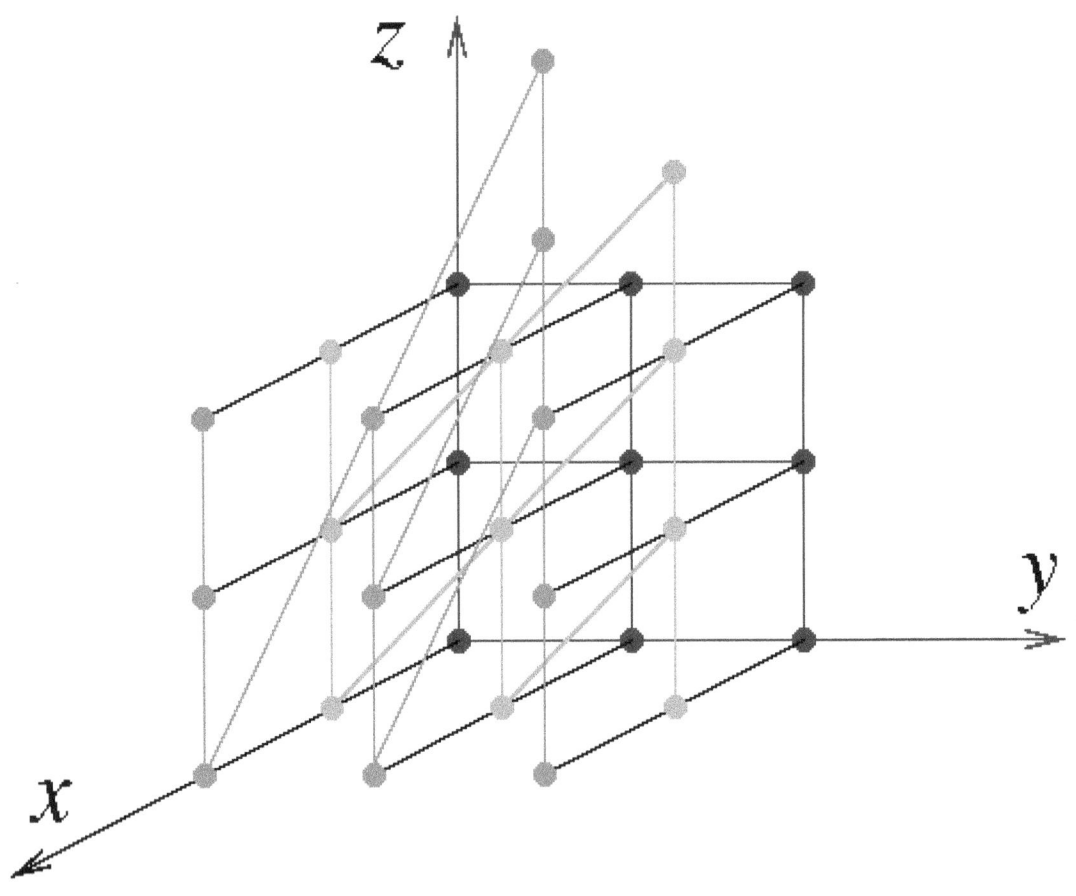

A portion of the discrete Heisenberg group, a discrete subgroup of the continuous Heisenberg Lie group. (The coloring and edges are only for visual aid.)

In Lie theory and related areas of mathematics, a **lattice** in a locally compact topological group is a discrete subgroup with the property that the quotient space has finite invariant measure. In the special case of subgroups of $\mathbf{R}^n$, this amounts to the usual geometric notion of a lattice, and both the algebraic structure of lattices and the geometry of the totality of all lattices are relatively well understood. Deep results of Borel, Harish-Chandra, Mostow, Tamagawa, M. S. Raghunathan,

141

Margulis, Zimmer obtained from the 1950s through the 1970s provided examples and generalized much of the theory to the setting of nilpotent Lie groups and semisimple algebraic groups over a local field. In the 1990s, Bass and Lubotzky initiated the study of *tree lattices*, which remains an active research area.

20.1   Definition

Let G be a locally compact topological group with the Haar measure μ. A discrete subgroup Γ is called a **lattice** in G if the quotient space G/Γ has finite invariant measure, that is, if G is a unimodular group and the volume $\mu(G/\Gamma)$ is finite. The lattice is **uniform** (or *cocompact*) if the quotient space is compact, and **nonuniform** otherwise.

20.2   Arithmetic lattices

An archetypical example of a nonuniform lattice is given by the group $SL(2,\mathbf{Z})$, which is a lattice in the special linear group $SL(2,\mathbf{R})$, and by the closely related modular group. This construction admits a far-reaching generalization to a class of lattices in all semisimple algebraic groups over a local field F called *arithmetic lattices*. For example, let $F = \mathbf{R}$ be the field of real numbers. Roughly speaking, the Lie group $G(\mathbf{R})$ is formed by all matrices with entries in $\mathbf{R}$ satisfying certain algebraic conditions, and by restricting the entries to the integers $\mathbf{Z}$, one obtains a lattice $G(\mathbf{Z})$. Conversely, Grigory Margulis proved that under certain assumptions on G, any lattice in it essentially arises in this way. This remarkable statement is known as *Arithmeticity of lattices* or *Margulis Arithmeticity Theorem*.

20.3   *S*-arithmetic lattices

Arithmetic lattices admit an important generalization, known as the *S-arithmetic lattices*. The first example is given by the diagonally embedded subgroup

$$SL\left(2, \mathbb{Z}\left[\frac{1}{p}\right]\right) \subset SL(2,\mathbb{R}) \times SL(2,\mathbb{Q}_p), S = \{p, \infty\}.$$

This is a lattice in the product of algebraic groups over *different* local fields, both real and p-adic. It is formed by the unimodular matrices of order 2 with entries in the localization of the ring of integers at the prime p. The set S is a finite set of places of **Q** which includes all archimedean places and the locally compact group is the direct product of the groups of points of a fixed linear algebraic group G defined over **Q** (or a more general global field) over the completions of **Q** at the places from S. To form the discrete subgroup, instead of matrices with integer entries, one considers matrices with entries in the localization over the primes (nonarchimedean places) in S. Under fairly general assumptions, this construction indeed produces a lattice. The class of S-arithmetic lattices is much wider than the class of arithmetic lattices, but they share many common features.

20.4   Adelic case

A lattice of fundamental importance for the theory of automorphic forms is given by the group $G(K)$ of K-points of a semisimple (or reductive) linear algebraic group G defined over a global field K. This group diagonally embeds into the adelic algebraic group $G(A)$, where A is the ring of adeles of K, and is a lattice there. Unlike arithmetic lattices, $G(K)$ is not finitely generated.

20.5 Rigidity

Another group of phenomena concerning lattices in semisimple algebraic groups is collectively known as *rigidity*. The Mostow rigidity theorem showed that the algebraic structure of a lattice in simple Lie group G of split rank at least two determines G. Thus any isomorphism of lattices in two such groups is essentially induced by an isomorphism between the groups themselves. *Superrigidity* provides a generalization dealing with homomorphisms from a lattice in an algebraic group G into another algebraic group H.

20.6 Tree lattices

Let X be a locally finite tree. Then the automorphism group G of X is a locally compact topological group, in which the basis of the topology is given by the stabilizers of finite sets of vertices. Vertex stabilizers Gx are thus compact open subgroups, and a subgroup Γ of G is discrete if Γx is finite for some (and hence, for any) vertex x. The subgroup Γ is an **X-lattice** if the suitably defined volume of X/Γ is finite, and a **uniform X-lattice** if this quotient is a finite graph. In case $G\backslash X$ is finite, this is equivalent to Γ being a lattice (respectively, a uniform lattice) in G.

20.7 See also

- Kazhdan's property (T)
- Graph of groups

20.8 References

- Hyman Bass and Alexander Lubotzky, *Tree lattices*. With appendices by H. Bass, L. Carbone, A. Lubotzky, G. Rosenberg, and J. Tits. Progress in Mathematics, vol 176, Birkhäuser Verlag, Boston, 2001 ISBN 0-8176-4120-3

- Grigory Margulis, *Discrete subgroups of semisimple Lie groups*, Ergebnisse der Mathematik und ihrer Grenzgebiete (3) [Results in Mathematics and Related Areas (3)], 17. Springer-Verlag, Berlin, 1991. x+388 pp. ISBN 3-540-12179-X MR 1090825

- Dave Witte Morris: Introduction to Arithmetic Groups, draft of a book

- Platonov, Vladimir; Rapinchuk, Andrei (1994), *Algebraic groups and number theory. (Translated from the 1991 Russian original by Rachel Rowen.)*, Pure and Applied Mathematics **139**, Boston, MA: Academic Press, Inc., ISBN 0-12-558180-7, MR 1278263

- M.S.Raghunathan, *Discrete subgroups of Lie groups*. Ergebnisse der Mathematik und ihrer Grenzgebiete, Band 68. Springer-Verlag, New York-Heidelberg, 1972 MR 0507234

Chapter 21

Continuous symmetry

In mathematics, **continuous symmetry** is an intuitive idea corresponding to the concept of viewing some symmetries as motions, as opposed to e.g. reflection symmetry, which is invariance under a kind of flip from one state to another.

21.1 Formalization

The notion of continuous symmetry has largely and successfully been formalised in the mathematical notions of topological group, Lie group and group action. For most practical purposes continuous symmetry is modelled by a *group action* of a topological group.

21.1.1 One-parameter subgroups

The simplest motions follow a one-parameter subgroup of a Lie group, such as the Euclidean group of three-dimensional space. For example translation parallel to the x-axis by u units, as u varies, is a one-parameter group of motions. Rotation around the z-axis is also a one-parameter group.

21.2 Noether's theorem

Continuous symmetry has a basic role in Noether's theorem in theoretical physics, in the derivation of conservation laws from symmetry principles, specifically for continuous symmetries. The search for continuous symmetries only intensified with the further developments of quantum field theory.

21.3 See also

- Goldstone's theorem

- Infinitesimal transformation

- Noether's theorem

- Sophus Lie

21.4 References

- William H. Barker, Roger Howe (2007), *Continuous Symmetry: from Euclid to Klein*

Chapter 22

Mathematical object

A **mathematical object** is an abstract object arising in philosophy of mathematics and mathematics itself.

Commonly encountered mathematical objects include numbers, permutations, partitions, matrices, sets, functions, and relations. Geometry as a branch of mathematics has such objects as hexagons, points, lines, triangles, circles, spheres, polyhedra, topological spaces and manifolds. Another branch—algebra—has groups, rings, fields, group-theoretic lattices, and order-theoretic lattices. Categories are simultaneously homes to mathematical objects and mathematical objects in their own right.

The ontological status of mathematical objects has been the subject of much investigation and debate by philosophers of mathematics.[1]

22.1 Cantorian framework

One view that emerged around the turn of the 20th century with the work of Cantor is that all mathematical objects can be defined as sets. The set $\{0,1\}$ is a relatively clear-cut example. On the face of it the group $\mathbf{Z}_2$ of integers mod 2 is also a set with two elements. However, it cannot simply be the set $\{0,1\}$, because this does not mention the additional structure imputed to $\mathbf{Z}_2$ by the operations of addition and negation mod 2: how are we to tell which of 0 or 1 is the additive identity, for example? To organize this group as a set it can first be coded as the quadruple $(\{0,1\},+,-,0)$, which in turn can be coded using one of several conventions as a set representing that quadruple, which in turn entails encoding the operations $+$ and $-$ and the constant 0 as sets.

Sets may include ordered denotation of the particular identities and operations that apply to them, indicating a group, abelian group, ring, field, or other mathematical object. These types of mathematical objects are commonly studied in abstract algebra.

22.2 Foundational paradoxes

If, however, the goal of mathematical ontology is taken to be the internal consistency of mathematics, it is more important that mathematical objects be definable in some uniform way (for example, as sets) regardless of actual practice, in order to lay bare the essence of its paradoxes. This has been the viewpoint taken by foundations of mathematics, which has traditionally accorded the management of paradox higher priority than the faithful reflection of the details of mathematical practice as a justification for defining mathematical objects to be sets.

Much of the tension created by this foundational identification of mathematical objects with sets can be relieved without unduly compromising the goals of foundations by allowing two kinds of objects into the mathematical universe, sets and relations, without requiring that either be considered merely an instance of the other. These form the basis of model theory as the domain of discourse of predicate logic. From this viewpoint, mathematical objects are entities satisfying the axioms of a formal theory expressed in the language of predicate logic.

22.3 Category theory

A variant of this approach replaces relations with operations, the basis of universal algebra. In this variant the axioms often take the form of equations, or implications between equations.

A more abstract variant is category theory, which abstracts sets as objects and the operations thereon as morphisms between those objects. At this level of abstraction mathematical objects reduce to mere vertices of a graph whose edges as the morphisms abstract the ways in which those objects can transform and whose structure is encoded in the composition law for morphisms. Categories may arise as the models of some axiomatic theory and the homomorphisms between them (in which case they are usually concrete, meaning equipped with a faithful forgetful functor to the category **Set** or more generally to a suitable topos), or they may be constructed from other more primitive categories, or they may be studied as abstract objects in their own right without regard for their provenance.

22.4 See also

- Abstract object

- Mathematical structure

22.5 References

[1] Burgess, John, and Rosen, Gideon, 1997. *A Subject with No Object: Strategies for Nominalistic Reconstrual of Mathematics.* Oxford University Press. ISBN 0198236158

- Azzouni, J., 1994. *Metaphysical Myths, Mathematical Practice.* Cambridge University Press.

- Burgess, John, and Rosen, Gideon, 1997. *A Subject with No Object.* Oxford Univ. Press.

- Davis, Philip and Reuben Hersh, 1999 [1981]. *The Mathematical Experience.* Mariner Books: 156-62.

- Gold, Bonnie, and Simons, Roger A., 2008. *Proof and Other Dilemmas: Mathematics and Philosophy.* Mathematical Association of America.

- Hersh, Reuben, 1997. *What is Mathematics, Really?* Oxford University Press.

- Sfard, A., 2000, "Symbolizing mathematical reality into being, Or how mathematical discourse and mathematical objects create each other," in Cobb, P., *et al.*, *Symbolizing and communicating in mathematics classrooms: Perspectives on discourse, tools and instructional design.* Lawrence Erlbaum.

- Stewart Shapiro, 2000. *Thinking about mathematics: The philosophy of mathematics.* Oxford University Press.

22.6 External links

- Stanford Encyclopedia of Philosophy: "Abstract Objects"—by Gideon Rosen.

- Wells, Charles, "Mathematical Objects."

- AMOF: The Amazing Mathematical Object Factory

- Mathematical Object Exhibit

Chapter 23

Semisimple Lie algebra

In mathematics, a Lie algebra is **semisimple** if it is a direct sum of simple Lie algebras, i.e., non-abelian Lie algebras $\mathfrak{g}$ whose only ideals are $\{0\}$ and $\mathfrak{g}$ itself.

Throughout the article, unless otherwise stated, $\mathfrak{g}$ is a finite-dimensional Lie algebra over a field of characteristic 0. The following conditions are equivalent:

- $\mathfrak{g}$ is semisimple

- the Killing form, $\kappa(x,y) = \mathrm{tr}(\mathrm{ad}(x)\mathrm{ad}(y))$, is non-degenerate,

- $\mathfrak{g}$ has no non-zero abelian ideals,

- $\mathfrak{g}$ has no non-zero solvable ideals,

- The radical (maximal solvable ideal) of $\mathfrak{g}$ is zero.

23.1 Examples

Examples of semisimple Lie algebras, with notation coming from classification by Dynkin diagrams, are:

- $A_n : \mathfrak{sl}_{n+1}$, the special linear Lie algebra.

- $B_n : \mathfrak{so}_{2n+1}$, the odd-dimensional special orthogonal Lie algebra.

- $C_n : \mathfrak{sp}_{2n}$, the symplectic Lie algebra.

- $D_n : \mathfrak{so}_{2n}$, the even-dimensional special orthogonal Lie algebra.

These Lie algebras are numbered so that n is the rank. Except certain exceptions in low dimensions, many of these are simple Lie algebras, which are *a fortiori* semisimple. These four families, together with five exceptions (E_6, E_7, E_8, F_4, and G_2), are in fact the *only* simple Lie algebras over the complex numbers.

23.2 Classification

See also: Root system

Every semisimple Lie algebra over an algebraically closed field is a direct sum of simple Lie algebras (by definition), and the finite-dimensional simple Lie algebras fall in four families – A_n, B_n, C_n, and D_n – with five exceptions E_6, E_7, E_8, F_4, and G_2. Simple Lie algebras are classified by the connected Dynkin diagrams, shown on the right, while semisimple Lie algebras correspond to not necessarily connected Dynkin diagrams, where each component of the diagram corresponds to a summand of the decomposition of the semisimple Lie algebra into simple Lie algebras.

The classification proceeds by considering a Cartan subalgebra (maximal abelian Lie algebra; corresponds to a maximal torus in a Lie group) and the adjoint action of the Lie algebra on this subalgebra. The root system of the action then both determines the original Lie algebra and must have a very constrained form, which can be classified by the Dynkin diagrams.

The classification is widely considered one of the most elegant results in mathematics – a brief list of axioms yields, via a relatively short proof, a complete but non-trivial classification with surprising structure. This should be compared to the classification of finite simple groups, which is significantly more complicated.

The enumeration of the four families is non-redundant and consists only of simple algebras if $n \geq 1$ for A_n, $n \geq 2$ for B_n, $n \geq 3$ for C_n, and $n \geq 4$ for D_n. If one starts numbering lower, the enumeration is redundant, and one has exceptional isomorphisms between simple Lie algebras, which are reflected in isomorphisms of Dynkin diagrams; the E_n can also be extended down, but below E_6 are isomorphic to other, non-exceptional algebras.

Over a non-algebraically closed field, the classification is more complicated – one classifies simple Lie algebras over the algebraic closure, then for each of these, one classifies simple Lie algebras over the original field which have this form (over the closure). For example, to classify simple real Lie algebras, one classifies real Lie algebras with a given complexification, which are known as real forms of the complex Lie algebra; this can be done by Satake diagrams, which are Dynkin diagrams with additional data ("decorations").

23.3   History

The semisimple Lie algebras over the complex numbers were first classified by Wilhelm Killing (1888–90), though his proof lacked rigor. His proof was made rigorous by Élie Cartan (1894) in his Ph.D. thesis, who also classified semisimple real Lie algebras. This was subsequently refined, and the present classification by Dynkin diagrams was given by then 22-year old Eugene Dynkin in 1947. Some minor modifications have been made (notably by J. P. Serre), but the proof is unchanged in its essentials and can be found in any standard reference, such as (Humphreys 1972).

23.4   Properties

23.4.1   Complete reducibility

A consequence of semisimplicity is a theorem due to Weyl: every finite-dimensional representation is completely reducible; that is for every invariant subspace of the representation there is an invariant complement. Infinite-dimensional representations of semisimple Lie algebras are not in general completely reducible.

23.4.2   Centerless

Since the center of a Lie algebra $\mathfrak{g}$ is an abelian ideal, if $\mathfrak{g}$ is semisimple, then its center is zero. (Note: since $\mathfrak{gl}_n$ has non-trivial center, it is not semisimple.) In other words, the adjoint representation ad is injective. Moreover, it can be shown that the dimension of the Lie algebra $\mathrm{Der}(\mathfrak{g})$ of derivations on $\mathfrak{g}$ is equal to the dimension of $\mathfrak{g}$. Hence, $\mathfrak{g}$ is Lie algebra isomorphic to $\mathrm{Der}(\mathfrak{g})$. (This is a special case of Whitehead's lemma.) Every ideal, quotient and product of semisimple Lie algebras is again semisimple.

23.4.3   Linear

The adjoint representation is injective, and so a semisimple Lie algebra is also a linear Lie algebra under the adjoint representation. This may lead to some ambiguity, as every Lie algebra is already linear with respect to some other vector space (Ado's theorem), although not necessarily via the adjoint representation. But in practice, such ambiguity rarely occurs.

23.4.4   Jordan decomposition

Any endomorphism x of a finite-dimensional vector space over an algebraically closed field can be decomposed uniquely into a diagonalizable (or semisimple) and nilpotent part

$$x = s + n$$

such that s and n commute with each other. Moreover, each of s and n is a polynomial in x. This is a consequence of the Jordan decomposition.

If $x \in \mathfrak{g}$, then the image of x under the adjoint map decomposes as

$$\mathrm{ad}(x) = \mathrm{ad}(s) + \mathrm{ad}(n).$$

The elements s and n are *unique* elements of $\mathfrak{g}$ such that n is nilpotent, s is semisimple, n and s commute, and for which such a decomposition holds. This abstract Jordan decomposition factors through any representation of $\mathfrak{g}$ in the sense that given any representation ρ,

$$\rho(x) = \rho(s) + \rho(n)$$

is the Jordan decomposition of $\rho(x)$ in the endomorphism ring of the representation space.

23.4.5   Rank

The **rank** of a complex semisimple Lie algebra is the dimension of any of its Cartan subalgebras.

23.5   Significance

The significance of semisimplicity comes firstly from the Levi decomposition, which states that every finite dimensional Lie algebra is the semidirect product of a solvable ideal (its radical) and a semisimple algebra. In particular, there is no nonzero Lie algebra that is both solvable and semisimple.

Semisimple Lie algebras have a very elegant classification, in stark contrast to solvable Lie algebras. Semisimple Lie algebras over an algebraically closed field are completely classified by their root system, which are in turn classified by Dynkin diagrams. Semisimple algebras over non-algebraically closed fields can be understood in terms of those over the algebraic closure, though the classification is somewhat more intricate; see real form for the case of real semisimple Lie algebras, which were classified by Élie Cartan.

Further, the representation theory of semisimple Lie algebras is much cleaner than that for general Lie algebras. For example, the Jordan decomposition in a semisimple Lie algebra coincides with the Jordan decomposition in its representation; this is not the case for Lie algebras in general.

If $\mathfrak{g}$ is semisimple, then $\mathfrak{g} = [\mathfrak{g}, \mathfrak{g}]$. In particular, every linear semisimple Lie algebra is a subalgebra of $\mathfrak{sl}$, the special linear Lie algebra. The study of the structure of $\mathfrak{sl}$ constitutes an important part of the representation theory for semisimple Lie algebras.

23.6 Generalizations

Main articles: Reductive Lie algebra and Split Lie algebra

Semisimple Lie algebras admit certain generalizations. Firstly, many statements that are true for semisimple Lie algebras are true more generally for reductive Lie algebras. Abstractly, a reductive Lie algebra is one whose adjoint representation is completely reducible, while concretely, a reductive Lie algebra is a direct sum of a semisimple Lie algebra and an abelian Lie algebra; for example, $\mathfrak{sl}_n$ is semisimple, and $\mathfrak{gl}_n$ is reductive. Many properties of semisimple Lie algebras depend only on reducibility.

Many properties of complex semisimple/reductive Lie algebras are true not only for semisimple/reductive Lie algebras over algebraically closed fields, but more generally for split semisimple/reductive Lie algebras over other fields: semisimple/reductive Lie algebras over algebraically closed fields are always split, but over other fields this is not always the case. Split Lie algebras have essentially the same representation theory as semsimple Lie algebras over algebraically closed fields, for instance, the splitting Cartan subalgebra playing the same role as the Cartan subalgebra plays over algebraically closed fields. This is the approach followed in (Bourbaki 2005), for instance, which classifies representations of split semisimple/reductive Lie algebras.

23.7 References

- Bourbaki, Nicolas (2005), "VIII: Split Semi-simple Lie Algebras", *Elements of Mathematics: Lie Groups and Lie Algebras: Chapters 7–9*

- Erdmann, Karin; Wildon, Mark (2006), *Introduction to Lie Algebras* (1st ed.), Springer, ISBN 1-84628-040-0.

- Humphreys, James E. (1972), *Introduction to Lie Algebras and Representation Theory*, Berlin, New York: Springer-Verlag, ISBN 978-0-387-90053-7.

- Varadarajan, V. S. (2004), *Lie Groups, Lie Algebras, and Their Representations* (1st ed.), Springer, ISBN 0-387-90969-9.

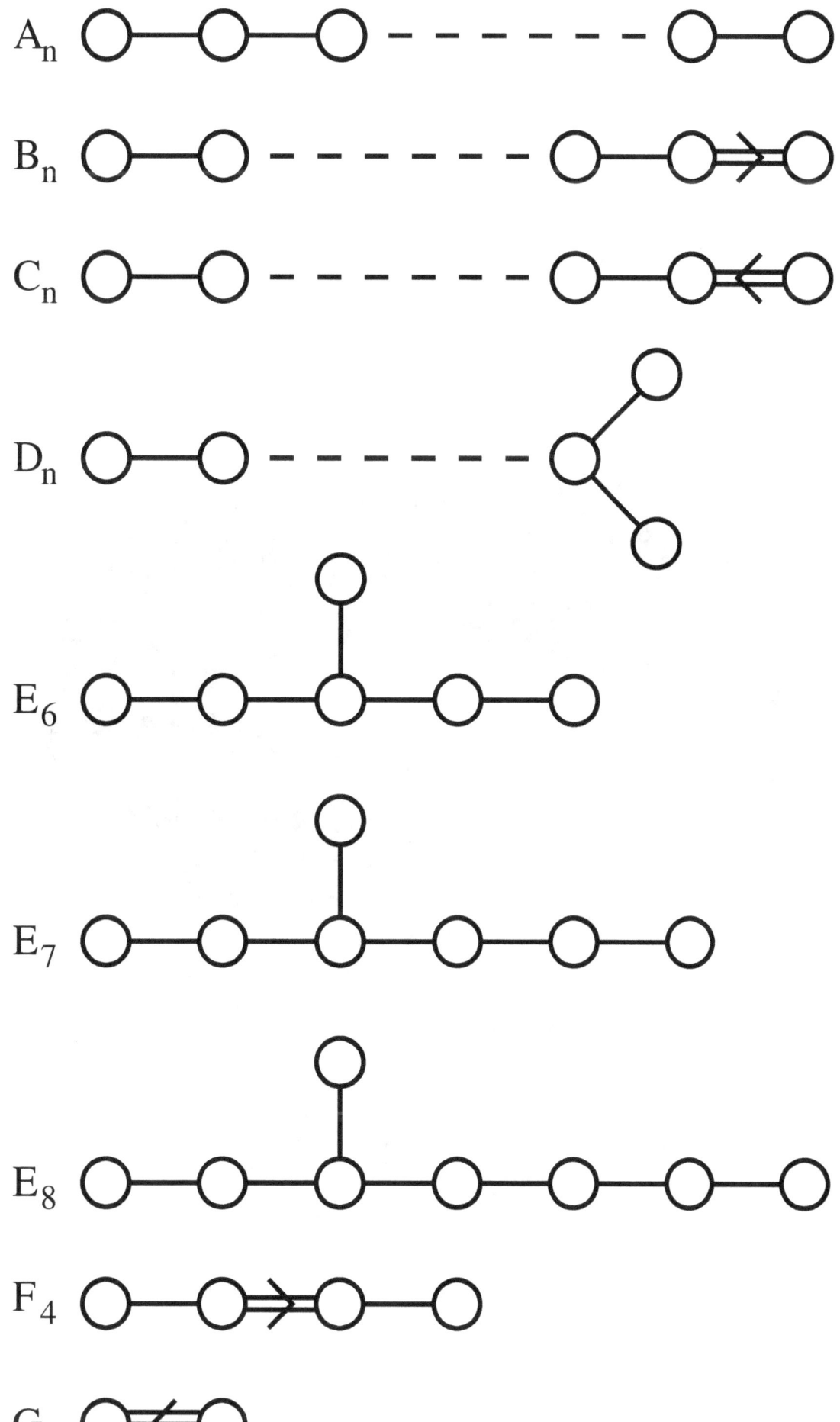

Chapter 24

Homogeneous space

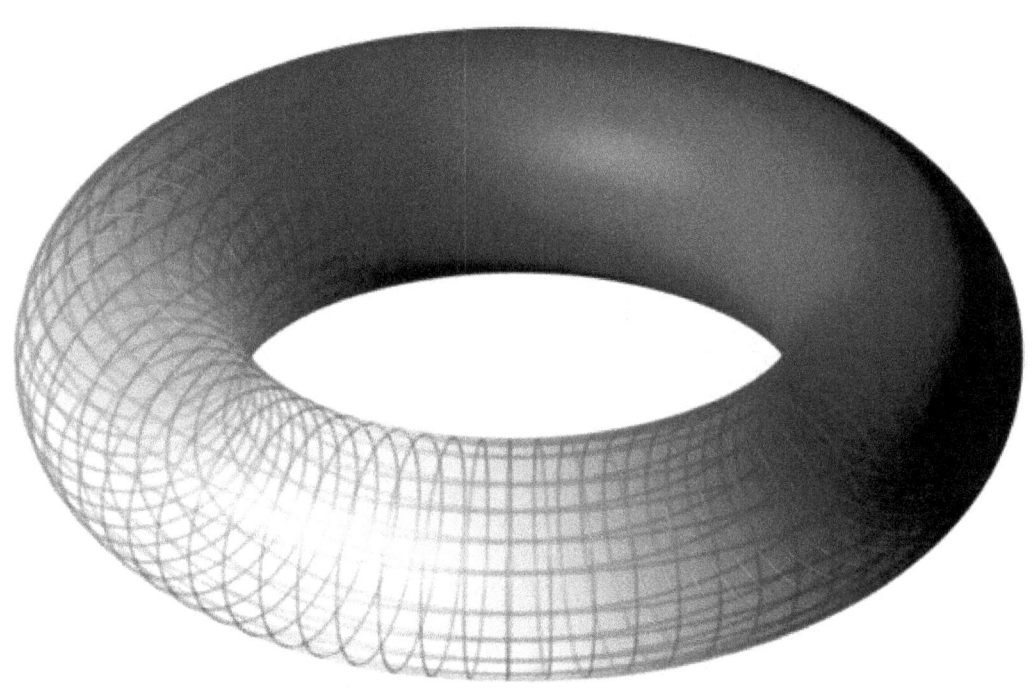

A torus. The standard torus is homogeneous under its diffeomorphism and homeomorphism groups, and the flat torus is homogeneous under its diffeomorphism, homeomorphism, and isometry groups.

In mathematics, particularly in the theories of Lie groups, algebraic groups and topological groups, a **homogeneous space** for a group G is a non-empty manifold or topological space X on which G acts transitively. The elements of G are called the **symmetries** of X. A special case of this is when the group G in question is the automorphism group of the space X – here "automorphism group" can mean isometry group, diffeomorphism group, or homeomorphism group. In this case X is homogeneous if intuitively X looks locally the same at each point, either in the sense of isometry (rigid geometry), diffeomorphism (differential geometry), or homeomorphism (topology). Some authors insist that the action of G be faithful (non-identity elements act non-trivially), although the present article does not. Thus there is a group action of G on X which can be thought of as preserving some "geometric structure" on X, and making X into a single G-orbit.

153

24.1   Formal definition

Let X be a non-empty set and G a group. Then X is called a G-space if it is equipped with an action of G on X.[1] Note that automatically G acts by automorphisms (bijections) on the set. If X in addition belongs to some category, then the elements of G are assumed to act as automorphisms in the same category. Thus the maps on X effected by G are structure preserving. A homogeneous space is a G-space on which G acts transitively.

Succinctly, if X is an object of the category $\mathbf{C}$, then the structure of a G-space is a homomorphism:

$$\rho : G \to \mathrm{Aut}_{\mathbf{C}}(X)$$

into the group of automorphisms of the object X in the category $\mathbf{C}$. The pair (X, ρ) defines a homogeneous space provided $\rho(G)$ is a transitive group of symmetries of the underlying set of X.

24.1.1   Examples

For example, if X is a topological space, then group elements are assumed to act as homeomorphisms on X. The structure of a G-space is a group homomorphism $\rho : G \to \mathrm{Homeo}(X)$ into the homeomorphism group of X.

Similarly, if X is a differentiable manifold, then the group elements are diffeomorphisms. The structure of a G-space is a group homomorphism $\rho : G \to \mathrm{Diffeo}(X)$ into the diffeomorphism group of X.

Riemannian symmetric spaces are an important class of homogeneous spaces, and include many of the examples listed below.

Concrete examples include:

Isometry groups

- Positive curvature:

1. Sphere (orthogonal group): $S^{n-1} \cong \mathrm{O}(n)/\mathrm{O}(n-1)$
2. Oriented sphere (special orthogonal group): $S^{n-1} \cong \mathrm{SO}(n)/\mathrm{SO}(n-1)$
3. Projective space (projective orthogonal group): $\mathrm{P}^{n-1} \cong \mathrm{PO}(n)/\mathrm{PO}(n-1)$

- Flat (zero curvature):

1. Euclidean space (Euclidean group, point stabilizer is orthogonal group): $\mathbf{A}^n \cong \mathrm{E}(n)/\mathrm{O}(n)$

- Negative curvature:

1. Hyperbolic space (orthochronous Lorentz group, point stabilizer orthogonal group, corresponding to hyperboloid model): $\mathbf{H}^n \cong \mathrm{O}^+(1, n)/\mathrm{O}(n)$
2. Oriented hyperbolic space: $\mathrm{SO}^+(1, n)/\mathrm{SO}(n)$
3. Anti-de Sitter space: $\mathrm{AdS}_{n+1} = \mathrm{O}(2, n)/\mathrm{O}(1, n)$

Others

- Affine space (for affine group, point stabilizer general linear group): $\mathbf{A}^n = \mathrm{Aff}(n, K)/\mathrm{GL}(n, k)$.
- Grassmannian: $\mathrm{Gr}(r, n) = \mathrm{O}(n)/(\mathrm{O}(r) \times \mathrm{O}(n-r))$

24.2 Geometry

From the point of view of the Erlangen program, one may understand that "all points are the same", in the geometry of X. This was true of essentially all geometries proposed before Riemannian geometry, in the middle of the nineteenth century.

Thus, for example, Euclidean space, affine space and projective space are all in natural ways homogeneous spaces for their respective symmetry groups. The same is true of the models found of non-Euclidean geometry of constant curvature, such as hyperbolic space.

A further classical example is the space of lines in projective space of three dimensions (equivalently, the space of two-dimensional subspaces of a four-dimensional vector space). It is simple linear algebra to show that GL_4 acts transitively on those. We can parameterize them by *line co-ordinates*: these are the 2×2 minors of the 4×2 matrix with columns two basis vectors for the subspace. The geometry of the resulting homogeneous space is the line geometry of Julius Plücker.

24.3 Homogeneous spaces as coset spaces

In general, if X is a homogeneous space, and Ho is the stabilizer of some marked point o in X (a choice of origin), the points of X correspond to the left cosets G/Ho, and the marked point o corresponds to the coset of the identity. Conversely, given a coset space G/H, it is a homogeneous space for G with a distinguished point, namely the coset of the identity. Thus a homogeneous space can be thought of as a coset space without a choice of origin.

In general, a different choice of origin o will lead to a quotient of G by a different subgroup Ho' which is related to Ho by an inner automorphism of G. Specifically,

$$H_{o'} = gH_og^{-1} \qquad (1)$$

where g is any element of G for which $go = o'$. Note that the inner automorphism (1) does not depend on which such g is selected; it depends only on g modulo Ho.

If the action of G on X is continuous, then H is a closed subgroup of G. In particular, if G is a Lie group, then H is a Lie subgroup by Cartan's theorem. Hence G/H is a smooth manifold and so X carries a unique smooth structure compatible with the group action.

If H is the identity subgroup $\{e\}$, then X is a principal homogeneous space.

One can go further to *double* coset spaces, notably Clifford–Klein forms $\Gamma\backslash G/H$, where Γ is a discrete subgroup (of G) acting properly discontinuously.

24.4 Example

For example in the line geometry case, we can identify H as a 12-dimensional subgroup of the 16-dimensional general linear group, GL(4), defined by conditions on the matrix entries

$$h_{13} = h_{14} = h_{23} = h_{24} = 0,$$

by looking for the stabilizer of the subspace spanned by the first two standard basis vectors. That shows that X has dimension 4.

Since the homogeneous coordinates given by the minors are 6 in number, this means that the latter are not independent of each other. In fact a single quadratic relation holds between the six minors, as was known to nineteenth-century geometers.

This example was the first known example of a Grassmannian, other than a projective space. There are many further homogeneous spaces of the classical linear groups in common use in mathematics.

24.5   Prehomogeneous vector spaces

The idea of a prehomogeneous vector space was introduced by Mikio Sato.

It is a finite-dimensional vector space V with a group action of an algebraic group G, such that there is an orbit of G that is open for the Zariski topology (and so, dense). An example is GL(1) acting on a one-dimensional space.

The definition is more restrictive than it initially appears: such spaces have remarkable properties, and there is a classification of irreducible prehomogeneous vector spaces, up to a transformation known as "castling".

24.6   Homogeneous spaces in physics

Cosmology using the general theory of relativity makes use of the Bianchi classification system. Homogeneous spaces in relativity represent the space part of background metrics for some cosmological models; for example, the three cases of the Friedmann–Lemaître–Robertson–Walker metric may be represented by subsets of the Bianchi I (flat), V (open), VII (flat or open) and IX (closed) types, while the Mixmaster universe represents an anisotropic example of a Bianchi IX cosmology.[2]

A homogeneous space of N dimensions admits a set of $\frac{1}{2}N(N+1)$ Killing vectors.[3] For three dimensions, this gives a total of six linearly independent Killing vector fields; homogeneous 3-spaces have the property that one may use linear combinations of these to find three everywhere non-vanishing Killing vector fields $\xi_i^{(a)}$,

$$\xi_{[i;k]}^{(a)} = C^a_{bc}\xi_i^{(b)}\xi_k^{(c)}$$

where the object C^a_{bc}, the "structure constants", form a constant order-three tensor antisymmetric in its lower two indices (on the left-hand side, the brackets denote antisymmetrisation and ";" represents the covariant differential operator). In the case of a flat isotropic universe, one possibility is $C^a_{bc} = 0$ (type I), but in the case of a closed FLRW universe, $C^a_{bc} = \varepsilon^a_{bc}$ where ε^a_{bc} is the Levi-Civita symbol.

24.7   See also

- Erlangen program

- Klein geometry

- Heap (mathematics)

- Homogeneous variety

24.8   References

[1] We assume that the action is on the *left*. The distinction is only important in the description of X as a coset space.

[2] Lev Landau and Evgeny Lifshitz (1980), *Course of Theoretical Physics vol. 2: The Classical Theory of Fields*, Butterworth-Heinemann, ISBN 978-0-7506-2768-9

[3] Steven Weinberg (1972), *Gravitation and Cosmology*, John Wiley and Sons

Chapter 25

Representation theory

This article is about the theory of representations of algebraic structures by linear transformations and matrices. For representation theory in other disciplines, see Representation.

Representation theory is a branch of mathematics that studies abstract algebraic structures by *representing* their elements as linear transformations of vector spaces, and studies modules over these abstract algebraic structures.[1] In essence, a representation makes an abstract algebraic object more concrete by describing its elements by matrices and the algebraic operations in terms of matrix addition and matrix multiplication. The algebraic objects amenable to such a description include groups, associative algebras and Lie algebras. The most prominent of these (and historically the first) is the representation theory of groups, in which elements of a group are represented by invertible matrices in such a way that the group operation is matrix multiplication.[2]

Representation theory is a useful method because it reduces problems in abstract algebra to problems in linear algebra, a subject that is well understood.[3] Furthermore, the vector space on which a group (for example) is represented can be infinite-dimensional, and by allowing it to be, for instance, a Hilbert space, methods of analysis can be applied to the theory of groups.[4] Representation theory is also important in physics because, for example, it describes how the symmetry group of a physical system affects the solutions of equations describing that system.[5]

A feature of representation theory is its pervasiveness in mathematics. There are two sides to this. First, the applications of representation theory are diverse:[6] in addition to its impact on algebra, representation theory:

- illuminates and generalizes Fourier analysis via harmonic analysis,[7]

- is connected to geometry via invariant theory and the Erlangen program,[8]

- has an impact in number theory via automorphic forms and the Langlands program.[9]

The second aspect is the diversity of approaches to representation theory. The same objects can be studied using methods from algebraic geometry, module theory, analytic number theory, differential geometry, operator theory, algebraic combinatorics and topology.[10]

The success of representation theory has led to numerous generalizations. One of the most general is in category theory.[11] The algebraic objects to which representation theory applies can be viewed as particular kinds of categories, and the representations as functors from the object category to the category of vector spaces. This description points to two obvious generalizations: first, the algebraic objects can be replaced by more general categories; second, the target category of vector spaces can be replaced by other well-understood categories.

A *representation* should not be confused with a *presentation*.

25.1 Definitions and concepts

Let V be a vector space over a field $\mathbf{F}$.[3] For instance, suppose V is $\mathbf{R}^n$ or $\mathbf{C}^n$, the standard n-dimensional space of column vectors over the real or complex numbers respectively. In this case, the idea of representation theory is to do abstract algebra concretely by using $n \times n$ matrices of real or complex numbers.

There are three main sorts of algebraic objects for which this can be done: groups, associative algebras and Lie algebras.[12]

- The set of all *invertible* $n \times n$ matrices is a group under matrix multiplication and the representation theory of groups analyses a group by describing ("representing") its elements in terms of invertible matrices.

- Matrix addition and multiplication make the set of *all* $n \times n$ matrices into an associative algebra and hence there is a corresponding representation theory of associative algebras.

- If we replace matrix multiplication MN by the matrix commutator $MN - NM$, then the $n \times n$ matrices become instead a Lie algebra, leading to a representation theory of Lie algebras.

This generalizes to any field $\mathbf{F}$ and any vector space V over $\mathbf{F}$, with linear maps replacing matrices and composition replacing matrix multiplication: there is a group $\mathrm{GL}(V,\mathbf{F})$ of automorphisms of V, an associative algebra $\mathrm{End}\mathbf{F}(V)$ of all endomorphisms of V, and a corresponding Lie algebra $\mathbf{gl}(V,\mathbf{F})$.

25.1.1 Definition

See also: group representation, algebra representation and Lie algebra representation

There are two ways to say what a representation is.[13] The first uses the idea of an action, generalizing the way that matrices act on column vectors by matrix multiplication. A representation of a group G or (associative or Lie) algebra A on a vector space V is a map

$$\Phi : G \times V \to V \quad \text{or} \quad \Phi : A \times V \to V$$

with two properties. First, for any g in G (or a in A), the map

$$\varphi(g) : V \to V$$
$$v \mapsto \Phi(g, v)$$

is linear (over $\mathbf{F}$). Second, if we introduce the notation $g \cdot v$ for $\Phi(g, v)$, then for any g_1, g_2 in G and v in V:

(1) $e \cdot v = v$

(2) $g_1 \cdot (g_2 \cdot v) = (g_1 g_2) \cdot v$

where e is the identity element of G and $g_1 g_2$ is product in G. The requirement for associative algebras is analogous, except that associative algebras do not always have an identity element, in which case equation (1) is ignored. Equation (2) is an abstract expression of the associativity of matrix multiplication. This doesn't hold for the matrix commutator and also there is no identity element for the commutator. Hence for Lie algebras, the only requirement is that for any x_1, x_2 in A and v in V:

(2′) $x_1 \cdot (x_2 \cdot v) - x_2 \cdot (x_1 \cdot v) = [x_1, x_2] \cdot v$

where $[x_1, x_2]$ is the Lie bracket, which generalizes the matrix commutator $MN - NM$.

The second way to define a representation focuses on the map φ sending g in G to a linear map $\varphi(g)$: $V \to V$, which satisfies

$$\varphi(g_1 g_2) = \varphi(g_1) \circ \varphi(g_2) \quad \text{all for } g_1, g_2 \in G$$

and similarly in the other cases. This approach is both more concise and more abstract. From this point of view:

- a representation of a group G on a vector space V is a group homomorphism φ: $G \to \mathrm{GL}(V, \mathbf{F})$;

- a representation of an associative algebra A on a vector space V is an algebra homomorphism φ: $A \to \mathrm{End}\mathbf{F}(V)$;

- a representation of a Lie algebra $\mathbf{a}$ on a vector space V is a Lie algebra homomorphism φ: $\mathbf{a} \to \mathbf{gl}(V, \mathbf{F})$.

25.1.2 Terminology

The vector space V is called the **representation space** of φ and its dimension (if finite) is called the **dimension** of the representation (sometimes *degree*, as in [14]). It is also common practice to refer to V itself as the representation when the homomorphism φ is clear from the context; otherwise the notation (V, φ) can be used to denote a representation.

When V is of finite dimension n, one can choose a basis for V to identify V with $\mathbf{F}^n$ and hence recover a matrix representation with entries in the field $\mathbf{F}$.

An effective or faithful representation is a representation (V, φ) for which the homomorphism φ is injective.

25.1.3 Equivariant maps and isomorphisms

See also: Equivariant map

If V and W are vector spaces over $\mathbf{F}$, equipped with representations φ and ψ of a group G, then an equivariant map from V to W is a linear map α: $V \to W$ such that

$$\alpha(g \cdot v) = g \cdot \alpha(v)$$

for all g in G and v in V. In terms of φ: $G \to \mathrm{GL}(V)$ and ψ: $G \to \mathrm{GL}(W)$, this means

$$\alpha \circ \phi(g) = \psi(g) \circ \alpha$$

for all g in G.

Equivariant maps for representations of an associative or Lie algebra are defined similarly. If α is invertible, then it is said to be an isomorphism, in which case V and W (or, more precisely, φ and ψ) are *isomorphic representations*.

Isomorphic representations are, for all practical purposes, "the same": they provide the same information about the group or algebra being represented. Representation theory therefore seeks to classify representations "up to isomorphism".

25.1.4 Subrepresentations, quotients, and irreducible representations

See also: Irreducible representation and simple module

If (W, ψ) is a representation of (say) a group G, and V is a linear subspace of W that is preserved by the action of G in the sense that $g \cdot v \in V$ for all $v \in V$ (Serre [14] calls these V *stable under G*), then V is called a *subrepresentation*: by defining

$\varphi(g)$ to be the restriction of $\psi(g)$ to V, (V, φ) is a representation of G and the inclusion of V into W is an equivariant map. The quotient space W/V can also be made into a representation of G.

If W has exactly two subrepresentations, namely the trivial subspace $\{0\}$ and W itself, then the representation is said to be *irreducible*; if W has a proper nontrivial subrepresentation, the representation is said to be *reducible*.[15]

The definition of an irreducible representation implies Schur's lemma: an equivariant map $\alpha: V \to W$ between irreducible representations is either the zero map or an isomorphism, since its kernel and image are subrepresentations. In particular, when $V = W$, this shows that the equivariant endomorphisms of V form an associative division algebra over the underlying field $\mathbf{F}$. If $\mathbf{F}$ is algebraically closed, the only equivariant endomorphisms of an irreducible representation are the scalar multiples of the identity.

Irreducible representations are the building blocks of representation theory: if a representation W is not irreducible then it is built from a subrepresentation and a quotient that are both "simpler" in some sense; for instance, if W is finite-dimensional, then both the subrepresentation and the quotient have smaller dimension.

25.1.5   Direct sums and indecomposable representations

See also: Direct sum, indecomposable module and semisimple module

If (V,φ) and (W,ψ) are representations of (say) a group G, then the direct sum of V and W is a representation, in a canonical way, via the equation

$$g \cdot (v, w) = (g \cdot v, g \cdot w).$$

The direct sum of two representations carries no more information about the group G than the two representations do individually. If a representation is the direct sum of two proper nontrivial subrepresentations, it is said to be decomposable. Otherwise, it is said to be indecomposable.

In favourable circumstances, every representation is a direct sum of irreducible representations: such representations are said to be semisimple. In this case, it suffices to understand only the irreducible representations. In other cases, one must understand how indecomposable representations can be built from irreducible representations as extensions of a quotient by a subrepresentation.

25.2   Branches and topics

See also: Group representation

Representation theory is notable for the number of branches it has, and the diversity of the approaches to studying representations of groups and algebras. Although, all the theories have in common the basic concepts discussed already, they differ considerably in detail. The differences are at least 3-fold:

1. Representation theory depends upon the type of algebraic object being represented. There are several different classes of groups, associative algebras and Lie algebras, and their representation theories all have an individual flavour.

2. Representation theory depends upon the nature of the vector space on which the algebraic object is represented. The most important distinction is between finite-dimensional representations and infinite-dimensional ones. In the infinite-dimensional case, additional structures are important (e.g. whether or not the space is a Hilbert space, Banach space, etc.). Additional algebraic structures can also be imposed in the finite-dimensional case.

3. Representation theory depends upon the type of field over which the vector space is defined. The most important case is the field of complex numbers. The other important cases are the field of real numbers, finite fields, and

fields of p-adic numbers. Additional difficulties arise for fields of positive characteristic and for fields that are not algebraically closed.

25.2.1 Finite groups

Main article: Representation of a finite group

Group representations are a very important tool in the study of finite groups.[16] They also arise in the applications of finite group theory to geometry and crystallography.[17] Representations of finite groups exhibit many of the features of the general theory and point the way to other branches and topics in representation theory.

Over a field of characteristic zero, the representation theory of a finite group G has a number of convenient properties. First, the representations of G are semisimple (completely reducible). This is a consequence of Maschke's theorem, which states that any subrepresentation V of a G-representation W has a G-invariant complement. One proof is to choose any projection π from W to V and replace it by its average πG defined by

$$\pi_G(x) = \frac{1}{|G|} \sum_{g \in G} g \cdot \pi(g^{-1} \cdot x).$$

πG is equivariant, and its kernel is the required complement.

The finite-dimensional G-representations can be understood using character theory: the character of a representation φ: $G \to \mathrm{GL}(V)$ is the class function $\chi\varphi$: $G \to \mathbf{F}$ defined by

$$\chi_\varphi(g) = \mathrm{Tr}(\varphi(g))$$

where Tr is the trace. An irreducible representation of G is completely determined by its character.

Maschke's theorem holds more generally for fields of positive characteristic p, such as the finite fields, as long as the prime p is coprime to the order of G. When p and $|G|$ have a common factor, there are G-representations that are not semisimple, which are studied in a subbranch called modular representation theory.

Averaging techniques also show that if $\mathbf{F}$ is the real or complex numbers, then any G-representation preserves an inner product $\langle \cdot, \cdot \rangle$ on V in the sense that

$$\langle g \cdot v, g \cdot w \rangle = \langle v, w \rangle$$

for all g in G and v, w in W. Hence any G-representation is unitary.

Unitary representations are automatically semisimple, since Maschke's result can be proven by taking the orthogonal complement of a subrepresentation. When studying representations of groups that are not finite, the unitary representations provide a good generalization of the real and complex representations of a finite group.

Results such as Maschke's theorem and the unitary property that rely on averaging can be generalized to more general groups by replacing the average with an integral, provided that a suitable notion of integral can be defined. This can be done for compact groups or locally compact groups, using Haar measure, and the resulting theory is known as abstract harmonic analysis.

Over arbitrary fields, another class of finite groups that have a good representation theory are the finite groups of Lie type. Important examples are linear algebraic groups over finite fields. The representation theory of linear algebraic groups and Lie groups extends these examples to infinite-dimensional groups, the latter being intimately related to Lie algebra representations. The importance of character theory for finite groups has an analogue in the theory of weights for representations of Lie groups and Lie algebras.

Representations of a finite group G are also linked directly to algebra representations via the group algebra $\mathbf{F}[G]$, which is a vector space over $\mathbf{F}$ with the elements of G as a basis, equipped with the multiplication operation defined by the group operation, linearity, and the requirement that the group operation and scalar multiplication commute.

25.2.2 Modular representations

Main article: Modular representation theory

Modular representations of a finite group G are representations over a field whose characteristic is not coprime to $|G|$, so that Maschke's theorem no longer holds (because $|G|$ is not invertible in $\mathbf{F}$ and so one cannot divide by it).[18] Nevertheless, Richard Brauer extended much of character theory to modular representations, and this theory played an important role in early progress towards the classification of finite simple groups, especially for simple groups whose characterization was not amenable to purely group-theoretic methods because their Sylow 2-subgroups were "too small".[19]

As well as having applications to group theory, modular representations arise naturally in other branches of mathematics, such as algebraic geometry, coding theory, combinatorics and number theory.

25.2.3 Unitary representations

Main article: Unitary representation

A unitary representation of a group G is a linear representation φ of G on a real or (usually) complex Hilbert space V such that $\varphi(g)$ is a unitary operator for every $g \in G$. Such representations have been widely applied in quantum mechanics since the 1920s, thanks in particular to the influence of Hermann Weyl,[20] and this has inspired the development of the theory, most notably through the analysis of representations of the Poincaré group by Eugene Wigner.[21] One of the pioneers in constructing a general theory of unitary representations (for any group G rather than just for particular groups useful in applications) was George Mackey, and an extensive theory was developed by Harish-Chandra and others in the 1950s and 1960s.[22]

A major goal is to describe the "unitary dual", the space of irreducible unitary representations of G.[23] The theory is most well-developed in the case that G is a locally compact (Hausdorff) topological group and the representations are strongly continuous.[7] For G abelian, the unitary dual is just the space of characters, while for G compact, the Peter–Weyl theorem shows that the irreducible unitary representations are finite-dimensional and the unitary dual is discrete.[24] For example, if G is the circle group S^1, then the characters are given by integers, and the unitary dual is $\mathbf{Z}$.

For non-compact G, the question of which representations are unitary is a subtle one. Although irreducible unitary representations must be "admissible" (as Harish-Chandra modules) and it is easy to detect which admissible representations have a nondegenerate invariant sesquilinear form, it is hard to determine when this form is positive definite. An effective description of the unitary dual, even for relatively well-behaved groups such as real reductive Lie groups (discussed below), remains an important open problem in representation theory. It has been solved for many particular groups, such as SL(2,$\mathbf{R}$) and the Lorentz group.[25]

25.2.4 Harmonic analysis

Main article: Abstract harmonic analysis

The duality between the circle group S^1 and the integers $\mathbf{Z}$, or more generally, between a torus T^n and $\mathbf{Z}^n$ is well known in analysis as the theory of Fourier series, and the Fourier transform similarly expresses the fact that the space of characters on a real vector space is the dual vector space. Thus unitary representation theory and harmonic analysis are intimately related, and abstract harmonic analysis exploits this relationship, by developing the analysis of functions on locally compact topological groups and related spaces.[7]

A major goal is to provide a general form of the Fourier transform and the Plancherel theorem. This is done by constructing a measure on the unitary dual and an isomorphism between the regular representation of G on the space $L^2(G)$ of square integrable functions on G and its representation on the space of L^2 functions on the unitary dual. Pontrjagin duality and the Peter–Weyl theorem achieve this for abelian and compact G respectively.[24][26]

Another approach involves considering all unitary representations, not just the irreducible ones. These form a category, and Tannaka–Krein duality provides a way to recover a compact group from its category of unitary representations.

If the group is neither abelian nor compact, no general theory is known with an analogue of the Plancherel theorem or Fourier inversion, although Alexander Grothendieck extended Tannaka–Krein duality to a relationship between linear algebraic groups and tannakian categories.

Harmonic analysis has also been extended from the analysis of functions on a group *G* to functions on homogeneous spaces for *G*. The theory is particularly well developed for symmetric spaces and provides a theory of automorphic forms (discussed below).

25.2.5 Lie groups

Main article: Representation of a Lie group

A Lie group is a group that is also a smooth manifold. Many classical groups of matrices over the real or complex numbers are Lie groups.[27] Many of the groups important in physics and chemistry are Lie groups, and their representation theory is crucial to the application of group theory in those fields.[5]

The representation theory of Lie groups can be developed first by considering the compact groups, to which results of compact representation theory apply.[23] This theory can be extended to finite-dimensional representations of semisimple Lie groups using Weyl's unitary trick: each semisimple real Lie group *G* has a complexification, which is a complex Lie group G^c, and this complex Lie group has a maximal compact subgroup *K*. The finite-dimensional representations of *G* closely correspond to those of *K*.

A general Lie group is a semidirect product of a solvable Lie group and a semisimple Lie group (the Levi decomposition).[28] The classification of representations of solvable Lie groups is intractable in general, but often easy in practical cases. Representations of semidirect products can then be analysed by means of general results called *Mackey theory*, which is a generalization of the methods used in Wigner's classification of representations of the Poincaré group.

25.2.6 Lie algebras

Main article: Lie algebra representation

A Lie algebra over a field **F** is a vector space over **F** equipped with a skew-symmetric bilinear operation called the Lie bracket, which satisfies the Jacobi identity. Lie algebras arise in particular as tangent spaces to Lie groups at the identity element, leading to their interpretation as "infinitesimal symmetries".[28] An important approach to the representation theory of Lie groups is to study the corresponding representation theory of Lie algebras, but representations of Lie algebras also have an intrinsic interest.[29]

Lie algebras, like Lie groups, have a Levi decomposition into semisimple and solvable parts, with the representation theory of solvable Lie algebras being intractable in general. In contrast, the finite-dimensional representations of semisimple Lie algebras are completely understood, after work of Élie Cartan. A representation of a semisimple Lie algebra **g** is analysed by choosing a Cartan subalgebra, which is essentially a generic maximal subalgebra **h** of **g** on which the Lie bracket is zero ("abelian"). The representation of **g** can be decomposed into weight spaces that are eigenspaces for the action of **h** and the infinitesimal analogue of characters. The structure of semisimple Lie algebras then reduces the analysis of representations to easily understood combinatorics of the possible weights that can occur.[28]

Infinite-dimensional Lie algebras

See also: Affine Lie algebra and Kac–Moody algebra

There are many classes of infinite-dimensional Lie algebras whose representations have been studied. Among these, an important class are the Kac–Moody algebras.[30] They are named after Victor Kac and Robert Moody, who independently discovered them. These algebras form a generalization of finite-dimensional semisimple Lie algebras, and share many of their combinatorial properties. This means that they have a class of representations that can be understood in the same way as representations of semisimple Lie algebras.

Affine Lie algebras are a special case of Kac–Moody algebras, which have particular importance in mathematics and theoretical physics, especially conformal field theory and the theory of exactly solvable models. Kac discovered an elegant proof of certain combinatorial identities, Macdonald identities, which is based on the representation theory of affine Kac–Moody algebras.

Lie superalgebras

Main article: Representation of a Lie superalgebra

Lie superalgebras are generalizations of Lie algebras in which the underlying vector space has a $\mathbf{Z}_2$-grading, and skew-symmetry and Jacobi identity properties of the Lie bracket are modified by signs. Their representation theory is similar to the representation theory of Lie algebras.[31]

25.2.7   Linear algebraic groups

See also: Linear algebraic group

Linear algebraic groups (or more generally, affine group schemes) are analogues in algebraic geometry of Lie groups, but over more general fields than just $\mathbf{R}$ or $\mathbf{C}$. In particular, over finite fields, they give rise to finite groups of Lie type. Although linear algebraic groups have a classification that is very similar to that of Lie groups, their representation theory is rather different (and much less well understood) and requires different techniques, since the Zariski topology is relatively weak, and techniques from analysis are no longer available.[32]

25.2.8   Invariant theory

Main article: Invariant theory

Invariant theory studies actions on algebraic varieties from the point of view of their effect on functions, which form representations of the group. Classically, the theory dealt with the question of explicit description of polynomial functions that do not change, or are *invariant*, under the transformations from a given linear group. The modern approach analyses the decomposition of these representations into irreducibles.[33]

Invariant theory of infinite groups is inextricably linked with the development of linear algebra, especially, the theories of quadratic forms and determinants. Another subject with strong mutual influence is projective geometry, where invariant theory can be used to organize the subject, and during the 1960s, new life was breathed into the subject by David Mumford in the form of his geometric invariant theory.[34]

The representation theory of semisimple Lie groups has its roots in invariant theory[27] and the strong links between representation theory and algebraic geometry have many parallels in differential geometry, beginning with Felix Klein's Erlangen program and Élie Cartan's connections, which place groups and symmetry at the heart of geometry.[35] Modern developments link representation theory and invariant theory to areas as diverse as holonomy, differential operators and the theory of several complex variables.

25.2.9   Automorphic forms and number theory

Main article: Automorphic form

Automorphic forms are a generalization of modular forms to more general analytic functions, perhaps of several complex variables, with similar transformation properties.[36] The generalization involves replacing the modular group PSL_2 (**R**) and a chosen congruence subgroup by a semisimple Lie group G and a discrete subgroup Γ. Just as modular forms can be viewed as differential forms on a quotient of the upper half space $H = PSL_2$ (**R**)$/SO(2)$, automorphic forms can be viewed as differential forms (or similar objects) on $\Gamma\backslash G/K$, where K is (typically) a maximal compact subgroup of G. Some care is required, however, as the quotient typically has singularities. The quotient of a semisimple Lie group by a compact subgroup is a symmetric space and so the theory of automorphic forms is intimately related to harmonic analysis on symmetric spaces.

Before the development of the general theory, many important special cases were worked out in detail, including the Hilbert modular forms and Siegel modular forms. Important results in the theory include the Selberg trace formula and the realization by Robert Langlands that the Riemann-Roch theorem could be applied to calculate the dimension of the space of automorphic forms. The subsequent notion of "automorphic representation" has proved of great technical value for dealing with the case that G is an algebraic group, treated as an adelic algebraic group. As a result an entire philosophy, the Langlands program has developed around the relation between representation and number theoretic properties of automorphic forms.[37]

25.2.10 Associative algebras

Main article: Algebra representation

In one sense, associative algebra representations generalize both representations of groups and Lie algebras. A representation of a group induces a representation of a corresponding group ring or group algebra, while representations of a Lie algebra correspond bijectively to representations of its universal enveloping algebra. However, the representation theory of general associative algebras does not have all of the nice properties of the representation theory of groups and Lie algebras.

Module theory

Main article: Module theory

When considering representations of an associative algebra, one can forget the underlying field, and simply regard the associative algebra as a ring, and its representations as modules. This approach is surprisingly fruitful: many results in representation theory can be interpreted as special cases of results about modules over a ring.

Hopf algebras and quantum groups

Main article: Representation theory of Hopf algebras

Hopf algebras provide a way to improve the representation theory of associative algebras, while retaining the representation theory of groups and Lie algebras as special cases. In particular, the tensor product of two representations is a representation, as is the dual vector space.

The Hopf algebras associated to groups have a commutative algebra structure, and so general Hopf algebras are known as quantum groups, although this term is often restricted to certain Hopf algebras arising as deformations of groups or their universal enveloping algebras. The representation theory of quantum groups has added surprising insights to the representation theory of Lie groups and Lie algebras, for instance through the crystal basis of Kashiwara.

25.3 Generalizations

25.3.1   Set-theoretic representations

Main article: Group action

A *set-theoretic representation* (also known as a group action or *permutation representation*) of a group G on a set X is given by a function ρ from G to X^X, the set of functions from X to X, such that for all g_1, g_2 in G and all x in X:

$$\rho(1)[x] = x$$

$$\rho(g_1 g_2)[x] = \rho(g_1)[\rho(g_2)[x]].$$

This condition and the axioms for a group imply that $\rho(g)$ is a bijection (or permutation) for all g in G. Thus we may equivalently define a permutation representation to be a group homomorphism from G to the symmetric group SX of X.

25.3.2   Representations in other categories

See also: Category theory

Every group G can be viewed as a category with a single object; morphisms in this category are just the elements of G. Given an arbitrary category C, a *representation* of G in C is a functor from G to C. Such a functor selects an object X in C and a group homomorphism from G to $\mathrm{Aut}(X)$, the automorphism group of X.

In the case where C is **VectF**, the category of vector spaces over a field **F**, this definition is equivalent to a linear representation. Likewise, a set-theoretic representation is just a representation of G in the category of sets.

For another example consider the category of topological spaces, **Top**. Representations in **Top** are homomorphisms from G to the homeomorphism group of a topological space X.

Two types of representations closely related to linear representations are:

- projective representations: in the category of projective spaces. These can be described as "linear representations up to scalar transformations".

- affine representations: in the category of affine spaces. For example, the Euclidean group acts affinely upon Euclidean space.

25.3.3   Representations of categories

See also: Quiver (mathematics)

Since groups are categories, one can also consider representation of other categories. The simplest generalization is to monoids, which are categories with one object. Groups are monoids for which every morphism is invertible. General monoids have representations in any category. In the category of sets, these are monoid actions, but monoid representations on vector spaces and other objects can be studied.

More generally, one can relax the assumption that the category being represented has only one object. In full generality, this is simply the theory of functors between categories, and little can be said.

One special case has had a significant impact on representation theory, namely the representation theory of quivers.[11] A quiver is simply a directed graph (with loops and multiple arrows allowed), but it can be made into a category (and also an algebra) by considering paths in the graph. Representations of such categories/algebras have illuminated several aspects of representation theory, for instance by allowing non-semisimple representation theory questions about a group to be reduced in some cases to semisimple representation theory questions about a quiver.

25.4 See also

- Philosophy of cusp forms
- Representation (mathematics)
- Representation theorem
- List of representation theory topics
- List of harmonic analysis topics
- Galois representation

25.5 Notes

[1] Classic texts on representation theory include Curtis & Reiner (1962) and Serre (1977). Other excellent sources are Fulton & Harris (1991) and Goodman & Wallach (1998).

[2] For the history of the representation theory of finite groups, see Lam (1998). For algebraic and Lie groups, see Borel (2001).

[3] There are many textbooks on vector spaces and linear algebra. For an advanced treatment, see Kostrikin & Manin (1997).

[4] Sally & Vogan 1989.

[5] Sternberg 1994.

[6] Lam 1998, p. 372.

[7] Folland 1995.

[8] Goodman & Wallach 1998, Olver 1999, Sharpe 1997.

[9] Borel & Casselman 1979, Gelbert 1984.

[10] See the previous footnotes and also Borel (2001).

[11] Simson, Skowronski & Assem 2007.

[12] Fulton & Harris 1991, Simson, Skowronski & Assem 2007, Humphreys 1972.

[13] This material can be found in standard textbooks, such as Curtis & Reiner (1962), Fulton & Harris (1991), Goodman & Wallach (1998), Gordon & Liebeck (1993), Humphreys (1972), Jantzen (2003), Knapp (2001) and Serre (1977).

[14] Serre 1977

[15] The representation {0} of dimension zero is considered to be neither reducible nor irreducible, just like the number 1 is considered to be neither composite nor prime.

[16] Alperin 1986, Lam 1998, Serre 1977.

[17] Kim 1999.

[18] Serre 1977, Part III

[19] Alperin 1986.

[20] See Weyl 1928.

[21] Wigner 1939.

[22] Borel 2001.

[23] Knapp 2001.

[24] Peter & Weyl 1927.

[25] Bargmann 1947.

[26] Pontrjagin 1934.

[27] Weyl 1946.

[28] Fulton & Harris 1991.

[29] Humphreys 1972a.

[30] Kac 1990.

[31] Kac 1977.

[32] Humphreys 1972b, Jantzen 2003.

[33] Olver 1999.

[34] Mumford, Fogarty & Kirwan 1994.

[35] Sharpe 1997.

[36] Borel & Casselman 1979.

[37] Gelbart 1984.

25.6 References

- Alperin, J. L. (1986), *Local Representation Theory: Modular Representations as an Introduction to the Local Representation Theory of Finite Groups*, Cambridge University Press, ISBN 978-0-521-44926-7.

- Bargmann, V. (1947), "Irreducible unitary representations of the Lorenz group", *Annals of Mathematics* (Annals of Mathematics) **48** (3): 568–640, doi:10.2307/1969129, JSTOR 1969129.

- Borel, Armand (2001), *Essays in the History of Lie Groups and Algebraic Groups*, American Mathematical Society, ISBN 978-0-8218-0288-5.

- Borel, Armand; Casselman, W. (1979), *Automorphic Forms, Representations, and L-functions*, American Mathematical Society, ISBN 978-0-8218-1435-2.

- Curtis, Charles W.; Reiner, Irving (1962), *Representation Theory of Finite Groups and Associative Algebras*, John Wiley & Sons (Reedition 2006 by AMS Bookstore), ISBN 978-0-470-18975-7.

- Gelbart, Stephen (1984), "An Elementary Introduction to the Langlands Program", *Bulletin of the American Mathematical Society* **10** (2): 177–219, doi:10.1090/S0273-0979-1984-15237-6.

- Folland, Gerald B. (1995), *A Course in Abstract Harmonic Analysis*, CRC Press, ISBN 978-0-8493-8490-5.

- Fulton, William; Harris, Joe (1991), *Representation theory. A first course*, Graduate Texts in Mathematics, Readings in Mathematics **129**, New York: Springer-Verlag, ISBN 978-0-387-97495-8, MR 1153249, ISBN 978-0-387-97527-6.

- Goodman, Roe; Wallach, Nolan R. (1998), *Representations and Invariants of the Classical Groups*, Cambridge University Press, ISBN 978-0-521-66348-9.

- Gordon, James; Liebeck, Martin (1993), *Representations and Characters of Finite Groups*, Cambridge: Cambridge University Press, ISBN 978-0-521-44590-0.

- Helgason, Sigurdur (1978), *Differential Geometry, Lie groups and Symmetric Spaces*, Academic Press, ISBN 978-0-12-338460-7

- Humphreys, James E. (1972a), *Introduction to Lie Algebras and Representation Theory*, Birkhäuser, ISBN 978-0-387-90053-7.

- Humphreys, James E. (1972b), *Linear Algebraic Groups*, Graduate Texts in Mathematics **21**, Berlin, New York: Springer-Verlag, ISBN 978-0-387-90108-4, MR 0396773

- Jantzen, Jens Carsten (2003), *Representations of Algebraic Groups*, American Mathematical Society, ISBN 978-0-8218-3527-2.

- Kac, Victor G. (1977), "Lie superalgebras", *Advances in Mathematics* **26** (1): 8–96, doi:10.1016/0001-8708(77)90017-2.

- Kac, Victor G. (1990), *Infinite Dimensional Lie Algebras* (3rd ed.), Cambridge University Press, ISBN 978-0-521-46693-6.

- Knapp, Anthony W. (2001), *Representation Theory of Semisimple Groups: An Overview Based on Examples*, Princeton University Press, ISBN 978-0-691-09089-4.

- Kim, Shoon Kyung (1999), *Group Theoretical Methods and Applications to Molecules and Crystals: And Applications to Molecules and Crystals*, Cambridge University Press, ISBN 978-0-521-64062-6.

- Kostrikin, A. I.; Manin, Yuri I. (1997), *Linear Algebra and Geometry*, Taylor & Francis, ISBN 978-90-5699-049-7.

- Lam, T. Y. (1998), "Representations of finite groups: a hundred years", *Notices of the AMS* (American Mathematical Society) **45** (3,4): 361–372 (Part I), 465–474 (Part II).

- Yurii I. Lyubich. *Introduction to the Theory of Banach Representations of Groups.* Translated from the 1985 Russian-language edition (Kharkov, Ukraine). Birkhäuser Verlag. 1988.

- Mumford, David; Fogarty, J.; Kirwan, F. (1994), *Geometric invariant theory*, Ergebnisse der Mathematik und ihrer Grenzgebiete (2) [Results in Mathematics and Related Areas (2)] **34** (3rd ed.), Berlin, New York: Springer-Verlag, ISBN 978-3-540-56963-3, MR 0214602(1st ed. 1965) [[Mathematical Reviews|MR]] [http://www.ams.org/mathscinet-getitem?mr=0719371 0719371] (2nd ed.) [[Mathematical Reviews|MR]] [http://www.ams.org/mathscinet-getitem?mr=1304906 1304906](3rd ed.)

- Olver, Peter J. (1999), *Classical invariant theory*, Cambridge: Cambridge University Press, ISBN 0-521-55821-2.

- Peter, F.; Weyl, Hermann (1927), "Die Vollständigkeit der primitiven Darstellungen einer geschlossenen kontinuierlichen Gruppe", *Mathematische Annalen* **97** (1): 737–755, doi:10.1007/BF01447892.

- Pontrjagin, Lev S. (1934), "The theory of topological commutative groups", *Annals of Mathematics* (Annals of Mathematics) **35** (2): 361–388, doi:10.2307/1968438, JSTOR 1968438.

- Sally, Paul; Vogan, David A. (1989), *Representation Theory and Harmonic Analysis on Semisimple Lie Groups*, American Mathematical Society, ISBN 978-0-8218-1526-7.

- Serre, Jean-Pierre (1977), *Linear Representations of Finite Groups*, Springer-Verlag, ISBN 978-0387901909.

- Sharpe, Richard W. (1997), *Differential Geometry: Cartan's Generalization of Klein's Erlangen Program*, Springer, ISBN 978-0-387-94732-7.

- Simson, Daniel; Skowronski, Andrzej; Assem, Ibrahim (2007), *Elements of the Representation Theory of Associative Algebras*, Cambridge University Press, ISBN 978-0-521-88218-7.

- Sternberg, Shlomo (1994), *Group Theory and Physics*, Cambridge University Press, ISBN 978-0-521-55885-3.

- Weyl, Hermann (1928), *Gruppentheorie und Quantenmechanik* (The Theory of Groups and Quantum Mechanics, translated H.P. Robertson, 1931 ed.), S. Hirzel, Leipzig (reprinted 1950, Dover), ISBN 978-0-486-60269-1.

- Weyl, Hermann (1946), *The Classical Groups: Their Invariants and Representations* (2nd ed.), Princeton University Press (reprinted 1997), ISBN 978-0-691-05756-9.

- Wigner, Eugene P. (1939), "On unitary representations of the inhomogeneous Lorentz group", *Annals of Mathematics* (Annals of Mathematics) **40** (1): 149–204, doi:10.2307/1968551, JSTOR 1968551.

25.7 External links

- Hazewinkel, Michiel, ed. (2001), "Representation theory", *Encyclopedia of Mathematics*, Springer, ISBN 978-1-55608-010-4

Chapter 26

Vector space

This article is about linear (vector) spaces. For the structure in incidence geometry, see Linear space (geometry).

A **vector space** (also called a **linear space**) is a collection of objects called **vectors**, which may be added together

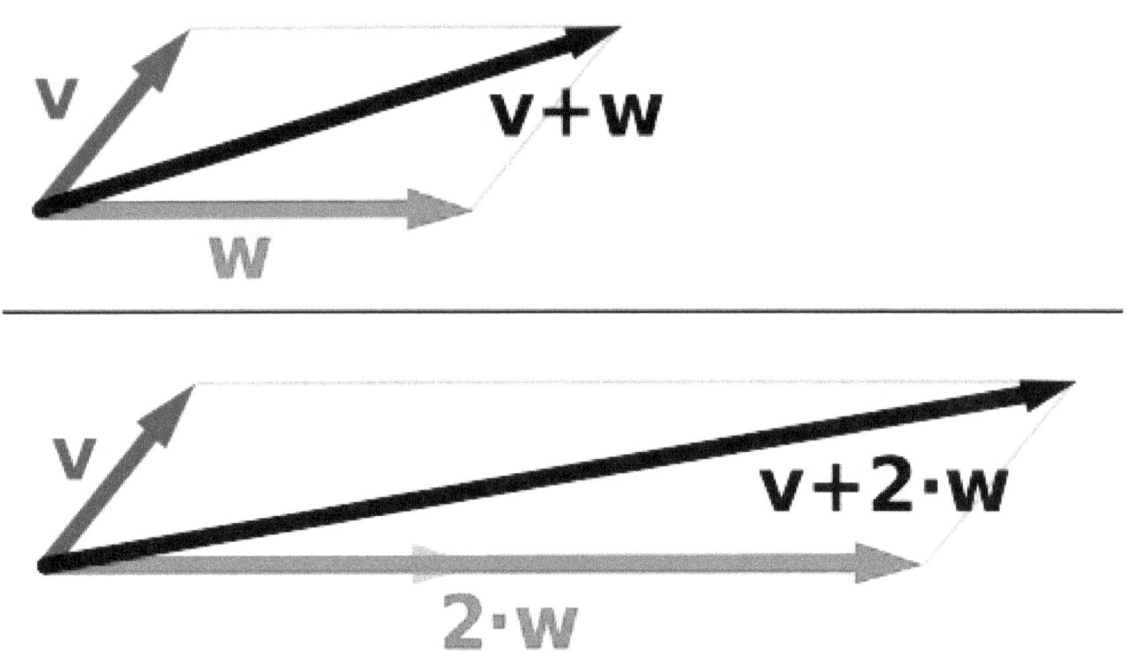

*Vector addition and scalar multiplication: a vector **v** (blue) is added to another vector **w** (red, upper illustration). Below, **w** is stretched by a factor of 2, yielding the sum **v** + 2**w**.*

and multiplied ("scaled") by numbers, called *scalars* in this context. Scalars are often taken to be real numbers, but there are also vector spaces with scalar multiplication by complex numbers, rational numbers, or generally any field. The operations of vector addition and scalar multiplication must satisfy certain requirements, called *axioms*, listed below. Euclidean vectors are an example of a vector space. They represent physical quantities such as forces: any two forces (of the same type) can be added to yield a third, and the multiplication of a force vector by a real multiplier is another force vector. In the same vein, but in a more geometric sense, vectors representing displacements in the plane or in three-dimensional space also form vector spaces. Vectors in vector spaces do not necessarily have to be arrow-like objects as they appear in the mentioned examples: vectors are regarded as abstract mathematical objects with particular properties, which in some cases can be visualized as arrows.

Vector spaces are the subject of linear algebra and are well understood from this point of view since vector spaces are characterized by their dimension, which, roughly speaking, specifies the number of independent directions in the space. A vector space may be endowed with additional structure, such as a norm or inner product. Such spaces arise naturally in mathematical analysis, mainly in the guise of infinite-dimensional function spaces whose vectors are functions. Analytical problems call for the ability to decide whether a sequence of vectors converges to a given vector. This is accomplished by considering vector spaces with additional structure, mostly spaces endowed with a suitable topology, thus allowing the consideration of proximity and continuity issues. These topological vector spaces, in particular Banach spaces and Hilbert spaces, have a richer theory.

Historically, the first ideas leading to vector spaces can be traced back as far as 17th century's analytic geometry, matrices, systems of linear equations, and Euclidean vectors. The modern, more abstract treatment, first formulated by Giuseppe Peano in 1888, encompasses more general objects than Euclidean space, but much of the theory can be seen as an extension of classical geometric ideas like lines, planes and their higher-dimensional analogs.

Today, vector spaces are applied throughout mathematics, science and engineering. They are the appropriate linear-algebraic notion to deal with systems of linear equations; offer a framework for Fourier expansion, which is employed in image compression routines; or provide an environment that can be used for solution techniques for partial differential equations. Furthermore, vector spaces furnish an abstract, coordinate-free way of dealing with geometrical and physical objects such as tensors. This in turn allows the examination of local properties of manifolds by linearization techniques. Vector spaces may be generalized in several ways, leading to more advanced notions in geometry and abstract algebra.

26.1 Introduction and definition

The concept of vector space will first be explained by describing two particular examples:

26.1.1 First example: arrows in the plane

The first example of a vector space consists of arrows in a fixed plane, starting at one fixed point. This is used in physics to describe forces or velocities. Given any two such arrows, $\mathbf{v}$ and $\mathbf{w}$, the parallelogram spanned by these two arrows contains one diagonal arrow that starts at the origin, too. This new arrow is called the *sum* of the two arrows and is denoted $\mathbf{v} + \mathbf{w}$. In the special case of two arrows on the same line, their sum is the arrow on this line whose length is the sum or the difference of the lengths, depending on whether the arrows have the same direction. Another operation that can be done with arrows is scaling: given any positive real number a, the arrow that has the same direction as $\mathbf{v}$, but is dilated or shrunk by multiplying its length by a, is called *multiplication* of $\mathbf{v}$ by a. It is denoted $a\mathbf{v}$. When a is negative, $a\mathbf{v}$ is defined as the arrow pointing in the opposite direction, instead.

The following shows a few examples: if $a = 2$, the resulting vector $a\mathbf{w}$ has the same direction as $\mathbf{w}$, but is stretched to the double length of $\mathbf{w}$ (right image below). Equivalently $2\mathbf{w}$ is the sum $\mathbf{w} + \mathbf{w}$. Moreover, $(-1)\mathbf{v} = -\mathbf{v}$ has the opposite direction and the same length as $\mathbf{v}$ (blue vector pointing down in the right image).

26.1.2 Second example: ordered pairs of numbers

A second key example of a vector space is provided by pairs of real numbers x and y. (The order of the components x and y is significant, so such a pair is also called an ordered pair.) Such a pair is written as (x, y). The sum of two such pairs and multiplication of a pair with a number is defined as follows:

$$(x_1, y_1) + (x_2, y_2) = (x_1 + x_2, y_1 + y_2)$$

and

$$a\,(x, y) = (ax, ay).$$

The first example above reduces to this one if the arrows are represented by the pair of Cartesian coordinates of their end points.

26.1.3 Definition

A vector space over a field F is a set V together with two operations that satisfy the eight axioms listed below. Elements of V are commonly called *vectors*. Elements of F are commonly called *scalars*. The first operation, called *vector addition* or simply *addition*, takes any two vectors **v** and **w** and assigns to them a third vector which is commonly written as **v** + **w**, and called the sum of these two vectors. The second operation, called *scalar multiplication* takes any scalar a and any vector **v** and gives another vector a**v**.

In this article, vectors are distinguished from scalars by boldface.[nb 1] In the two examples above, the field is the field of the real numbers and the set of the vectors consists of the planar arrows with fixed starting point and of pairs of real numbers, respectively.

To qualify as a vector space, the set V and the operations of addition and multiplication must adhere to a number of requirements called axioms.[1] In the list below, let **u**, **v** and **w** be arbitrary vectors in V, and a and b scalars in F.

These axioms generalize properties of the vectors introduced in the above examples. Indeed, the result of addition of two ordered pairs (as in the second example above) does not depend on the order of the summands:

$$(x_\mathbf{v}, y_\mathbf{v}) + (x_\mathbf{w}, y_\mathbf{w}) = (x_\mathbf{w}, y_\mathbf{w}) + (x_\mathbf{v}, y_\mathbf{v}).$$

Likewise, in the geometric example of vectors as arrows, **v** + **w** = **w** + **v** since the parallelogram defining the sum of the vectors is independent of the order of the vectors. All other axioms can be checked in a similar manner in both examples. Thus, by disregarding the concrete nature of the particular type of vectors, the definition incorporates these two and many more examples in one notion of vector space.

Subtraction of two vectors and division by a (non-zero) scalar can be defined as

$$\mathbf{v} - \mathbf{w} = \mathbf{v} + (-\mathbf{w}),$$
$$\mathbf{v}/a = (1/a)\mathbf{v}.$$

When the scalar field F is the real numbers **R**, the vector space is called a *real vector space*. When the scalar field is the complex numbers, it is called a *complex vector space*. These two cases are the ones used most often in engineering. The general definition of a vector space allows scalars to be elements of any fixed field F. The notion is then known as an *F-vector spaces* or a *vector space over F*. A field is, essentially, a set of numbers possessing addition, subtraction, multiplication and division operations.[nb 3] For example, rational numbers also form a field.

In contrast to the intuition stemming from vectors in the plane and higher-dimensional cases, there is, in general vector spaces, no notion of nearness, angles or distances. To deal with such matters, particular types of vector spaces are introduced; see below.

26.1.4 Alternative formulations and elementary consequences

The requirement that vector addition and scalar multiplication be binary operations includes (by definition of binary operations) a property called closure: that **u** + **v** and a**v** are in V for all a in F, and **u**, **v** in V. Some older sources mention these properties as separate axioms.[2]

In the parlance of abstract algebra, the first four axioms can be subsumed by requiring the set of vectors to be an abelian group under addition. The remaining axioms give this group an F-module structure. In other words there is a ring homomorphism f from the field F into the endomorphism ring of the group of vectors. Then scalar multiplication a**v** is defined as $(f(a))(\mathbf{v})$.[3]

There are a number of direct consequences of the vector space axioms. Some of them derive from elementary group theory, applied to the additive group of vectors: for example the zero vector **0** of V and the additive inverse −**v** of any vector **v** are unique. Other properties follow from the distributive law, for example a**v** equals **0** if and only if a equals 0 or **v** equals **0**.

26.2 History

Vector spaces stem from affine geometry via the introduction of coordinates in the plane or three-dimensional space. Around 1636, Descartes and Fermat founded analytic geometry by equating solutions to an equation of two variables with points on a plane curve.[4] To achieve geometric solutions without using coordinates, Bolzano introduced, in 1804, certain operations on points, lines and planes, which are predecessors of vectors.[5] This work was made use of in the conception of barycentric coordinates by Möbius in 1827.[6] The foundation of the definition of vectors was Bellavitis' notion of the bipoint, an oriented segment one of whose ends is the origin and the other one a target. Vectors were reconsidered with the presentation of complex numbers by Argand and Hamilton and the inception of quaternions and biquaternions by the latter.[7] They are elements in $\mathbf{R}^2$, $\mathbf{R}^4$, and $\mathbf{R}^8$; treating them using linear combinations goes back to Laguerre in 1867, who also defined systems of linear equations.

In 1857, Cayley introduced the matrix notation which allows for a harmonization and simplification of linear maps. Around the same time, Grassmann studied the barycentric calculus initiated by Möbius. He envisaged sets of abstract objects endowed with operations.[8] In his work, the concepts of linear independence and dimension, as well as scalar products, are present. Actually Grassmann's 1844 work exceeds the framework of vector spaces, since his considering multiplication, too, led him to what are today called algebras. Peano was the first to give the modern definition of vector spaces and linear maps in 1888.[9]

An important development of vector spaces is due to the construction of function spaces by Lebesgue. This was later formalized by Banach and Hilbert, around 1920.[10] At that time, algebra and the new field of functional analysis began to interact, notably with key concepts such as spaces of p-integrable functions and Hilbert spaces.[11] Vector spaces, including infinite-dimensional ones, then became a firmly established notion, and many mathematical branches started making use of this concept.

26.3 Examples

Main article: Examples of vector spaces

26.3.1 Coordinate spaces

Main article: Coordinate space

The most simple example of a vector space over a field F is the field itself, equipped with its standard addition and multiplication. More generally, a vector space can be composed of n-tuples (sequences of length n) of elements of F, such as

$(a_1, a_2, ..., an)$, where each ai is an element of F.[12]

A vector space composed of all the n-tuples of a field F is known as a *coordinate space*, usually denoted F^n. The case $n = 1$ is the above-mentioned simplest example, in which the field F is also regarded as a vector space over itself. The case $F = \mathbf{R}$ and $n = 2$ was discussed in the introduction above.

26.3.2 Complex numbers and other field extensions

The set of complex numbers $\mathbf{C}$, i.e., numbers that can be written in the form $x + iy$ for real numbers x and y where i is the imaginary unit, form a vector space over the reals with the usual addition and multiplication: $(x + iy) + (a + ib) = (x$

+ a) + i(y + b)$ and $c \cdot (x + iy) = (c \cdot x) + i(c \cdot y)$ for real numbers x, y, a, b and c. The various axioms of a vector space follow from the fact that the same rules hold for complex number arithmetic.

In fact, the example of complex numbers is essentially the same (i.e., it is *isomorphic*) to the vector space of ordered pairs of real numbers mentioned above: if we think of the complex number $x + i y$ as representing the ordered pair (x, y) in the complex plane then we see that the rules for sum and scalar product correspond exactly to those in the earlier example.

More generally, field extensions provide another class of examples of vector spaces, particularly in algebra and algebraic number theory: a field F containing a smaller field E is an E-vector space, by the given multiplication and addition operations of F.[13] For example, the complex numbers are a vector space over **R**, and the field extension $\mathbf{Q}(i\sqrt{5})$ is a vector space over **Q**.

26.3.3 Function spaces

Functions from any fixed set Ω to a field F also form vector spaces, by performing addition and scalar multiplication pointwise. That is, the sum of two functions f and g is the function $(f + g)$ given by

$$(f + g)(w) = f(w) + g(w),$$

and similarly for multiplication. Such function spaces occur in many geometric situations, when Ω is the real line or an interval, or other subsets of **R**. Many notions in topology and analysis, such as continuity, integrability or differentiability are well-behaved with respect to linearity: sums and scalar multiples of functions possessing such a property still have that property.[14] Therefore, the set of such functions are vector spaces. They are studied in greater detail using the methods of functional analysis, see below. Algebraic constraints also yield vector spaces: the vector space $F[x]$ is given by polynomial functions:

$$f(x) = r_0 + r_1 x + ... + r_{n-1}x^{n-1} + r_n x^n, \text{ where the coefficients } r_0, ..., r_n \text{ are in } F.[15]$$

26.3.4 Linear equations

Main articles: Linear equation, Linear differential equation and Systems of linear equations

Systems of homogeneous linear equations are closely tied to vector spaces.[16] For example, the solutions of

are given by triples with arbitrary a, $b = a/2$, and $c = -5a/2$. They form a vector space: sums and scalar multiples of such triples still satisfy the same ratios of the three variables; thus they are solutions, too. Matrices can be used to condense multiple linear equations as above into one vector equation, namely

$$A\mathbf{x} = \mathbf{0},$$

where $A = \begin{bmatrix} 1 & 3 & 1 \\ 4 & 2 & 2 \end{bmatrix}$ is the matrix containing the coefficients of the given equations, $\mathbf{x}$ is the vector (a, b, c), $A\mathbf{x}$ denotes the matrix product, and $\mathbf{0} = (0, 0)$ is the zero vector. In a similar vein, the solutions of homogeneous *linear differential equations* form vector spaces. For example

$$f''(x) + 2f'(x) + f(x) = 0$$

yields $f(x) = a\,e^{-x} + bx\,e^{-x}$, where a and b are arbitrary constants, and e^x is the natural exponential function.

26.4   Basis and dimension

Main articles: Basis and Dimension
Bases allow to represent vectors by a sequence of scalars called *coordinates* or *components*. A basis is a (finite or infinite)

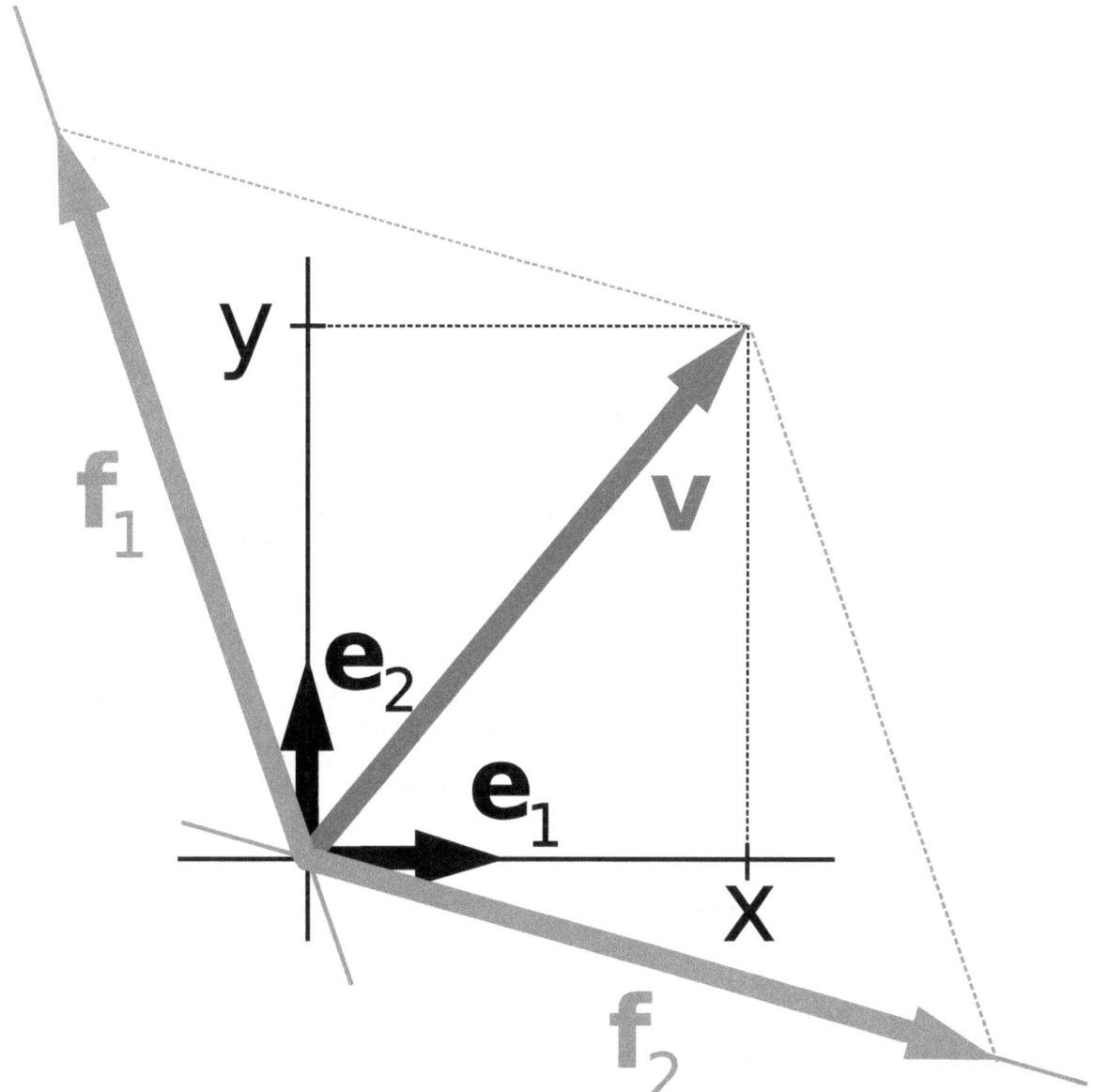

A vector v in $\mathbf{R}^2$ (blue) expressed in terms of different bases: using the standard basis of $\mathbf{R}^2$ $v = xe_1 + ye_2$ (black), and using a different, non-orthogonal basis: $v = f_1 + f_2$ (red).

set $B = \{\mathbf{b}i\}i \in I$ of vectors $\mathbf{b}i$, for convenience often indexed by some index set I, that spans the whole space and is linearly independent. "Spanning the whole space" means that any vector $\mathbf{v}$ can be expressed as a finite sum (called a *linear combination*) of the basis elements:

where the ak are scalars, called the coordinates (or the components) of the vector $\mathbf{v}$ with respect to the basis B, and $\mathbf{b}ik$ ($k = 1, ..., n$) elements of B. Linear independence means that the coordinates ak are uniquely determined for any vector in the vector space.

For example, the coordinate vectors $\mathbf{e}_1 = (1, 0, ..., 0)$, $\mathbf{e}_2 = (0, 1, 0, ..., 0)$, to $\mathbf{e}n = (0, 0, ..., 0, 1)$, form a basis of F^n, called the standard basis, since any vector $(x_1, x_2, ..., xn)$ can be uniquely expressed as a linear combination of these vectors:

$$(x_1, x_2, ..., xn) = x_1(1, 0, ..., 0) + x_2(0, 1, 0, ..., 0) + ... + xn(0, ..., 0, 1) = x_1\mathbf{e}_1 + x_2\mathbf{e}_2 + ... + xn\mathbf{e}n.$$

The corresponding coordinates $x_1, x_2, ..., xn$ are just the Cartesian coordinates of the vector.

Every vector space has a basis. This follows from Zorn's lemma, an equivalent formulation of the Axiom of Choice.[17] Given the other axioms of Zermelo–Fraenkel set theory, the existence of bases is equivalent to the axiom of choice.[18] The ultrafilter lemma, which is weaker than the axiom of choice, implies that all bases of a given vector space have the same number of elements, or cardinality (cf. *Dimension theorem for vector spaces*).[19] It is called the *dimension* of the vector space, denoted dim V. If the space is spanned by finitely many vectors, the above statements can be proven without such fundamental input from set theory.[20]

The dimension of the coordinate space F^n is n, by the basis exhibited above. The dimension of the polynomial ring $F[x]$ introduced above is countably infinite, a basis is given by $1, x, x^2, ...$ A fortiori, the dimension of more general function spaces, such as the space of functions on some (bounded or unbounded) interval, is infinite.[nb 4] Under suitable regularity assumptions on the coefficients involved, the dimension of the solution space of a homogeneous ordinary differential equation equals the degree of the equation.[21] For example, the solution space for the above equation is generated by e^{-x} and xe^{-x}. These two functions are linearly independent over $\mathbf{R}$, so the dimension of this space is two, as is the degree of the equation.

A field extension over the rationals $\mathbf{Q}$ can be thought of as a vector space over $\mathbf{Q}$ (by defining vector addition as field addition, defining scalar multiplication as field multiplication by elements of $\mathbf{Q}$, and otherwise ignoring the field multiplication). The dimension (or degree) of the field extension $\mathbf{Q}(\alpha)$ over $\mathbf{Q}$ depends on α. If α satisfies some polynomial equation

$$qn\alpha^n + qn_{-1}\alpha^{n-1} + ... + q_0 = 0, \text{ with rational coefficients } qn, ..., q_0.$$

("α is algebraic"), the dimension is finite. More precisely, it equals the degree of the minimal polynomial having α as a root.[22] For example, the complex numbers $\mathbf{C}$ are a two-dimensional real vector space, generated by 1 and the imaginary unit i. The latter satisfies $i^2 + 1 = 0$, an equation of degree two. Thus, $\mathbf{C}$ is a two-dimensional $\mathbf{R}$-vector space (and, as any field, one-dimensional as a vector space over itself, $\mathbf{C}$). If α is not algebraic, the dimension of $\mathbf{Q}(\alpha)$ over $\mathbf{Q}$ is infinite. For instance, for $\alpha = \pi$ there is no such equation, in other words π is transcendental.[23]

26.5 Linear maps and matrices

Main article: Linear map

The relation of two vector spaces can be expressed by *linear map* or *linear transformation*. They are functions that reflect the vector space structure—i.e., they preserve sums and scalar multiplication:

$$f(\mathbf{x} + \mathbf{y}) = f(\mathbf{x}) + f(\mathbf{y}) \text{ and } f(a \cdot \mathbf{x}) = a \cdot f(\mathbf{x}) \text{ for all } \mathbf{x} \text{ and } \mathbf{y} \text{ in } V, \text{ all } a \text{ in } F.[24]$$

An *isomorphism* is a linear map $f : V \to W$ such that there exists an inverse map $g : W \to V$, which is a map such that the two possible compositions $f \circ g : W \to W$ and $g \circ f : V \to V$ are identity maps. Equivalently, f is both one-to-one (injective) and onto (surjective).[25] If there exists an isomorphism between V and W, the two spaces are said to be *isomorphic*; they are then essentially identical as vector spaces, since all identities holding in V are, via f, transported to similar ones in W, and vice versa via g.

For example, the "arrows in the plane" and "ordered pairs of numbers" vector spaces in the introduction are isomorphic: a planar arrow $\mathbf{v}$ departing at the origin of some (fixed) coordinate system can be expressed as an ordered pair by considering the x- and y-component of the arrow, as shown in the image at the right. Conversely, given a pair (x, y), the arrow going by x to the right (or to the left, if x is negative), and y up (down, if y is negative) turns back the arrow $\mathbf{v}$.

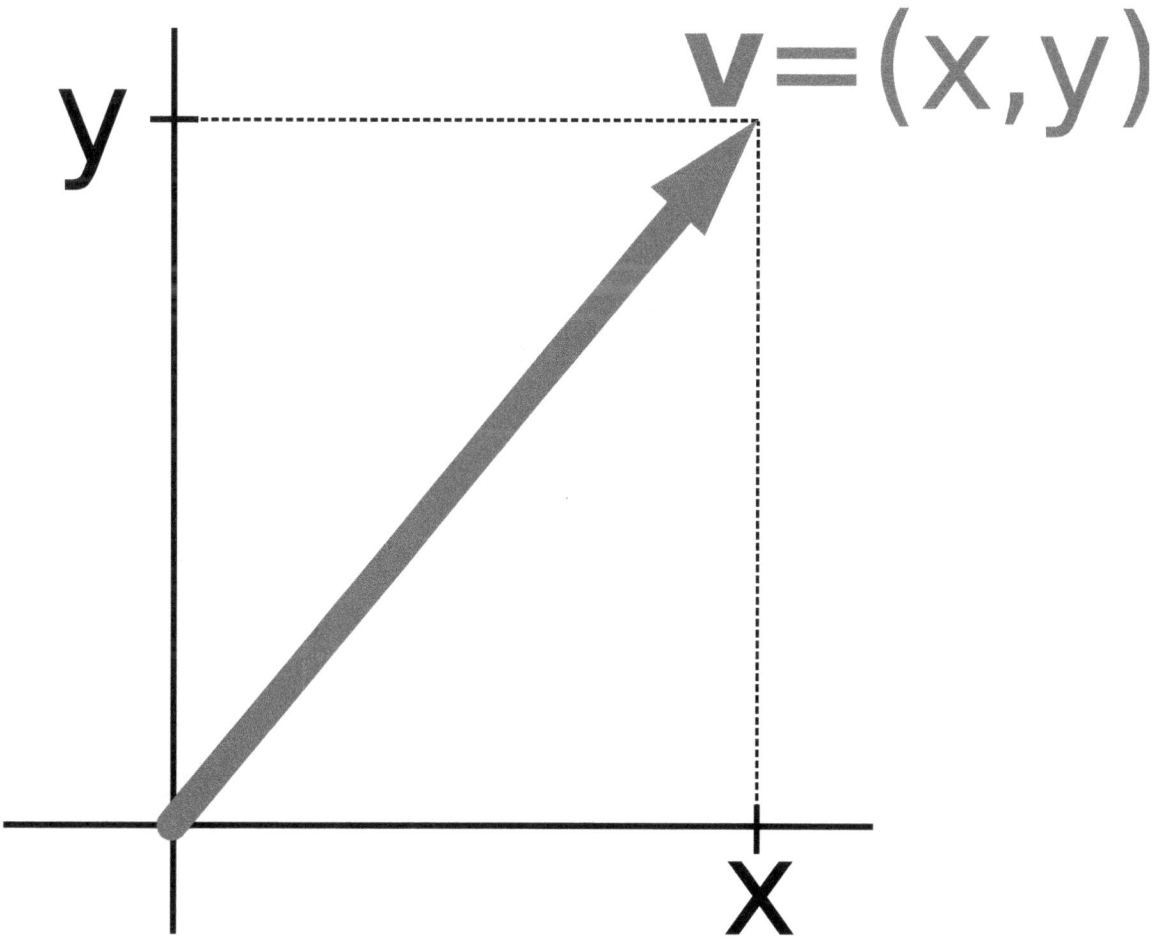

*Describing an arrow vector **v** by its coordinates* x *and* y *yields an isomorphism of vector spaces.*

Linear maps $V \to W$ between two vector spaces form a vector space $\mathrm{Hom}_F(V, W)$, also denoted $\mathrm{L}(V, W)$.[26] The space of linear maps from V to F is called the *dual vector space*, denoted V^*.[27] Via the injective natural map $V \to V^{**}$, any vector space can be embedded into its *bidual*; the map is an isomorphism if and only if the space is finite-dimensional.[28]

Once a basis of V is chosen, linear maps $f : V \to W$ are completely determined by specifying the images of the basis vectors, because any element of V is expressed uniquely as a linear combination of them.[29] If dim V = dim W, a 1-to-1 correspondence between fixed bases of V and W gives rise to a linear map that maps any basis element of V to the corresponding basis element of W. It is an isomorphism, by its very definition.[30] Therefore, two vector spaces are isomorphic if their dimensions agree and vice versa. Another way to express this is that any vector space is *completely classified* (up to isomorphism) by its dimension, a single number. In particular, any n-dimensional F-vector space V is isomorphic to F^n. There is, however, no "canonical" or preferred isomorphism; actually an isomorphism $\varphi : F^n \to V$ is equivalent to the choice of a basis of V, by mapping the standard basis of F^n to V, via φ. The freedom of choosing a convenient basis is particularly useful in the infinite-dimensional context, see below.

26.5.1   Matrices

Main articles: Matrix and Determinant

Matrices are a useful notion to encode linear maps.[31] They are written as a rectangular array of scalars as in the image at the right. Any m-by-n matrix A gives rise to a linear map from F^n to F^m, by the following

$$\mathbf{x} = (x_1, x_2, \cdots, x_n) \mapsto \left(\sum_{j=1}^{n} a_{1j}x_j, \sum_{j=1}^{n} a_{2j}x_j, \cdots, \sum_{j=1}^{n} a_{mj}x_j \right), \text{ where } \sum \text{ denotes summation,}$$

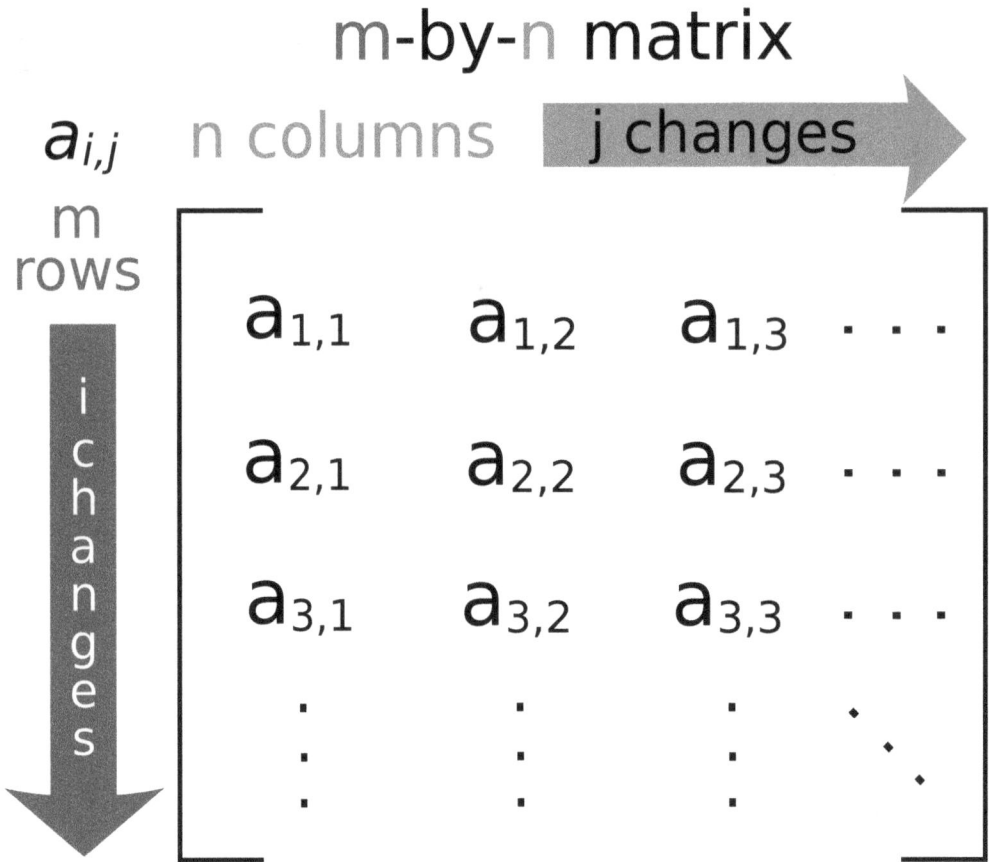

A typical matrix

or, using the matrix multiplication of the matrix A with the coordinate vector $\mathbf{x}$:

$$\mathbf{x} \mapsto A\mathbf{x}.$$

Moreover, after choosing bases of V and W, *any* linear map $f : V \to W$ is uniquely represented by a matrix via this assignment.[32]

The determinant det (A) of a square matrix A is a scalar that tells whether the associated map is an isomorphism or not: to be so it is sufficient and necessary that the determinant is nonzero.[33] The linear transformation of $\mathbf{R}^n$ corresponding to a real n-by-n matrix is orientation preserving if and only if its determinant is positive.

26.5.2 Eigenvalues and eigenvectors

Main article: Eigenvalues and eigenvectors

Endomorphisms, linear maps $f : V \to V$, are particularly important since in this case vectors $\mathbf{v}$ can be compared with their image under f, $f(\mathbf{v})$. Any nonzero vector $\mathbf{v}$ satisfying $\lambda\mathbf{v} = f(\mathbf{v})$, where λ is a scalar, is called an *eigenvector* of f with *eigenvalue* λ.[nb 5][34] Equivalently, $\mathbf{v}$ is an element of the kernel of the difference $f - \lambda \cdot \mathrm{Id}$ (where Id is the identity map $V \to V$). If V is finite-dimensional, this can be rephrased using determinants: f having eigenvalue λ is equivalent to

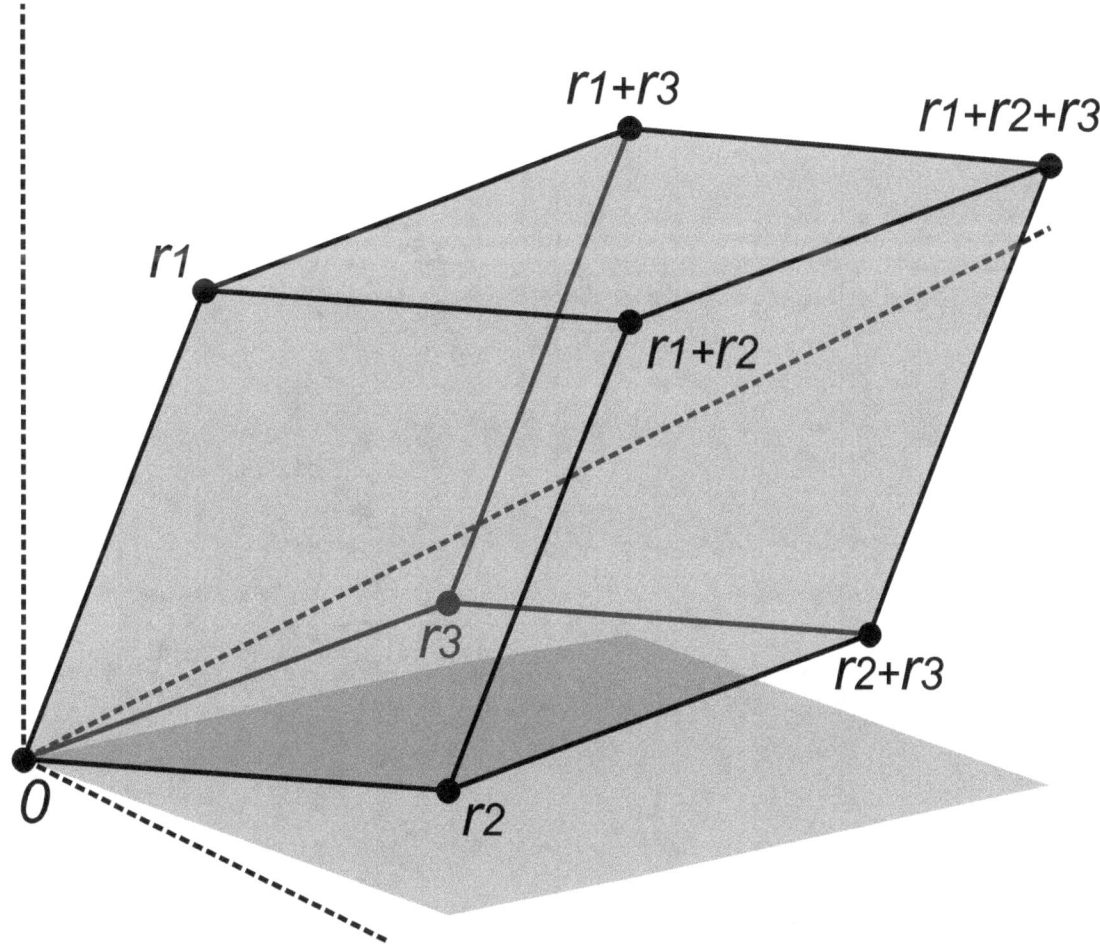

The volume of this parallelepiped is the absolute value of the determinant of the 3-by-3 matrix formed by the vectors r_1, r_2, *and* r_3.

$$\det(f - \lambda \cdot \mathrm{Id}) = 0.$$

By spelling out the definition of the determinant, the expression on the left hand side can be seen to be a polynomial function in λ, called the characteristic polynomial of f.[35] If the field F is large enough to contain a zero of this polynomial (which automatically happens for F algebraically closed, such as $F = \mathbf{C}$) any linear map has at least one eigenvector. The vector space V may or may not possess an eigenbasis, a basis consisting of eigenvectors. This phenomenon is governed by the Jordan canonical form of the map.[nb 6] The set of all eigenvectors corresponding to a particular eigenvalue of f forms a vector space known as the *eigenspace* corresponding to the eigenvalue (and f) in question. To achieve the spectral theorem, the corresponding statement in the infinite-dimensional case, the machinery of functional analysis is needed, see below.

26.6   Basic constructions

In addition to the above concrete examples, there are a number of standard linear algebraic constructions that yield vector spaces related to given ones. In addition to the definitions given below, they are also characterized by universal properties, which determine an object X by specifying the linear maps from X to any other vector space.

26.6.1 Subspaces and quotient spaces

Main articles: Linear subspace and Quotient vector space

A nonempty subset W of a vector space V that is closed under addition and scalar multiplication (and therefore contains

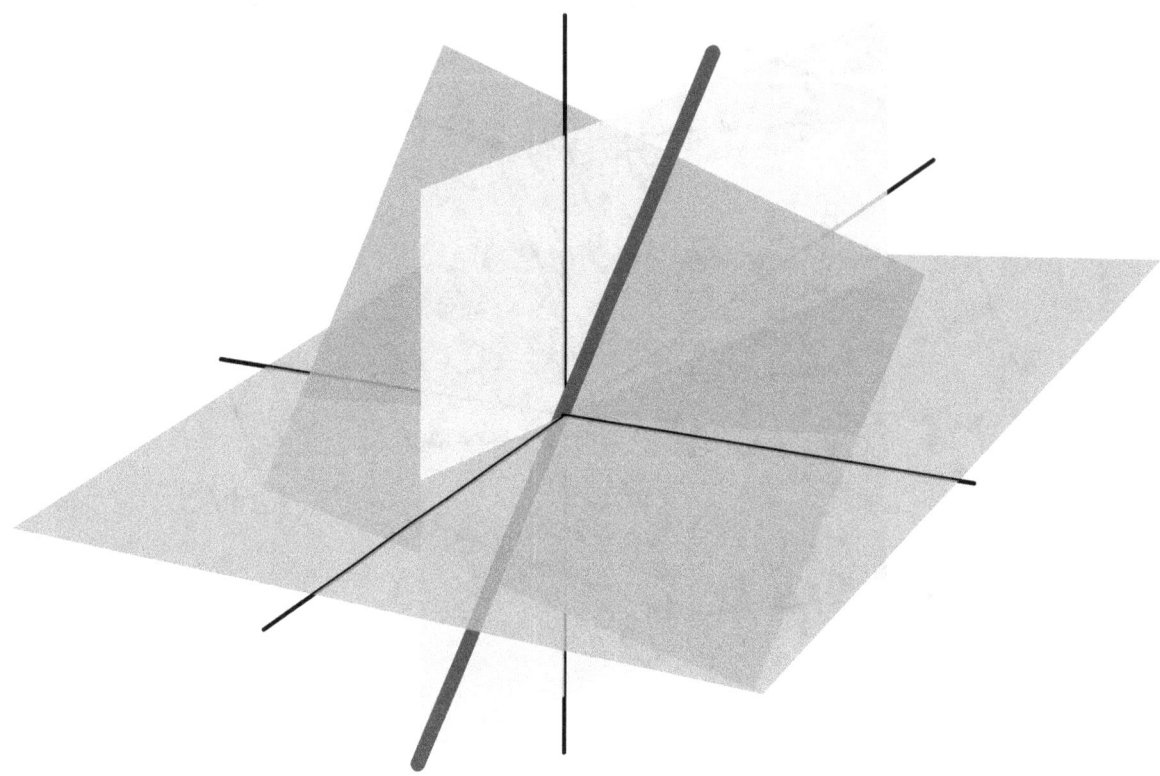

A line passing through the origin (blue, thick) in $\mathbf{R}^3$ is a linear subspace. It is the intersection of two planes (green and yellow).

the **0**-vector of V) is called a *subspace* of V.[36] Subspaces of V are vector spaces (over the same field) in their own right. The intersection of all subspaces containing a given set S of vectors is called its span, and it is the smallest subspace of V containing the set S. Expressed in terms of elements, the span is the subspace consisting of all the linear combinations of elements of S.[37]

The counterpart to subspaces are *quotient vector spaces*.[38] Given any subspace $W \subset V$, the quotient space V/W ("V modulo W") is defined as follows: as a set, it consists of $\mathbf{v} + W = \{\mathbf{v} + \mathbf{w} : \mathbf{w} \in W\}$, where $\mathbf{v}$ is an arbitrary vector in V. The sum of two such elements $\mathbf{v}_1 + W$ and $\mathbf{v}_2 + W$ is $(\mathbf{v}_1 + \mathbf{v}_2) + W$, and scalar multiplication is given by $a \cdot (\mathbf{v} + W) = (a \cdot \mathbf{v}) + W$. The key point in this definition is that $\mathbf{v}_1 + W = \mathbf{v}_2 + W$ if and only if the difference of $\mathbf{v}_1$ and $\mathbf{v}_2$ lies in W.[nb 7] This way, the quotient space "forgets" information that is contained in the subspace W.

The kernel $\ker(f)$ of a linear map $f : V \to W$ consists of vectors $\mathbf{v}$ that are mapped to $\mathbf{0}$ in W.[39] Both kernel and image $\operatorname{im}(f) = \{f(\mathbf{v}) : \mathbf{v} \in V\}$ are subspaces of V and W, respectively.[40] The existence of kernels and images is part of the statement that the category of vector spaces (over a fixed field F) is an abelian category, i.e. a corpus of mathematical objects and structure-preserving maps between them (a category) that behaves much like the category of abelian groups.[41] Because of this, many statements such as the first isomorphism theorem (also called rank–nullity theorem in matrix-related terms)

$$V \,/\, \ker(f) \equiv \operatorname{im}(f).$$

and the second and third isomorphism theorem can be formulated and proven in a way very similar to the corresponding statements for groups.

An important example is the kernel of a linear map $\mathbf{x} \mapsto A\mathbf{x}$ for some fixed matrix A, as above. The kernel of this map is the subspace of vectors $\mathbf{x}$ such that $A\mathbf{x} = 0$, which is precisely the set of solutions to the system of homogeneous linear equations belonging to A. This concept also extends to linear differential equations

$$a_0 f + a_1 \frac{df}{dx} + a_2 \frac{d^2 f}{dx^2} + \cdots + a_n \frac{d^n f}{dx^n} = 0 \,, \text{ where the coefficients } a_i \text{ are functions in } x, \text{ too.}$$

In the corresponding map

$$f \mapsto D(f) = \sum_{i=0}^{n} a_i \frac{d^i f}{dx^i}$$

the derivatives of the function f appear linearly (as opposed to $f''(x)^2$, for example). Since differentiation is a linear procedure (i.e., $(f + g)' = f' + g'$ and $(c \cdot f)' = c \cdot f'$ for a constant c) this assignment is linear, called a linear differential operator. In particular, the solutions to the differential equation $D(f) = 0$ form a vector space (over $\mathbf{R}$ or $\mathbf{C}$).

26.6.2   Direct product and direct sum

Main articles: Direct product and Direct sum of modules

The *direct product* of vector spaces and the *direct sum* of vector spaces are two ways of combining an indexed family of vector spaces into a new vector space.

The *direct product* $\prod_{i \in I} V_i$ of a family of vector spaces V_i consists of the set of all tuples $(v_i)_i \in I$, which specify for each index i in some index set I an element $\mathbf{v}_i$ of V_i.[42] Addition and scalar multiplication is performed componentwise. A variant of this construction is the *direct sum* $\oplus_{i \in I} V_i$ (also called coproduct and denoted $\coprod_{i \in I} V_i$), where only tuples with finitely many nonzero vectors are allowed. If the index set I is finite, the two constructions agree, but in general they are different.

26.6.3   Tensor product

Main article: Tensor product of vector spaces

The *tensor product* $V \otimes_F W$, or simply $V \otimes W$, of two vector spaces V and W is one of the central notions of multilinear algebra which deals with extending notions such as linear maps to several variables. A map $g : V \times W \to X$ is called bilinear if g is linear in both variables $\mathbf{v}$ and $\mathbf{w}$. That is to say, for fixed $\mathbf{w}$ the map $\mathbf{v} \mapsto g(\mathbf{v}, \mathbf{w})$ is linear in the sense above and likewise for fixed $\mathbf{v}$.

The tensor product is a particular vector space that is a *universal* recipient of bilinear maps g, as follows. It is defined as the vector space consisting of finite (formal) sums of symbols called tensors

$$\mathbf{v}_1 \otimes \mathbf{w}_1 + \mathbf{v}_2 \otimes \mathbf{w}_2 + \ldots + \mathbf{v}n \otimes \mathbf{w}n,$$

subject to the rules

$$a \cdot (\mathbf{v} \otimes \mathbf{w}) = (a \cdot \mathbf{v}) \otimes \mathbf{w} = \mathbf{v} \otimes (a \cdot \mathbf{w}), \text{ where } a \text{ is a scalar,}$$

$$(\mathbf{v}_1 + \mathbf{v}_2) \otimes \mathbf{w} = \mathbf{v}_1 \otimes \mathbf{w} + \mathbf{v}_2 \otimes \mathbf{w}, \text{ and}$$

$$\mathbf{v} \otimes (\mathbf{w}_1 + \mathbf{w}_2) = \mathbf{v} \otimes \mathbf{w}_1 + \mathbf{v} \otimes \mathbf{w}_2.[43]$$

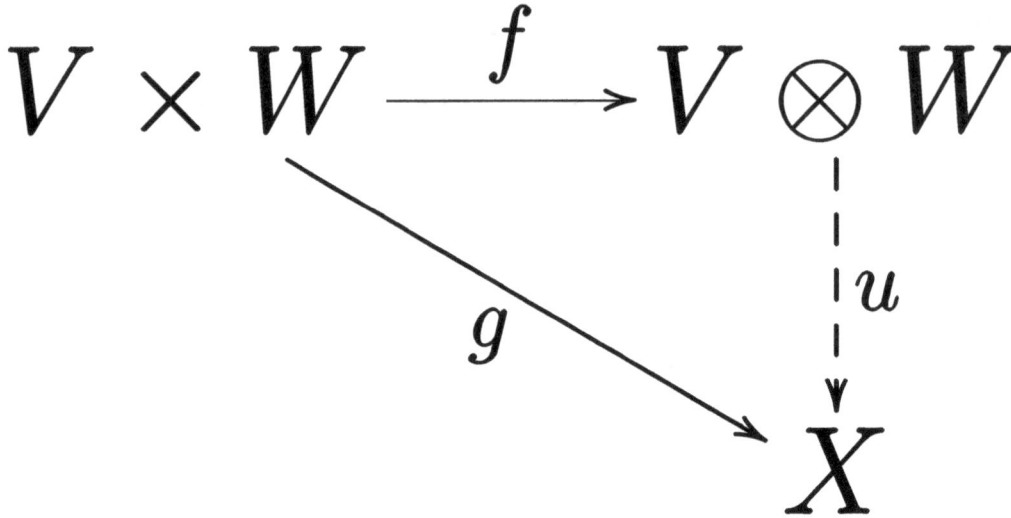

Commutative diagram depicting the universal property of the tensor product.

These rules ensure that the map f from the $V \times W$ to $V \otimes W$ that maps a tuple $(\mathbf{v}, \mathbf{w})$ to $\mathbf{v} \otimes \mathbf{w}$ is bilinear. The universality states that given *any* vector space X and *any* bilinear map $g : V \times W \rightarrow X$, there exists a unique map u, shown in the diagram with a dotted arrow, whose composition with f equals g: $u(\mathbf{v} \otimes \mathbf{w}) = g(\mathbf{v}, \mathbf{w})$.[44] This is called the universal property of the tensor product, an instance of the method—much used in advanced abstract algebra—to indirectly define objects by specifying maps from or to this object.

26.7   Vector spaces with additional structure

From the point of view of linear algebra, vector spaces are completely understood insofar as any vector space is characterized, up to isomorphism, by its dimension. However, vector spaces *per se* do not offer a framework to deal with the question—crucial to analysis—whether a sequence of functions converges to another function. Likewise, linear algebra is not adapted to deal with infinite series, since the addition operation allows only finitely many terms to be added. Therefore, the needs of functional analysis require considering additional structures.

A vector space may be given a partial order $\leq$, under which some vectors can be compared.[45] For example, n-dimensional real space $\mathbf{R}^n$ can be ordered by comparing its vectors componentwise. Ordered vector spaces, for example Riesz spaces, are fundamental to Lebesgue integration, which relies on the ability to express a function as a difference of two positive functions

$$f = f^+ - f^-,$$

where f^+ denotes the positive part of f and f^- the negative part.[46]

26.7.1   Normed vector spaces and inner product spaces

Main articles: Normed vector space and Inner product space

"Measuring" vectors is done by specifying a norm, a datum which measures lengths of vectors, or by an inner product, which measures angles between vectors. Norms and inner products are denoted $|\mathbf{v}|$ and $\langle \mathbf{v}, \mathbf{w} \rangle$, respectively. The datum of an inner product entails that lengths of vectors can be defined too, by defining the associated norm $|\mathbf{v}| := \sqrt{\langle \mathbf{v}, \mathbf{v} \rangle}$. Vector spaces endowed with such data are known as *normed vector spaces* and *inner product spaces*, respectively.[47]

Coordinate space F^n can be equipped with the standard dot product:

$$\langle \mathbf{x}, \mathbf{y} \rangle = \mathbf{x} \cdot \mathbf{y} = x_1 y_1 + \cdots + x_n y_n.$$

In $\mathbf{R}^2$, this reflects the common notion of the angle between two vectors $\mathbf{x}$ and $\mathbf{y}$, by the law of cosines:

$$\mathbf{x} \cdot \mathbf{y} = \cos\left(\angle(\mathbf{x}, \mathbf{y})\right) \cdot |\mathbf{x}| \cdot |\mathbf{y}|.$$

Because of this, two vectors satisfying $\langle \mathbf{x}, \mathbf{y} \rangle = 0$ are called orthogonal. An important variant of the standard dot product is used in Minkowski space: $\mathbf{R}^4$ endowed with the Lorentz product

$$\langle \mathbf{x} | \mathbf{y} \rangle = x_1 y_1 + x_2 y_2 + x_3 y_3 - x_4 y_4. \text{ [48]}$$

In contrast to the standard dot product, it is not positive definite: $\langle \mathbf{x} | \mathbf{x} \rangle$ also takes negative values, for example for $\mathbf{x} = (0, 0, 0, 1)$. Singling out the fourth coordinate—corresponding to time, as opposed to three space-dimensions—makes it useful for the mathematical treatment of special relativity.

26.7.2 Topological vector spaces

Main article: Topological vector space

Convergence questions are treated by considering vector spaces V carrying a compatible topology, a structure that allows one to talk about elements being close to each other.[49][50] Compatible here means that addition and scalar multiplication have to be continuous maps. Roughly, if $\mathbf{x}$ and $\mathbf{y}$ in V, and a in F vary by a bounded amount, then so do $\mathbf{x} + \mathbf{y}$ and $a\mathbf{x}$.[nb 8] To make sense of specifying the amount a scalar changes, the field F also has to carry a topology in this context; a common choice are the reals or the complex numbers.

In such *topological vector spaces* one can consider series of vectors. The infinite sum

$$\sum_{i=0}^{\infty} f_i$$

denotes the limit of the corresponding finite partial sums of the sequence $(f_i)_{i \in \mathbf{N}}$ of elements of V. For example, the f_i could be (real or complex) functions belonging to some function space V, in which case the series is a function series. The mode of convergence of the series depends on the topology imposed on the function space. In such cases, pointwise convergence and uniform convergence are two prominent examples.

A way to ensure the existence of limits of certain infinite series is to restrict attention to spaces where any Cauchy sequence has a limit; such a vector space is called complete. Roughly, a vector space is complete provided that it contains all necessary limits. For example, the vector space of polynomials on the unit interval [0,1], equipped with the topology of uniform convergence is not complete because any continuous function on [0,1] can be uniformly approximated by a sequence of polynomials, by the Weierstrass approximation theorem.[51] In contrast, the space of *all* continuous functions on [0,1] with the same topology is complete.[52] A norm gives rise to a topology by defining that a sequence of vectors $\mathbf{v}_n$ converges to $\mathbf{v}$ if and only if

$$\lim_{n \to \infty} |\mathbf{v}_n - \mathbf{v}| = 0.$$

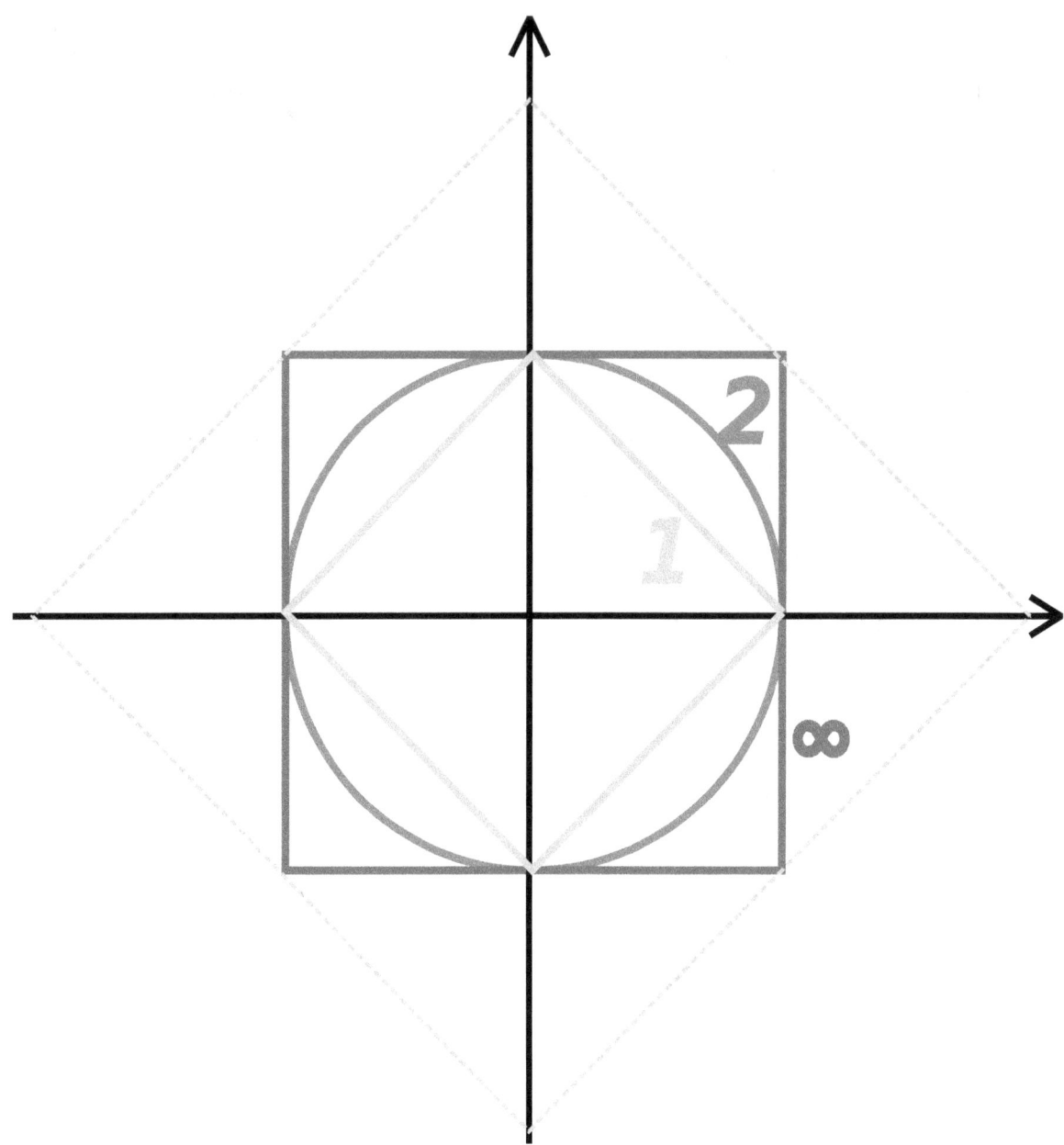

Unit "spheres" in $\mathbf{R}^2$ consist of plane vectors of norm 1. Depicted are the unit spheres in different p-norms, for p = 1, 2, and ∞. The bigger diamond depicts points of 1-norm equal to $\sqrt{2}$.

Banach and Hilbert spaces are complete topological vector spaces whose topologies are given, respectively, by a norm and an inner product. Their study—a key piece of functional analysis—focusses on infinite-dimensional vector spaces, since all norms on finite-dimensional topological vector spaces give rise to the same notion of convergence.[53] The image at the right shows the equivalence of the 1-norm and ∞-norm on $\mathbf{R}^2$: as the unit "balls" enclose each other, a sequence converges to zero in one norm if and only if it so does in the other norm. In the infinite-dimensional case, however, there will generally be inequivalent topologies, which makes the study of topological vector spaces richer than that of vector spaces without additional data.

From a conceptual point of view, all notions related to topological vector spaces should match the topology. For example, instead of considering all linear maps (also called functionals) $V \to W$, maps between topological vector spaces are required to be continuous.[54] In particular, the (topological) dual space V^* consists of continuous functionals $V \to \mathbf{R}$ (or to **C**). The fundamental Hahn–Banach theorem is concerned with separating subspaces of appropriate topological vector spaces by continuous functionals.[55]

Banach spaces

Main article: Banach space

Banach spaces, introduced by Stefan Banach, are complete normed vector spaces.[56] A first example is the vector space ℓ^p consisting of infinite vectors with real entries $\mathbf{x} = (x_1, x_2, ...)$ whose p-norm ($1 \leq p \leq \infty$) given by

$$|\mathbf{x}|_p := \left(\sum_i |x_i|^p\right)^{1/p} \text{ for } p < \infty \text{ and } |\mathbf{x}|_\infty := \sup_i |x_i|$$

is finite. The topologies on the infinite-dimensional space ℓ^p are inequivalent for different p. E.g. the sequence of vectors $\mathbf{x}n = (2^{-n}, 2^{-n}, ..., 2^{-n}, 0, 0, ...)$, i.e. the first 2^n components are 2^{-n}, the following ones are 0, converges to the zero vector for $p = \infty$, but does not for $p = 1$:

$$|x_n|_\infty = \sup(2^{-n}, 0) = 2^{-n} \to 0 \text{ , but } |x_n|_1 = \sum_{i=1}^{2^n} 2^{-n} = 2^n \cdot 2^{-n} = 1.$$

More generally than sequences of real numbers, functions $f \colon \Omega \to \mathbf{R}$ are endowed with a norm that replaces the above sum by the Lebesgue integral

$$|f|_p := \left(\int_\Omega |f(x)|^p \, dx\right)^{1/p}.$$

The space of integrable functions on a given domain Ω (for example an interval) satisfying $|f|p < \infty$, and equipped with this norm are called Lebesgue spaces, denoted $L^p(\Omega)$.[nb 9] These spaces are complete.[57] (If one uses the Riemann integral instead, the space is *not* complete, which may be seen as a justification for Lebesgue's integration theory.[nb 10]) Concretely this means that for any sequence of Lebesgue-integrable functions $f_1, f_2, ...$ with $|fn|p < \infty$, satisfying the condition

$$\lim_{k,\,n\to\infty} \int_\Omega |f_k(x) - f_n(x)|^p \, dx = 0$$

there exists a function $f(x)$ belonging to the vector space $L^p(\Omega)$ such that

$$\lim_{k\to\infty} \int_\Omega |f(x) - f_k(x)|^p \, dx = 0.$$

Imposing boundedness conditions not only on the function, but also on its derivatives leads to Sobolev spaces.[58]

Hilbert spaces

Main article: Hilbert space
 Complete inner product spaces are known as *Hilbert spaces*, in honor of David Hilbert.[59] The Hilbert space $L^2(\Omega)$, with inner product given by

$$\langle f, g \rangle = \int_\Omega f(x)\overline{g(x)} \, dx,$$

where $\overline{g(x)}$ denotes the complex conjugate of $g(x)$,[60][nb 11] is a key case.

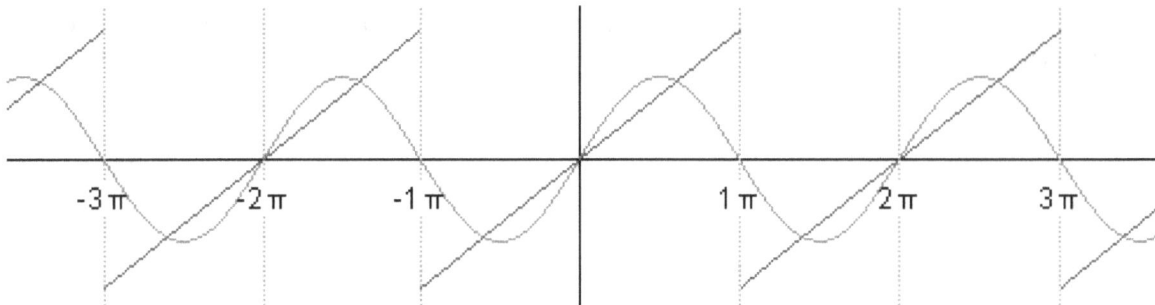

The succeeding snapshots show summation of 1 to 5 terms in approximating a periodic function (blue) by finite sum of sine functions (red).

By definition, in a Hilbert space any Cauchy sequence converges to a limit. Conversely, finding a sequence of functions fn with desirable properties that approximates a given limit function, is equally crucial. Early analysis, in the guise of the Taylor approximation, established an approximation of differentiable functions f by polynomials.[61] By the Stone–Weierstrass theorem, every continuous function on $[a, b]$ can be approximated as closely as desired by a polynomial.[62] A similar approximation technique by trigonometric functions is commonly called Fourier expansion, and is much applied in engineering, see below. More generally, and more conceptually, the theorem yields a simple description of what "basic functions", or, in abstract Hilbert spaces, what basic vectors suffice to generate a Hilbert space H, in the sense that the *closure* of their span (i.e., finite linear combinations and limits of those) is the whole space. Such a set of functions is called a *basis* of H, its cardinality is known as the Hilbert space dimension.[nb 12] Not only does the theorem exhibit suitable basis functions as sufficient for approximation purposes, but together with the Gram-Schmidt process, it enables one to construct a basis of orthogonal vectors.[63] Such orthogonal bases are the Hilbert space generalization of the coordinate axes in finite-dimensional Euclidean space.

The solutions to various differential equations can be interpreted in terms of Hilbert spaces. For example, a great many fields in physics and engineering lead to such equations and frequently solutions with particular physical properties are used as basis functions, often orthogonal.[64] As an example from physics, the time-dependent Schrödinger equation in quantum mechanics describes the change of physical properties in time by means of a partial differential equation, whose solutions are called wavefunctions.[65] Definite values for physical properties such as energy, or momentum, correspond to eigenvalues of a certain (linear) differential operator and the associated wavefunctions are called eigenstates. The spectral theorem decomposes a linear compact operator acting on functions in terms of these eigenfunctions and their eigenvalues.[66]

26.7.3 Algebras over fields

Main articles: Algebra over a field and Lie algebra

General vector spaces do not possess a multiplication between vectors. A vector space equipped with an additional bilinear operator defining the multiplication of two vectors is an *algebra over a field*.[67] Many algebras stem from functions on some geometrical object: since functions with values in a given field can be multiplied pointwise, these entities form algebras. The Stone–Weierstrass theorem mentioned above, for example, relies on Banach algebras which are both Banach spaces and algebras.

Commutative algebra makes great use of rings of polynomials in one or several variables, introduced above. Their multiplication is both commutative and associative. These rings and their quotients form the basis of algebraic geometry, because they are rings of functions of algebraic geometric objects.[68]

Another crucial example are *Lie algebras*, which are neither commutative nor associative, but the failure to be so is limited by the constraints ($[x, y]$ denotes the product of x and y):

- $[x, y] = -[y, x]$ (anticommutativity), and

- $[x, [y, z]] + [y, [z, x]] + [z, [x, y]] = 0$ (Jacobi identity).[69]

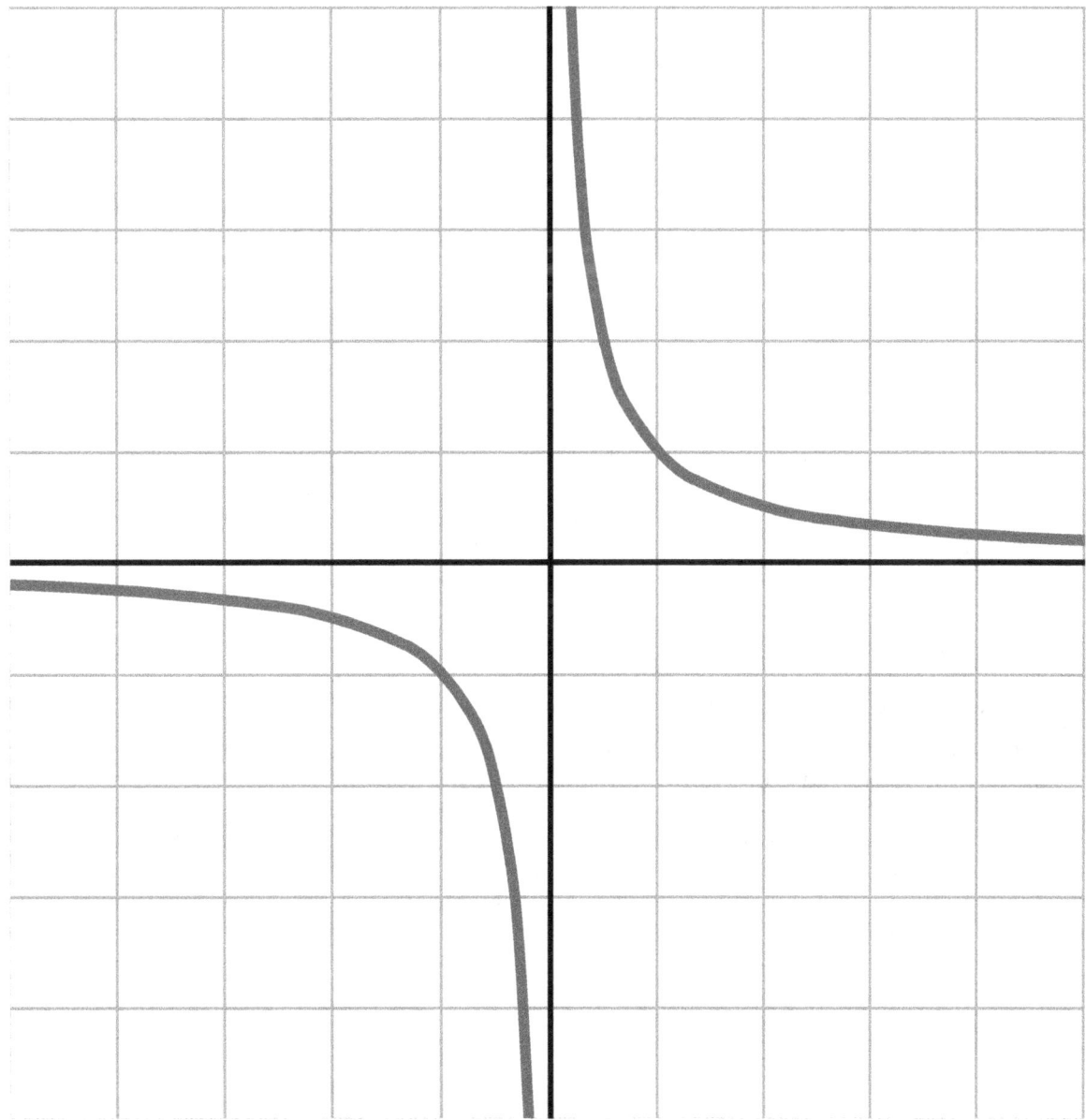

A hyperbola, given by the equation x ⬚ y = 1. *The coordinate ring of functions on this hyperbola is given by* **R***[x, y] / (x · y − 1), an infinite-dimensional vector space over* **R***.*

Examples include the vector space of *n*-by-*n* matrices, with $[x, y] = xy − yx$, the commutator of two matrices, and **R**^{3}, endowed with the cross product.

The tensor algebra T(V) is a formal way of adding products to any vector space V to obtain an algebra.[70] As a vector space, it is spanned by symbols, called simple tensors

$\mathbf{v}_1 \otimes \mathbf{v}_2 \otimes ... \otimes \mathbf{v}n$, where the degree *n* varies.

The multiplication is given by concatenating such symbols, imposing the distributive law under addition, and requiring that scalar multiplication commute with the tensor product ⊗, much the same way as with the tensor product of two vector spaces introduced above. In general, there are no relations between $\mathbf{v}_1 \otimes \mathbf{v}_2$ and $\mathbf{v}_2 \otimes \mathbf{v}_1$. Forcing two such elements to be equal leads to the symmetric algebra, whereas forcing $\mathbf{v}_1 \otimes \mathbf{v}_2 = − \mathbf{v}_2 \otimes \mathbf{v}_1$ yields the exterior algebra.[71]

When a field, F is explicitly stated, a common term used is F-algebra.

26.8 Applications

Vector spaces have manifold applications as they occur in many circumstances, namely wherever functions with values in some field are involved. They provide a framework to deal with analytical and geometrical problems, or are used in the Fourier transform. This list is not exhaustive: many more applications exist, for example in optimization. The minimax theorem of game theory stating the existence of a unique payoff when all players play optimally can be formulated and proven using vector spaces methods.[72] Representation theory fruitfully transfers the good understanding of linear algebra and vector spaces to other mathematical domains such as group theory.[73]

26.8.1 Distributions

Main article: Distribution

A *distribution* (or *generalized function*) is a linear map assigning a number to each "test" function, typically a smooth function with compact support, in a continuous way: in the above terminology the space of distributions is the (continuous) dual of the test function space.[74] The latter space is endowed with a topology that takes into account not only f itself, but also all its higher derivatives. A standard example is the result of integrating a test function f over some domain Ω:

$$I(f) = \int_\Omega f(x)\, dx.$$

When $\Omega = \{p\}$, the set consisting of a single point, this reduces to the Dirac distribution, denoted by δ, which associates to a test function f its value at the p: $\delta(f) = f(p)$. Distributions are a powerful instrument to solve differential equations. Since all standard analytic notions such as derivatives are linear, they extend naturally to the space of distributions. Therefore the equation in question can be transferred to a distribution space, which is bigger than the underlying function space, so that more flexible methods are available for solving the equation. For example, Green's functions and fundamental solutions are usually distributions rather than proper functions, and can then be used to find solutions of the equation with prescribed boundary conditions. The found solution can then in some cases be proven to be actually a true function, and a solution to the original equation (e.g., using the Lax–Milgram theorem, a consequence of the Riesz representation theorem).[75]

26.8.2 Fourier analysis

Main article: Fourier analysis

Resolving a periodic function into a sum of trigonometric functions forms a *Fourier series*, a technique much used in physics and engineering.[nb 13][76] The underlying vector space is usually the Hilbert space $L^2(0, 2\pi)$, for which the functions $\sin mx$ and $\cos mx$ (m an integer) form an orthogonal basis.[77] The Fourier expansion of an L^2 function f is

$$\frac{a_0}{2} + \sum_{m=1}^{\infty} \left[a_m \cos(mx) + b_m \sin(mx) \right].$$

The coefficients a_m and b_m are called Fourier coefficients of f, and are calculated by the formulas[78]

$$a_m = \tfrac{1}{\pi} \int_0^{2\pi} f(t) \cos(mt)\, dt \, , \, b_m = \tfrac{1}{\pi} \int_0^{2\pi} f(t) \sin(mt)\, dt.$$

In physical terms the function is represented as a superposition of sine waves and the coefficients give information about the function's frequency spectrum.[79] A complex-number form of Fourier series is also commonly used.[78] The concrete formulae above are consequences of a more general mathematical duality called Pontryagin duality.[80] Applied to the group $\mathbf{R}$, it yields the classical Fourier transform; an application in physics are reciprocal lattices, where the underlying

The heat equation describes the dissipation of physical properties over time, such as the decline of the temperature of a hot body placed in a colder environment (yellow depicts colder regions than red).

group is a finite-dimensional real vector space endowed with the additional datum of a lattice encoding positions of atoms in crystals.[81]

Fourier series are used to solve boundary value problems in partial differential equations.[82] In 1822, Fourier first used this technique to solve the heat equation.[83] A discrete version of the Fourier series can be used in sampling applications where the function value is known only at a finite number of equally spaced points. In this case the Fourier series is finite and its value is equal to the sampled values at all points.[84] The set of coefficients is known as the discrete Fourier transform (DFT) of the given sample sequence. The DFT is one of the key tools of digital signal processing, a field whose applications include radar, speech encoding, image compression.[85] The JPEG image format is an application of the closely related discrete cosine transform.[86]

The fast Fourier transform is an algorithm for rapidly computing the discrete Fourier transform.[87] It is used not only for calculating the Fourier coefficients but, using the convolution theorem, also for computing the convolution of two finite sequences.[88] They in turn are applied in digital filters[89] and as a rapid multiplication algorithm for polynomials and large integers (Schönhage-Strassen algorithm).[90][91]

26.8.3   Differential geometry

Main article: Tangent space

 The tangent plane to a surface at a point is naturally a vector space whose origin is identified with the point of contact. The tangent plane is the best linear approximation, or linearization, of a surface at a point.[nb 14] Even in a three-dimensional Euclidean space, there is typically no natural way to prescribe a basis of the tangent plane, and so it is conceived of as an abstract vector space rather than a real coordinate space. The *tangent space* is the generalization to higher-dimensional differentiable manifolds.[92]

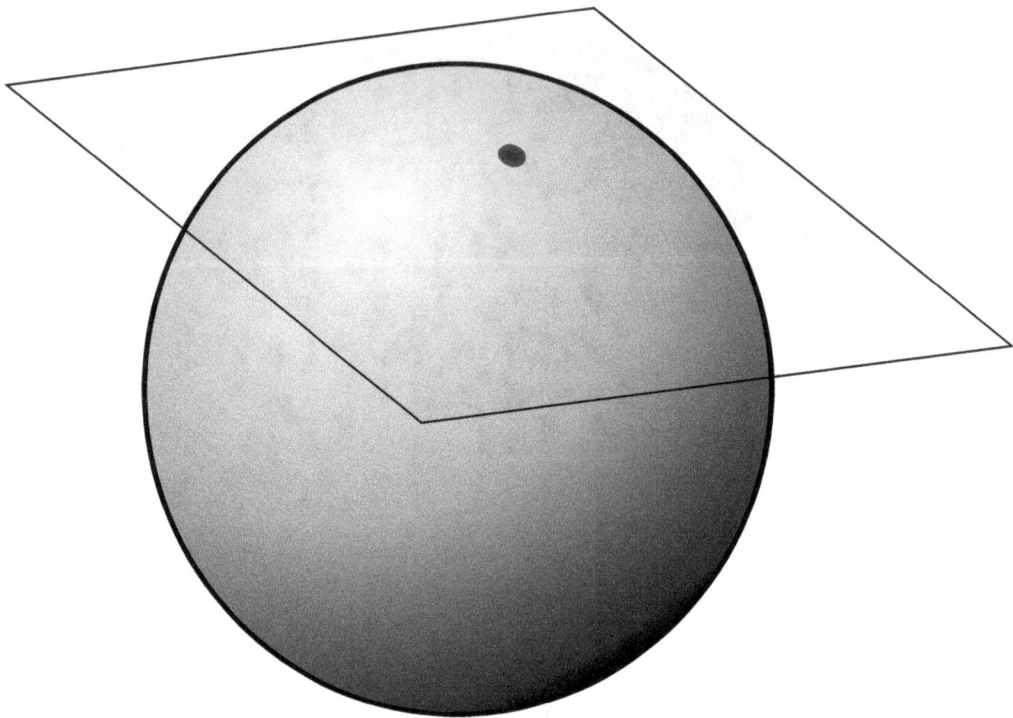

The tangent space to the 2-sphere at some point is the infinite plane touching the sphere in this point.

Riemannian manifolds are manifolds whose tangent spaces are endowed with a suitable inner product.[93] Derived therefrom, the Riemann curvature tensor encodes all curvatures of a manifold in one object, which finds applications in general relativity, for example, where the Einstein curvature tensor describes the matter and energy content of space-time.[94][95] The tangent space of a Lie group can be given naturally the structure of a Lie algebra and can be used to classify compact Lie groups.[96]

26.9 Generalizations

26.9.1 Vector bundles

Main articles: Vector bundle and Tangent bundle

A *vector bundle* is a family of vector spaces parametrized continuously by a topological space X.[92] More precisely, a vector bundle over X is a topological space E equipped with a continuous map

$$\pi : E \to X$$

such that for every x in X, the fiber $\pi^{-1}(x)$ is a vector space. The case dim $V = 1$ is called a line bundle. For any vector space V, the projection $X \times V \to X$ makes the product $X \times V$ into a "trivial" vector bundle. Vector bundles over X are required to be locally a product of X and some (fixed) vector space V: for every x in X, there is a neighborhood U of x such that the restriction of π to $\pi^{-1}(U)$ is isomorphic[nb 15] to the trivial bundle $U \times V \to U$. Despite their locally trivial character, vector bundles may (depending on the shape of the underlying space X) be "twisted" in the large (i.e., the bundle need not be (globally isomorphic to) the trivial bundle $X \times V$). For example, the Möbius strip can be seen as a line bundle over the circle S^1 (by identifying open intervals with the real line). It is, however, different from the cylinder $S^1 \times \mathbf{R}$, because the latter is orientable whereas the former is not.[97]

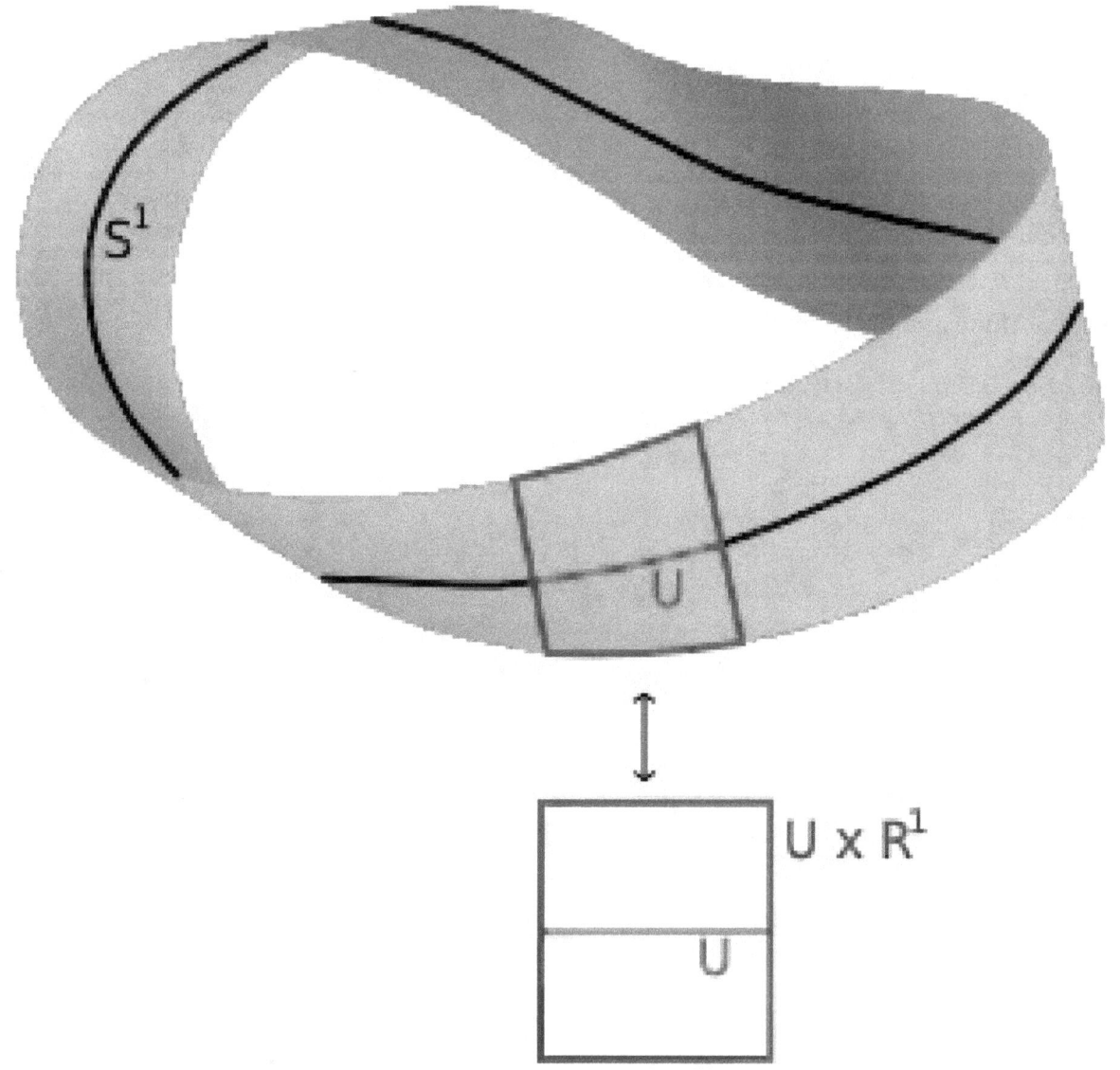

A Möbius strip. Locally, it looks like U × **R**.

Properties of certain vector bundles provide information about the underlying topological space. For example, the tangent bundle consists of the collection of tangent spaces parametrized by the points of a differentiable manifold. The tangent bundle of the circle S^1 is globally isomorphic to $S^1 \times \mathbf{R}$, since there is a global nonzero vector field on S^1.[nb 16] In contrast, by the hairy ball theorem, there is no (tangent) vector field on the 2-sphere S^2 which is everywhere nonzero.[98] K-theory studies the isomorphism classes of all vector bundles over some topological space.[99] In addition to deepening topological and geometrical insight, it has purely algebraic consequences, such as the classification of finite-dimensional real division algebras: **R**, **C**, the quaternions **H** and the octonions.

The cotangent bundle of a differentiable manifold consists, at every point of the manifold, of the dual of the tangent space, the cotangent space. Sections of that bundle are known as differential one-forms.

26.9.2 Modules

Main article: Module

Modules are to rings what vector spaces are to fields. The very same axioms, applied to a ring R instead of a field F yield modules.[100] The theory of modules, compared to that of vector spaces, is complicated by the presence of ring elements that do not have multiplicative inverses. For example, modules need not have bases, as the **Z**-module (i.e., abelian group) **Z**/2**Z** shows; those modules that do (including all vector spaces) are known as free modules. Nevertheless, a vector space can be compactly defined as a module over a ring which is a field with the elements being called vectors. Some authors use the term *vector space* to mean modules over a division ring.[101] The algebro-geometric interpretation of commutative rings via their spectrum allows the development of concepts such as locally free modules, the algebraic counterpart to vector bundles.

26.9.3 Affine and projective spaces

Main articles: Affine space and Projective space

Roughly, *affine spaces* are vector spaces whose origins are not specified.[102] More precisely, an affine space is a set with

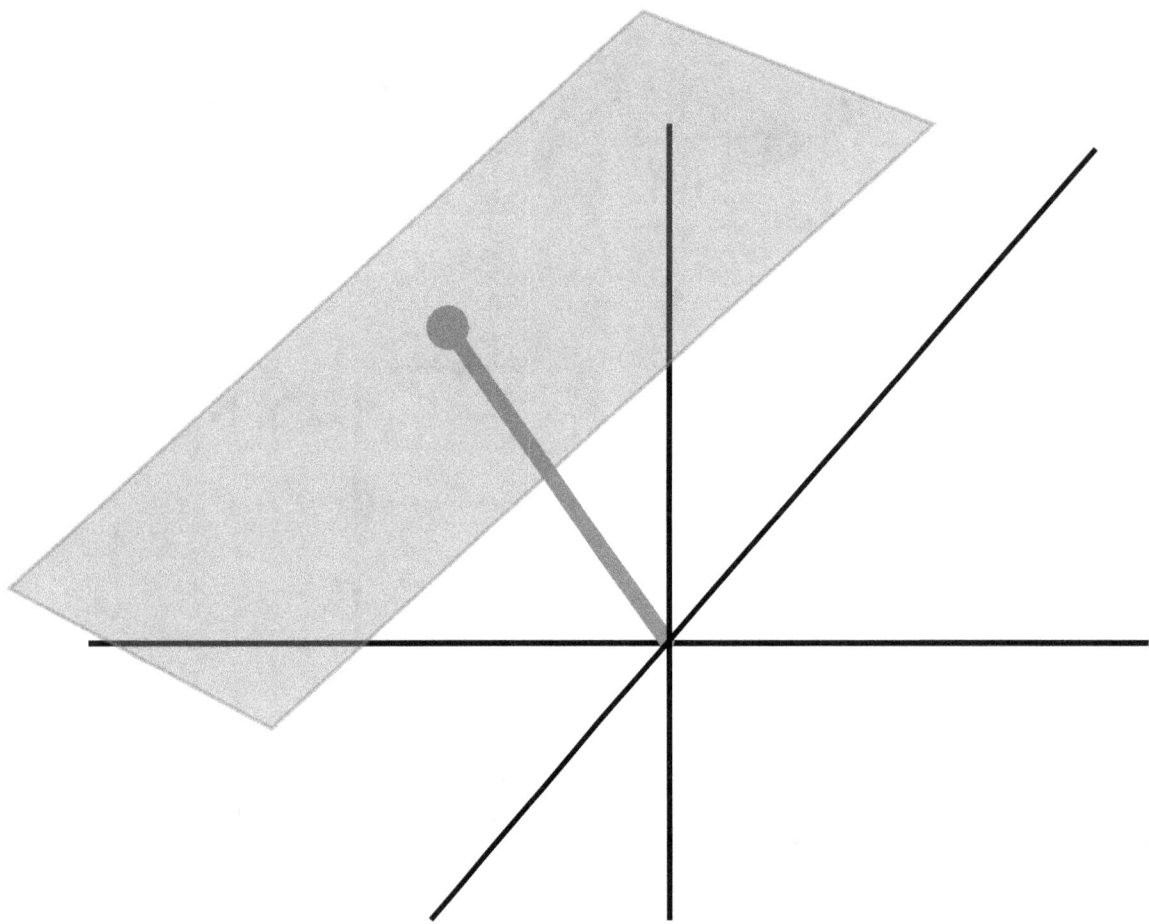

An affine plane (light blue) in $\mathbf{R}^3$. It is a two-dimensional subspace shifted by a vector $\mathbf{x}$ (red).

a free transitive vector space action. In particular, a vector space is an affine space over itself, by the map

$$V \times V \to V, (\mathbf{v}, \mathbf{a}) \mapsto \mathbf{a} + \mathbf{v}.$$

If W is a vector space, then an affine subspace is a subset of W obtained by translating a linear subspace V by a fixed vector $\mathbf{x} \in W$; this space is denoted by $\mathbf{x} + V$ (it is a coset of V in W) and consists of all vectors of the form $\mathbf{x} + \mathbf{v}$ for $\mathbf{v} \in V$. An important example is the space of solutions of a system of inhomogeneous linear equations

$$A\mathbf{x} = \mathbf{b}$$

generalizing the homogeneous case $\mathbf{b} = 0$ above.[103] The space of solutions is the affine subspace $\mathbf{x} + V$ where $\mathbf{x}$ is a particular solution of the equation, and V is the space of solutions of the homogeneous equation (the nullspace of A).

The set of one-dimensional subspaces of a fixed finite-dimensional vector space V is known as *projective space*; it may be used to formalize the idea of parallel lines intersecting at infinity.[104] Grassmannians and flag manifolds generalize this by parametrizing linear subspaces of fixed dimension k and flags of subspaces, respectively.

26.10 See also

- Vector (mathematics and physics), for a list of various kinds of vectors

26.11 Notes

[1] It is also common, especially in physics, to denote vectors with an arrow on top: $\vec{v}$.

[2] This axiom refers to two different operations: scalar multiplication: $b\mathbf{v}$; and field multiplication: ab. It does not assert the associativity of either operation. More formally, scalar multiplication is the *semigroup action* of the scalars on the vector space. Combined with the axiom of the identity element of scalar multiplication, it is a *monoid action*.

[3] Some authors (such as Brown 1991) restrict attention to the fields **R** or **C**, but most of the theory is unchanged for an arbitrary field.

[4] The indicator functions of intervals (of which there are infinitely many) are linearly independent, for example.

[5] The nomenclature derives from German "eigen", which means own or proper.

[6] Roman 2005, ch. 8, p. 140. See also Jordan–Chevalley decomposition.

[7] Some authors (such as Roman 2005) choose to start with this equivalence relation and derive the concrete shape of V/W from this.

[8] This requirement implies that the topology gives rise to a uniform structure, Bourbaki 1989, ch. II

[9] The triangle inequality for $|-|_p$ is provided by the Minkowski inequality. For technical reasons, in the context of functions one has to identify functions that agree almost everywhere to get a norm, and not only a seminorm.

[10] "Many functions in L^2 of Lebesgue measure, being unbounded, cannot be integrated with the classical Riemann integral. So spaces of Riemann integrable functions would not be complete in the L^2 norm, and the orthogonal decomposition would not apply to them. This shows one of the advantages of Lebesgue integration.", Dudley 1989, §5.3, p. 125

[11] For $p \neq 2$, $L^p(\Omega)$ is not a Hilbert space.

[12] A basis of a Hilbert space is not the same thing as a basis in the sense of linear algebra above. For distinction, the latter is then called a Hamel basis.

[13] Although the Fourier series is periodic, the technique can be applied to any L^2 function on an interval by considering the function to be continued periodically outside the interval. See Kreyszig 1988, p. 601

[14] That is to say (BSE-3 2001), the plane passing through the point of contact P such that the distance from a point P_1 on the surface to the plane is infinitesimally small compared to the distance from P_1 to P in the limit as P_1 approaches P along the surface.

[15] That is, there is a homeomorphism from $\pi^{-1}(U)$ to $V \times U$ which restricts to linear isomorphisms between fibers.

[16] A line bundle, such as the tangent bundle of S^1 is trivial if and only if there is a section that vanishes nowhere, see Husemoller 1994, Corollary 8.3. The sections of the tangent bundle are just vector fields.

26.12 Footnotes

[1] Roman 2005, ch. 1, p. 27

[2] van der Waerden 1993, Ch. 19

[3] Bourbaki 1998, §II.1.1. Bourbaki calls the group homomorphisms $f(a)$ *homotheties*.

[4] Bourbaki 1969, ch. "Algèbre linéaire et algèbre multilinéaire", pp. 78–91

[5] Bolzano 1804

[6] Möbius 1827

[7] Hamilton 1853

[8] Grassmann 2000

[9] Peano 1888, ch. IX

[10] Banach 1922

[11] Dorier 1995, Moore 1995

[12] Lang 1987, ch. I.1

[13] Lang 2002, ch. V.1

[14] e.g. Lang 1993, ch. XII.3., p. 335

[15] Lang 1987, ch. IX.1

[16] Lang 1987, ch. VI.3.

[17] Roman 2005, Theorem 1.9, p. 43

[18] Blass 1984

[19] Halpern 1966, pp. 670–673

[20] Artin 1991, Theorem 3.3.13

[21] Braun 1993, Th. 3.4.5, p. 291

[22] Stewart 1975, Proposition 4.3, p. 52

[23] Stewart 1975, Theorem 6.5, p. 74

[24] Roman 2005, ch. 2, p. 45

[25] Lang 1987, ch. IV.4, Corollary, p. 106

[26] Lang 1987, Example IV.2.6

[27] Lang 1987, ch. VI.6

[28] Halmos 1974, p. 28, Ex. 9

[29] Lang 1987, Theorem IV.2.1, p. 95

[30] Roman 2005, Th. 2.5 and 2.6, p. 49

[31] Lang 1987, ch. V.1

[32] Lang 1987, ch. V.3., Corollary, p. 106

[33] Lang 1987, Theorem VII.9.8, p. 198

[34] Roman 2005, ch. 8, p. 135–156

[35] Lang 1987, ch. IX.4

[36] Roman 2005, ch. 1, p. 29

[37] Roman 2005, ch. 1, p. 35

[38] Roman 2005, ch. 3, p. 64

[39] Lang 1987, ch. IV.3.

[40] Roman 2005, ch. 2, p. 48

[41] Mac Lane 1998

[42] Roman 2005, ch. 1, pp. 31–32

[43] Lang 2002, ch. XVI.1

[44] Roman 2005, Th. 14.3. See also Yoneda lemma.

[45] Schaefer & Wolff 1999, pp. 204–205

[46] Bourbaki 2004, ch. 2, p. 48

[47] Roman 2005, ch. 9

[48] Naber 2003, ch. 1.2

[49] Treves 1967

[50] Bourbaki 1987

[51] Kreyszig 1989, §4.11-5

[52] Kreyszig 1989, §1.5-5

[53] Choquet 1966, Proposition III.7.2

[54] Treves 1967, p. 34–36

[55] Lang 1983, Cor. 4.1.2, p. 69

[56] Treves 1967, ch. 11

[57] Treves 1967, Theorem 11.2, p. 102

[58] Evans 1998, ch. 5

[59] Treves 1967, ch. 12

[60] Dennery 1996, p.190

[61] Lang 1993, Th. XIII.6, p. 349

[62] Lang 1993, Th. III.1.1

[63] Choquet 1966, Lemma III.16.11

[64] Kreyszig 1999, Chapter 11

[65] Griffiths 1995, Chapter 1

[66] Lang 1993, ch. XVII.3

[67] Lang 2002, ch. III.1, p. 121

[68] Eisenbud 1995, ch. 1.6

[69] Varadarajan 1974

[70] Lang 2002, ch. XVI.7

[71] Lang 2002, ch. XVI.8

[72] Luenberger 1997, §7.13

[73] See representation theory and group representation.

[74] Lang 1993, Ch. XI.1

[75] Evans 1998, Th. 6.2.1

[76] Folland 1992, p. 349 *ff*

[77] Gasquet & Witomski 1999, p. 150

[78] Gasquet & Witomski 1999, §4.5

[79] Gasquet & Witomski 1999, p. 57

[80] Loomis 1953, Ch. VII

[81] Ashcroft & Mermin 1976, Ch. 5

[82] Kreyszig 1988, p. 667

[83] Fourier 1822

[84] Gasquet & Witomski 1999, p. 67

[85] Ifeachor & Jervis 2002, pp. 3–4, 11

[86] Wallace Feb 1992

[87] Ifeachor & Jervis 2002, p. 132

[88] Gasquet & Witomski 1999, §10.2

[89] Ifeachor & Jervis 2002, pp. 307–310

[90] Gasquet & Witomski 1999, §10.3

[91] Schönhage & Strassen 1971

[92] Spivak 1999, ch. 3

[93] Jost 2005. See also Lorentzian manifold.

[94] Misner, Thorne & Wheeler 1973, ch. 1.8.7, p. 222 and ch. 2.13.5, p. 325

[95] Jost 2005, ch. 3.1

[96] Varadarajan 1974, ch. 4.3, Theorem 4.3.27

[97] Kreyszig 1991, §34, p. 108

[98] Eisenberg & Guy 1979

[99] Atiyah 1989

[100] Artin 1991, ch. 12

[101] Grillet, Pierre Antoine. Abstract algebra. Vol. 242. Springer Science & Business Media, 2007.

[102] Meyer 2000, Example 5.13.5, p. 436

[103] Meyer 2000, Exercise 5.13.15–17, p. 442

[104] Coxeter 1987

26.13   References

26.13.1   Algebra

- Artin, Michael (1991), *Algebra*, Prentice Hall, ISBN 978-0-89871-510-1

- Blass, Andreas (1984), "Existence of bases implies the axiom of choice", *Axiomatic set theory (Boulder, Colorado, 1983)*, Contemporary Mathematics **31**, Providence, R.I.: American Mathematical Society, pp. 31–33, MR 763890

- Brown, William A. (1991), *Matrices and vector spaces*, New York: M. Dekker, ISBN 978-0-8247-8419-5

- Lang, Serge (1987), *Linear algebra*, Berlin, New York: Springer-Verlag, ISBN 978-0-387-96412-6

- Lang, Serge (2002), *Algebra*, Graduate Texts in Mathematics **211** (Revised third ed.), New York: Springer-Verlag, ISBN 978-0-387-95385-4, MR 1878556

- Mac Lane, Saunders (1999), *Algebra* (3rd ed.), pp. 193–222, ISBN 0-8218-1646-2

- Meyer, Carl D. (2000), *Matrix Analysis and Applied Linear Algebra*, SIAM, ISBN 978-0-89871-454-8

- Roman, Steven (2005), *Advanced Linear Algebra*, Graduate Texts in Mathematics **135** (2nd ed.), Berlin, New York: Springer-Verlag, ISBN 978-0-387-24766-3

- Spindler, Karlheinz (1993), *Abstract Algebra with Applications: Volume 1: Vector spaces and groups*, CRC, ISBN 978-0-8247-9144-5

- van der Waerden, Bartel Leendert (1993), *Algebra* (in German) (9th ed.), Berlin, New York: Springer-Verlag, ISBN 978-3-540-56799-8

26.13.2   Analysis

- Bourbaki, Nicolas (1987), *Topological vector spaces*, Elements of mathematics, Berlin, New York: Springer-Verlag, ISBN 978-3-540-13627-9

- Bourbaki, Nicolas (2004), *Integration I*, Berlin, New York: Springer-Verlag, ISBN 978-3-540-41129-1

- Braun, Martin (1993), *Differential equations and their applications: an introduction to applied mathematics*, Berlin, New York: Springer-Verlag, ISBN 978-0-387-97894-9

- BSE-3 (2001), "Tangent plane", in Hazewinkel, Michiel, *Encyclopedia of Mathematics*, Springer, ISBN 978-1-55608-010-4

- Choquet, Gustave (1966), *Topology*, Boston, MA: Academic Press

- Dennery, Philippe; Krzywicki, Andre (1996), *Mathematics for Physicists*, Courier Dover Publications, ISBN 978-0-486-69193-0

- Dudley, Richard M. (1989), *Real analysis and probability*, The Wadsworth & Brooks/Cole Mathematics Series, Pacific Grove, CA: Wadsworth & Brooks/Cole Advanced Books & Software, ISBN 978-0-534-10050-6

- Dunham, William (2005), *The Calculus Gallery*, Princeton University Press, ISBN 978-0-691-09565-3

- Evans, Lawrence C. (1998), *Partial differential equations*, Providence, R.I.: American Mathematical Society, ISBN 978-0-8218-0772-9

- Folland, Gerald B. (1992), *Fourier Analysis and Its Applications*, Brooks-Cole, ISBN 978-0-534-17094-3

- Gasquet, Claude; Witomski, Patrick (1999), *Fourier Analysis and Applications: Filtering, Numerical Computation, Wavelets*, Texts in Applied Mathematics, New York: Springer-Verlag, ISBN 0-387-98485-2

- Ifeachor, Emmanuel C.; Jervis, Barrie W. (2001), *Digital Signal Processing: A Practical Approach* (2nd ed.), Harlow, Essex, England: Prentice-Hall (published 2002), ISBN 0-201-59619-9

- Krantz, Steven G. (1999), *A Panorama of Harmonic Analysis*, Carus Mathematical Monographs, Washington, DC: Mathematical Association of America, ISBN 0-88385-031-1

- Kreyszig, Erwin (1988), *Advanced Engineering Mathematics* (6th ed.), New York: John Wiley & Sons, ISBN 0-471-85824-2

- Kreyszig, Erwin (1989), *Introductory functional analysis with applications*, Wiley Classics Library, New York: John Wiley & Sons, ISBN 978-0-471-50459-7, MR 992618

- Lang, Serge (1983), *Real analysis*, Addison-Wesley, ISBN 978-0-201-14179-5

- Lang, Serge (1993), *Real and functional analysis*, Berlin, New York: Springer-Verlag, ISBN 978-0-387-94001-4

- Loomis, Lynn H. (1953), *An introduction to abstract harmonic analysis*, Toronto-New York–London: D. Van Nostrand Company, Inc., pp. x+190

- Schaefer, Helmut H.; Wolff, M.P. (1999), *Topological vector spaces* (2nd ed.), Berlin, New York: Springer-Verlag, ISBN 978-0-387-98726-2

- Treves, François (1967), *Topological vector spaces, distributions and kernels*, Boston, MA: Academic Press

26.13.3 Historical references

- Banach, Stefan (1922), "Sur les opérations dans les ensembles abstraits et leur application aux équations intégrales (On operations in abstract sets and their application to integral equations)" (PDF), *Fundamenta Mathematicae* (in French) **3**, ISSN 0016-2736

- Bolzano, Bernard (1804), *Betrachtungen über einige Gegenstände der Elementargeometrie (Considerations of some aspects of elementary geometry)* (in German)

- Bourbaki, Nicolas (1969), *Éléments d'histoire des mathématiques (Elements of history of mathematics)* (in French), Paris: Hermann

- Dorier, Jean-Luc (1995), "A general outline of the genesis of vector space theory", *Historia Mathematica* **22** (3): 227–261, doi:10.1006/hmat.1995.1024, MR 1347828

- Fourier, Jean Baptiste Joseph (1822), *Théorie analytique de la chaleur* (in French), Chez Firmin Didot, père et fils

- Grassmann, Hermann (1844), *Die Lineale Ausdehnungslehre - Ein neuer Zweig der Mathematik* (in German), O. Wigand, reprint: Hermann Grassmann. Translated by Lloyd C. Kannenberg. (2000), Kannenberg, L.C., ed., *Extension Theory*, Providence, R.I.: American Mathematical Society, ISBN 978-0-8218-2031-5

- Hamilton, William Rowan (1853), *Lectures on Quaternions*, Royal Irish Academy

- Möbius, August Ferdinand (1827), *Der Barycentrische Calcul : ein neues Hülfsmittel zur analytischen Behandlung der Geometrie (Barycentric calculus: a new utility for an analytic treatment of geometry)* (in German)

- Moore, Gregory H. (1995), "The axiomatization of linear algebra: 1875–1940", *Historia Mathematica* **22** (3): 262–303, doi:10.1006/hmat.1995.1025

- Peano, Giuseppe (1888), *Calcolo Geometrico secondo l'Ausdehnungslehre di H. Grassmann preceduto dalle Operazioni della Logica Deduttiva* (in Italian), Turin

26.13.4   Further references

- Ashcroft, Neil; Mermin, N. David (1976), *Solid State Physics*, Toronto: Thomson Learning, ISBN 978-0-03-083993-1

- Atiyah, Michael Francis (1989), *K-theory*, Advanced Book Classics (2nd ed.), Addison-Wesley, ISBN 978-0-201-09394-0, MR 1043170

- Bourbaki, Nicolas (1998), *Elements of Mathematics : Algebra I Chapters 1-3*, Berlin, New York: Springer-Verlag, ISBN 978-3-540-64243-5

- Bourbaki, Nicolas (1989), *General Topology. Chapters 1-4*, Berlin, New York: Springer-Verlag, ISBN 978-3-540-64241-1

- Coxeter, Harold Scott MacDonald (1987), *Projective Geometry* (2nd ed.), Berlin, New York: Springer-Verlag, ISBN 978-0-387-96532-1

- Eisenberg, Murray; Guy, Robert (1979), "A proof of the hairy ball theorem", *The American Mathematical Monthly* (Mathematical Association of America) **86** (7): 572–574, doi:10.2307/2320587, JSTOR 2320587

- Eisenbud, David (1995), *Commutative algebra*, Graduate Texts in Mathematics **150**, Berlin, New York: Springer-Verlag, ISBN 978-0-387-94269-8, MR 1322960

- Goldrei, Derek (1996), *Classic Set Theory: A guided independent study* (1st ed.), London: Chapman and Hall, ISBN 0-412-60610-0

- Griffiths, David J. (1995), *Introduction to Quantum Mechanics*, Upper Saddle River, NJ: Prentice Hall, ISBN 0-13-124405-1

- Halmos, Paul R. (1974), *Finite-dimensional vector spaces*, Berlin, New York: Springer-Verlag, ISBN 978-0-387-90093-3

- Halpern, James D. (Jun 1966), "Bases in Vector Spaces and the Axiom of Choice", *Proceedings of the American Mathematical Society* (American Mathematical Society) **17** (3): 670–673, doi:10.2307/2035388, JSTOR 2035388

- Husemoller, Dale (1994), *Fibre Bundles* (3rd ed.), Berlin, New York: Springer-Verlag, ISBN 978-0-387-94087-8

- Jost, Jürgen (2005), *Riemannian Geometry and Geometric Analysis* (4th ed.), Berlin, New York: Springer-Verlag, ISBN 978-3-540-25907-7

- Kreyszig, Erwin (1991), *Differential geometry*, New York: Dover Publications, pp. xiv+352, ISBN 978-0-486-66721-8

- Kreyszig, Erwin (1999), *Advanced Engineering Mathematics* (8th ed.), New York: John Wiley & Sons, ISBN 0-471-15496-2

- Luenberger, David (1997), *Optimization by vector space methods*, New York: John Wiley & Sons, ISBN 978-0-471-18117-0

- Mac Lane, Saunders (1998), *Categories for the Working Mathematician* (2nd ed.), Berlin, New York: Springer-Verlag, ISBN 978-0-387-98403-2

- Misner, Charles W.; Thorne, Kip; Wheeler, John Archibald (1973), *Gravitation*, W. H. Freeman, ISBN 978-0-7167-0344-0

- Naber, Gregory L. (2003), *The geometry of Minkowski spacetime*, New York: Dover Publications, ISBN 978-0-486-43235-9, MR 2044239

- Schönhage, A.; Strassen, Volker (1971), "Schnelle Multiplikation großer Zahlen (Fast multiplication of big numbers)" (PDF), *Computing* (in German) **7**: 281–292, doi:10.1007/bf02242355, ISSN 0010-485X

- Spivak, Michael (1999), *A Comprehensive Introduction to Differential Geometry (Volume Two)*, Houston, TX: Publish or Perish

- Stewart, Ian (1975), *Galois Theory*, Chapman and Hall Mathematics Series, London: Chapman and Hall, ISBN 0-412-10800-3

- Varadarajan, V. S. (1974), *Lie groups, Lie algebras, and their representations*, Prentice Hall, ISBN 978-0-13-535732-3

- Wallace, G.K. (Feb 1992), "The JPEG still picture compression standard", *IEEE Transactions on Consumer Electronics* **38** (1): xviii–xxxiv, doi:10.1109/30.125072, ISSN 0098-3063

- Weibel, Charles A. (1994), *An introduction to homological algebra*, Cambridge Studies in Advanced Mathematics **38**, Cambridge University Press, ISBN 978-0-521-55987-4, OCLC 36131259, MR 1269324

26.14 External links

- Hazewinkel, Michiel, ed. (2001), "Vector space", *Encyclopedia of Mathematics*, Springer, ISBN 978-1-55608-010-4

- A lecture about fundamental concepts related to vector spaces (given at MIT)

- A graphical simulator for the concepts of span, linear dependency, base and dimension

Chapter 27

Group homomorphism

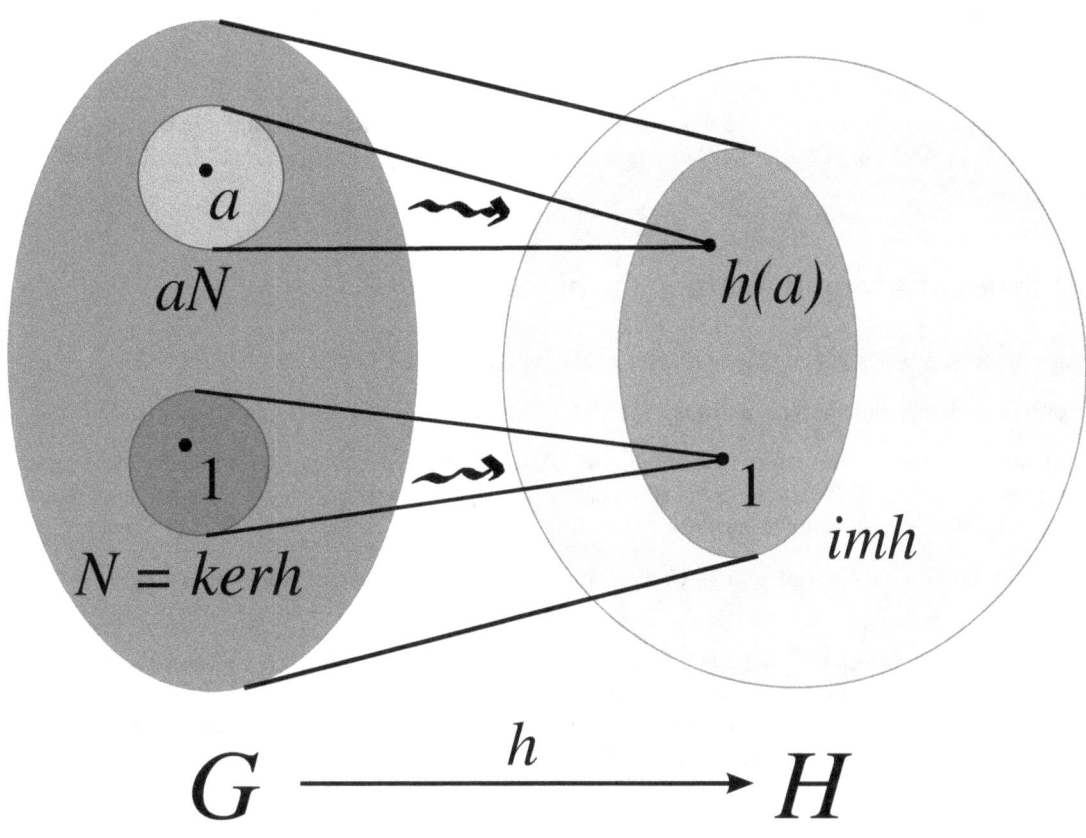

*Image of a group homomorphism (**h**) from **G** (left) to **H** (right). The smaller oval inside **H** is the image of **h**. **N** is the kernel of **h** and **aN** is a coset of **N**.*

In mathematics, given two groups, $(G, *)$ and $(H, \cdot)$, a **group homomorphism** from $(G, *)$ to $(H, \cdot)$ is a function $h : G \rightarrow H$ such that for all u and v in G it holds that

$$h(u * v) = h(u) \cdot h(v)$$

where the group operation on the left hand side of the equation is that of G and on the right hand side that of H.

From this property, one can deduce that h maps the identity element eG of G to the identity element eH of H, and it also maps inverses to inverses in the sense that

$$h\left(u^{-1}\right) = h(u)^{-1}.$$

Hence one can say that h "is compatible with the group structure".

Older notations for the homomorphism $h(x)$ may be xh, though this may be confused as an index or a general subscript. A more recent trend is to write group homomorphisms on the right of their arguments, omitting brackets, so that $h(x)$ becomes simply $x\,h$. This approach is especially prevalent in areas of group theory where automata play a role, since it accords better with the convention that automata read words from left to right.

In areas of mathematics where one considers groups endowed with additional structure, a *homomorphism* sometimes means a map which respects not only the group structure (as above) but also the extra structure. For example, a homomorphism of topological groups is often required to be continuous.

27.1 Intuition

The purpose of defining a group homomorphism is to create functions that preserve the algebraic structure. An equivalent definition of group homomorphism is: The function $h : G \rightarrow H$ is a group homomorphism if whenever $a * b = c$ we have $h(a) \cdot h(b) = h(c)$. In other words, the group H in some sense has a similar algebraic structure as G and the homomorphism h preserves that.

27.2 Types of group homomorphism

Monomorphism A group homomorphism that is injective (or, one-to-one); i.e., preserves distinctness.

Epimorphism A group homomorphism that is surjective (or, onto); i.e., reaches every point in the codomain.

Isomorphism A group homomorphism that is bijective; i.e, injective and surjective. Its inverse is also a group homomorphism. In this case, the groups G and H are called *isomorphic*; they differ only in the notation of their elements and are identical for all practical purposes.

Endomorphism A homomorphism, $h\colon G \rightarrow G$; the domain and codomain are the same. Also called an *endomorphism of* G.

Automorphism An endomorphism that is bijective, and hence an isomorphism. The set of all automorphisms of a group G, with functional composition as operation, forms itself a group, the *automorphism group* of G. It is denoted by $\text{Aut}(G)$. As an example, the automorphism group of $(\mathbf{Z}, +)$ contains only two elements, the identity transformation and multiplication with -1; it is isomorphic to $\mathbf{Z}/2\mathbf{Z}$.

27.3 Image and kernel

We define the *kernel of h* to be the set of elements in G which are mapped to the identity in H

$$\ker(h) \equiv \{u \in G\colon h(u) = e_H\}.$$

and the *image of h* to be

$$\text{im}(h) \equiv h(G) \equiv \{h(u)\colon u \in G\}.$$

The kernel and image of a homomorphism can be interpreted as measuring how close it is to being an isomorphism. The First Isomorphism Theorem states that the image of a group homomorphism, $h(G)$ is isomorphic to the quotient group $G/\ker h$.

The kernel of h is a normal subgroup of G and the image of h is a subgroup of H:

$$
\begin{aligned}
h\left(g^{-1} \circ u \circ g\right) &= h(g)^{-1} \cdot h(u) \cdot h(g) \\
&= h(g)^{-1} \cdot e_H \cdot h(g) \\
&= h(g)^{-1} \cdot h(g) = e_H.
\end{aligned}
$$

If and only if $\ker(h) = \{eG\}$, the homomorphism, h, is a *group monomorphism*; i.e., h is injective (one-to-one). Injection directly gives that there is a unique element in the kernel, and a unique element in the kernel gives injection:

$$
\begin{aligned}
& h(g_1) = h(g_2) \\
\Leftrightarrow \quad & h(g_1) \cdot h(g_2)^{-1} = e_H \\
\Leftrightarrow \quad & h\left(g_1 \circ g_2^{-1}\right) = e_H, \ \ker(h) = \{e_G\} \\
\Rightarrow \quad & g_1 \circ g_2^{-1} = e_G \\
\Leftrightarrow \quad & g_1 = g_2
\end{aligned}
$$

27.4 Examples

- Consider the cyclic group $\mathbf{Z}/3\mathbf{Z} = \{0, 1, 2\}$ and the group of integers $\mathbf{Z}$ with addition. The map $h : \mathbf{Z} \to \mathbf{Z}/3\mathbf{Z}$ with $h(u) = u \bmod 3$ is a group homomorphism. It is surjective and its kernel consists of all integers which are divisible by 3.

- Consider the group

$$
G \equiv \left\{ \begin{pmatrix} a & b \\ 0 & 1 \end{pmatrix} \middle| a > 0, b \in \mathbf{R} \right\}
$$

  For any complex number u the function $fu : G \to \mathbf{C}$ defined by:

$$
\begin{pmatrix} a & b \\ 0 & 1 \end{pmatrix} \mapsto a^u
$$

  is a group homomorphism.

- Consider multiplicative group of positive real numbers $(\mathbf{R}^+, \cdot)$ for any complex number u the function $fu : \mathbf{R}^+ \to \mathbf{C}$ defined by:

$$
f_u(a) = a^u
$$

  is a group homomorphism.

- The exponential map yields a group homomorphism from the group of real numbers $\mathbf{R}$ with addition to the group of non-zero real numbers $\mathbf{R}^*$ with multiplication. The kernel is $\{0\}$ and the image consists of the positive real numbers.

- The exponential map also yields a group homomorphism from the group of complex numbers $\mathbf{C}$ with addition to the group of non-zero complex numbers $\mathbf{C}^*$ with multiplication. This map is surjective and has the kernel $\{2\pi ki : k \in \mathbf{Z}\}$, as can be seen from Euler's formula. Fields like $\mathbf{R}$ and $\mathbf{C}$ that have homomorphisms from their additive group to their multiplicative group are thus called exponential fields.

27.5 The category of groups

If $h : G \to H$ and $k : H \to K$ are group homomorphisms, then so is $k \circ h : G \to K$. This shows that the class of all groups, together with group homomorphisms as morphisms, forms a category.

27.6 Homomorphisms of abelian groups

If G and H are abelian (i.e., commutative) groups, then the set $\mathrm{Hom}(G, H)$ of all group homomorphisms from G to H is itself an abelian group: the sum $h + k$ of two homomorphisms is defined by

$$(h + k)(u) = h(u) + k(u) \text{ for all } u \text{ in } G.$$

The commutativity of H is needed to prove that $h + k$ is again a group homomorphism.

The addition of homomorphisms is compatible with the composition of homomorphisms in the following sense: if f is in $\mathrm{Hom}(K, G)$, h, k are elements of $\mathrm{Hom}(G, H)$, and g is in $\mathrm{Hom}(H, L)$, then

$$(h + k) \circ f = (h \circ f) + (k \circ f) \text{ and } g \circ (h + k) = (g \circ h) + (g \circ k).$$

Since the composition is associative, this shows that the set $\mathrm{End}(G)$ of all endomorphisms of an abelian group forms a ring, the *endomorphism ring* of G. For example, the endomorphism ring of the abelian group consisting of the direct sum of m copies of $\mathbf{Z}/n\mathbf{Z}$ is isomorphic to the ring of m-by-m matrices with entries in $\mathbf{Z}/n\mathbf{Z}$. The above compatibility also shows that the category of all abelian groups with group homomorphisms forms a preadditive category; the existence of direct sums and well-behaved kernels makes this category the prototypical example of an abelian category.

27.7 See also

- Fundamental theorem on homomorphisms

- Ring homomorphism

27.8 References

- Dummit, D. S.; Foote, R. (2004). *Abstract Algebra* (3 ed.). Wiley. pp. 71–72. ISBN 9780471433347.

- Lang, Serge (2002), *Algebra*, Graduate Texts in Mathematics **211** (Revised third ed.), New York: Springer-Verlag, ISBN 978-0-387-95385-4, Zbl 0984.00001, MR 1878556

27.9 External links

- Group Homomorphism at PlanetMath.org.

Chapter 28

Automorphism

In mathematics, an **automorphism** is an isomorphism from a mathematical object to itself. It is, in some sense, a symmetry of the object, and a way of mapping the object to itself while preserving all of its structure. The set of all automorphisms of an object forms a group, called the **automorphism group**. It is, loosely speaking, the symmetry group of the object.

28.1 Definition

The exact definition of an automorphism depends on the type of "mathematical object" in question and what, precisely, constitutes an "isomorphism" of that object. The most general setting in which these words have meaning is an abstract branch of mathematics called category theory. Category theory deals with abstract objects and morphisms between those objects.

In category theory, an automorphism is an endomorphism (i.e. a morphism from an object to itself) which is also an isomorphism (in the categorical sense of the word).

This is a very abstract definition since, in category theory, morphisms aren't necessarily functions and objects aren't necessarily sets. In most concrete settings, however, the objects will be sets with some additional structure and the morphisms will be functions preserving that structure.

In the context of abstract algebra, for example, a mathematical object is an algebraic structure such as a group, ring, or vector space. An isomorphism is simply a bijective homomorphism. (The definition of a homomorphism depends on the type of algebraic structure; see, for example: group homomorphism, ring homomorphism, and linear operator).

The identity morphism (identity mapping) is called the **trivial automorphism** in some contexts. Respectively, other (non-identity) automorphisms are called **nontrivial automorphisms**.

28.2 Automorphism group

If the automorphisms of an object X form a set (instead of a proper class), then they form a group under composition of morphisms. This group is called the **automorphism group** of X. That this is indeed a group is simple to see:

- Closure: composition of two endomorphisms is another endomorphism.

- Associativity: composition of morphisms is *always* associative.

- Identity: the identity is the identity morphism from an object to itself, which exists by definition.

- Inverses: by definition every isomorphism has an inverse which is also an isomorphism, and since the inverse is also an endomorphism of the same object it is an automorphism.

The automorphism group of an object X in a category C is denoted Aut$C(X)$, or simply Aut(X) if the category is clear from context.

28.3 Examples

- In set theory, an arbitrary permutation of the elements of a set X is an automorphism. The automorphism group of X is also called the symmetric group on X.

- In elementary arithmetic, the set of integers, $\mathbf{Z}$, considered as a group under addition, has a unique nontrivial automorphism: negation. Considered as a ring, however, it has only the trivial automorphism. Generally speaking, negation is an automorphism of any abelian group, but not of a ring or field.

- A group automorphism is a group isomorphism from a group to itself. Informally, it is a permutation of the group elements such that the structure remains unchanged. For every group G there is a natural group homomorphism $G \to$ Aut(G) whose image is the group Inn(G) of inner automorphisms and whose kernel is the center of G. Thus, if G has trivial center it can be embedded into its own automorphism group.[1]

- In linear algebra, an endomorphism of a vector space V is a linear operator $V \to V$. An automorphism is an invertible linear operator on V. When the vector space is finite-dimensional, the automorphism group of V is the same as the general linear group, GL(V).

- A field automorphism is a bijective ring homomorphism from a field to itself. In the cases of the rational numbers ($\mathbf{Q}$) and the real numbers ($\mathbf{R}$) there are no nontrivial field automorphisms. Some subfields of $\mathbf{R}$ have nontrivial field automorphisms, which however do not extend to all of $\mathbf{R}$ (because they cannot preserve the property of a number having a square root in $\mathbf{R}$). In the case of the complex numbers, $\mathbf{C}$, there is a unique nontrivial automorphism that sends $\mathbf{R}$ into $\mathbf{R}$: complex conjugation, but there are infinitely (uncountably) many "wild" automorphisms (assuming the axiom of choice).[2][3] Field automorphisms are important to the theory of field extensions, in particular Galois extensions. In the case of a Galois extension L/K the subgroup of all automorphisms of L fixing K pointwise is called the Galois group of the extension.

- The field $\mathbf{Q}p$ of p-adic numbers has no nontrivial automorphisms.

- In graph theory an automorphism of a graph is a permutation of the nodes that preserves edges and non-edges. In particular, if two nodes are joined by an edge, so are their images under the permutation.

- For relations, see relation-preserving automorphism.

  - In order theory, see order automorphism.

- In geometry, an automorphism may be called a motion of the space. Specialized terminology is also used:

  - In metric geometry an automorphism is a self-isometry. The automorphism group is also called the isometry group.

  - In the category of Riemann surfaces, an automorphism is a bijective biholomorphic map (also called a conformal map), from a surface to itself. For example, the automorphisms of the Riemann sphere are Möbius transformations.

  - An automorphism of a differentiable manifold M is a diffeomorphism from M to itself. The automorphism group is sometimes denoted Diff(M).

  - In topology, morphisms between topological spaces are called continuous maps, and an automorphism of a topological space is a homeomorphism of the space to itself, or self-homeomorphism (see homeomorphism group). In this example it is *not sufficient* for a morphism to be bijective to be an isomorphism.

28.4 History

One of the earliest group automorphisms (automorphism of a group, not simply a group of automorphisms of points) was given by the Irish mathematician William Rowan Hamilton in 1856, in his icosian calculus, where he discovered an order two automorphism,[4] writing:

> so that μ is a new fifth root of unity, connected with the former fifth root λ by relations of perfect reciprocity.

28.5 Inner and outer automorphisms

In some categories—notably groups, rings, and Lie algebras—it is possible to separate automorphisms into two types, called "inner" and "outer" automorphisms.

In the case of groups, the inner automorphisms are the conjugations by the elements of the group itself. For each element a of a group G, conjugation by a is the operation $\varphi a : G \to G$ given by $\varphi a(g) = aga^{-1}$ (or $a^{-1}ga$; usage varies). One can easily check that conjugation by a is a group automorphism. The inner automorphisms form a normal subgroup of $\mathrm{Aut}(G)$, denoted by $\mathrm{Inn}(G)$; this is called Goursat's lemma.

The other automorphisms are called outer automorphisms. The quotient group $\mathrm{Aut}(G)$ / $\mathrm{Inn}(G)$ is usually denoted by $\mathrm{Out}(G)$; the non-trivial elements are the cosets that contain the outer automorphisms.

The same definition holds in any unital ring or algebra where a is any invertible element. For Lie algebras the definition is slightly different.

28.6 See also

- Endomorphism ring

- Antiautomorphism

- Frobenius automorphism

- Morphism

- Characteristic subgroup

28.7 References

[1] PJ Pahl, R Damrath (2001). "§7.5.5 Automorphisms". *Mathematical foundations of computational engineering* (Felix Pahl translation ed.). Springer. p. 376. ISBN 3-540-67995-2.

[2] Yale, Paul B. (May 1966). "Automorphisms of the Complex Numbers" (PDF). *Mathematics Magazine* **39** (3): 135–141. doi:10.2307/2689301. JSTOR 2689301.

[3] Lounesto, Pertti (2001), *Clifford Algebras and Spinors* (2nd ed.), Cambridge University Press, pp. 22–23, ISBN 0-521-00551-5

[4] Sir William Rowan Hamilton (1856). "Memorandum respecting a new System of Roots of Unity" (PDF). *Philosophical Magazine* **12**: 446.

28.8 External links

- *Automorphism* at Encyclopaedia of Mathematics

- Weisstein, Eric W., "Automorphism", *MathWorld*.

Chapter 29

Isomorphism

This article is about mathematics. For other uses, see Isomorphism (disambiguation).

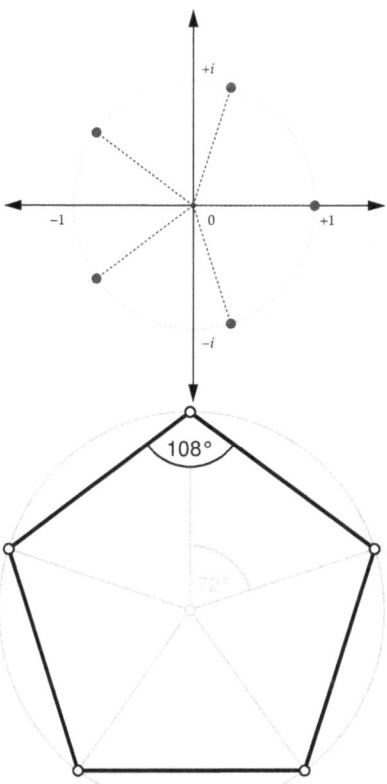

The group of fifth roots of unity under multiplication is isomorphic to the group of rotations of the regular pentagon under composition.

In mathematics, an **isomorphism** (from the Ancient Greek: ἴσος *isos* "equal", and μορφή *morphe* "shape") is a homomorphism (or more generally a morphism) that admits an inverse.[note 1] Two mathematical objects are **isomorphic** if an isomorphism exists between them. An *automorphism* is an isomorphism whose source and target coincide. The interest of isomorphisms lies in the fact that two isomorphic objects cannot be distinguished by using only the properties used to define morphisms; thus isomorphic objects may be considered the same as long as one considers only these properties and their consequences.

For most algebraic structures, including groups and rings, a homomorphism is an isomorphism if and only if it is bijective.

In topology, where the morphisms are continuous functions, isomorphisms are also called *homeomorphisms* or *bicontinuous functions*. In mathematical analysis, where the morphisms are differentiable functions, isomorphisms are also called *diffeomorphisms*.

A **canonical isomorphism** is a canonical map that is an isomorphism. Two objects are said to be **canonically isomorphic** if there is a canonical isomorphism between them. For example, the canonical map from a finite-dimensional vector space V to its second dual space is a canonical isomorphism; on the other hand, V is isomorphic to its dual space but not canonically in general.

Isomorphisms are formalized using category theory. A morphism $f : X \to Y$ in a category is an isomorphism if it admits a two-sided inverse, meaning that there is another morphism $g : Y \to X$ in that category such that $gf = 1X$ and $fg = 1Y$, where $1X$ and $1Y$ are the identity morphisms of X and Y, respectively.[1]

29.1 Examples

29.1.1 Logarithm and exponential

Let $\mathbb{R}^+$ be the multiplicative group of positive real numbers, and let $\mathbb{R}$ be the additive group of real numbers.

The logarithm function $\log : \mathbb{R}^+ \to \mathbb{R}$ satisfies $\log(xy) = \log x + \log y$ for all $x, y \in \mathbb{R}^+$, so it is a group homomorphism. The exponential function $\exp : \mathbb{R} \to \mathbb{R}^+$ satisfies $\exp(x + y) = (\exp x)(\exp y)$ for all $x, y \in \mathbb{R}$, so it too is a homomorphism.

The identities $\log \exp x = x$ and $\exp \log y = y$ show that log and exp are inverses of each other. Since log is a homomorphism that has an inverse that is also a homomorphism, log is an isomorphism of groups.

Because log is an isomorphism, it translates multiplication of positive real numbers into addition of real numbers. This facility makes it possible to multiply real numbers using a ruler and a table of logarithms, or using a slide rule with a logarithmic scale.

29.1.2 Integers modulo 6

Consider the group $(\mathbb{Z}_6, +)$, the integers from 0 to 5 with addition modulo 6. Also consider the group $(\mathbb{Z}_2 \times \mathbb{Z}_3, +)$, the ordered pairs where the x coordinates can be 0 or 1, and the y coordinates can be 0, 1, or 2, where addition in the x-coordinate is modulo 2 and addition in the y-coordinate is modulo 3.

These structures are isomorphic under addition, if you identify them using the following scheme:

$(0,0) \to 0$

$(1,1) \to 1$

$(0,2) \to 2$

$(1,0) \to 3$

$(0,1) \to 4$

$(1,2) \to 5$

or in general $(a,b) \to (3a + 4b) \bmod 6$.

For example note that $(1,1) + (1,0) = (0,1)$, which translates in the other system as $1 + 3 = 4$.

Even though these two groups "look" different in that the sets contain different elements, they are indeed **isomorphic**: their structures are exactly the same. More generally, the direct product of two cyclic groups $\mathbb{Z}_m$ and $\mathbb{Z}_n$ is isomorphic to $(\mathbb{Z}_{mn}, +)$ if and only if m and n are coprime.

29.1.3   Relation-preserving isomorphism

If one object consists of a set X with a binary relation R and the other object consists of a set Y with a binary relation S then an isomorphism from X to Y is a bijective function $f \colon X \to Y$ such that:[2]

$$S(f(u), f(v)) \iff R(u, v)$$

S is reflexive, irreflexive, symmetric, antisymmetric, asymmetric, transitive, total, trichotomous, a partial order, total order, strict weak order, total preorder (weak order), an equivalence relation, or a relation with any other special properties, if and only if R is.

For example, R is an ordering $\le$ and S an ordering $\sqsubseteq$, then an isomorphism from X to Y is a bijective function $f \colon X \to Y$ such that

$$f(u) \sqsubseteq f(v) \iff u \le v.$$

Such an isomorphism is called an *order isomorphism* or (less commonly) an *isotone isomorphism*.

If $X = Y$, then this is a relation-preserving automorphism.

29.2   Isomorphism vs. bijective morphism

In a concrete category (that is, roughly speaking, a category whose objects are sets and morphisms are mappings between sets), such as the category of topological spaces or categories of algebraic objects like groups, rings, and modules, an isomorphism must be bijective on the underlying sets. In algebraic categories (specifically, categories of varieties in the sense of universal algebra), an isomorphism is the same as a homomorphism which is bijective on underlying sets. However, there are concrete categories in which bijective morphisms are not necessarily isomorphisms (such as the category of topological spaces), and there are categories in which each object admits an underlying set but in which isomorphisms need not be bijective (such as the homotopy category of CW-complexes).

29.3   Applications

In abstract algebra, two basic isomorphisms are defined:

- Group isomorphism, an isomorphism between groups

- Ring isomorphism, an isomorphism between rings. (Note that isomorphisms between fields are actually ring isomorphisms)

Just as the automorphisms of an algebraic structure form a group, the isomorphisms between two algebras sharing a common structure form a heap. Letting a particular isomorphism identify the two structures turns this heap into a group.

In mathematical analysis, the Laplace transform is an isomorphism mapping hard differential equations into easier algebraic equations.

In category theory, let the category C consist of two classes, one of *objects* and the other of morphisms. Then a general definition of isomorphism that covers the previous and many other cases is: an isomorphism is a morphism $f \colon a \to b$ that has an inverse, i.e. there exists a morphism $g \colon b \to a$ with $fg = 1b$ and $gf = 1a$. For example, a bijective linear map is an isomorphism between vector spaces, and a bijective continuous function whose inverse is also continuous is an isomorphism between topological spaces, called a homeomorphism.

In graph theory, an isomorphism between two graphs G and H is a bijective map f from the vertices of G to the vertices of H that preserves the "edge structure" in the sense that there is an edge from vertex u to vertex v in G if and only if there is an edge from $f(u)$ to $f(v)$ in H. See graph isomorphism.

In mathematical analysis, an isomorphism between two Hilbert spaces is a bijection preserving addition, scalar multiplication, and inner product.

In early theories of logical atomism, the formal relationship between facts and true propositions was theorized by Bertrand Russell and Ludwig Wittgenstein to be isomorphic. An example of this line of thinking can be found in Russell's Introduction to Mathematical Philosophy.

In cybernetics, the Good Regulator or Conant-Ashby theorem is stated "Every Good Regulator of a system must be a model of that system". Whether regulated or self-regulating an isomorphism is required between regulator part and the processing part of the system.

29.4   Relation with equality

See also: Equality (mathematics)

In certain areas of mathematics, notably category theory, it is valuable to distinguish between *equality* on the one hand and *isomorphism* on the other.[3] Equality is when two objects are exactly the same, and everything that's true about one object is true about the other, while an isomorphism implies everything that's true about a designated part of one object's structure is true about the other's. For example, the sets

$$A = \{x \in \mathbb{Z} \mid x^2 < 2\} \text{ and } B = \{-1, 0, 1\}$$

are *equal*; they are merely different presentations—the first an intensional one (in set builder notation), and the second extensional (by explicit enumeration)—of the same subset of the integers. By contrast, the sets $\{A,B,C\}$ and $\{1,2,3\}$ are not *equal*—the first has elements that are letters, while the second has elements that are numbers. These are isomorphic as sets, since finite sets are determined up to isomorphism by their cardinality (number of elements) and these both have three elements, but there are many choices of isomorphism—one isomorphism is

$$A \mapsto 1, B \mapsto 2, C \mapsto 3, \text{ while another is } A \mapsto 3, B \mapsto 2, C \mapsto 1,$$

and no one isomorphism is intrinsically better than any other.[note 2][note 3] On this view and in this sense, these two sets are not equal because one cannot consider them *identical*: one can choose an isomorphism between them, but that is a weaker claim than identity—and valid only in the context of the chosen isomorphism.

Sometimes the isomorphisms can seem obvious and compelling, but are still not equalities. As a simple example, the genealogical relationships among Joe, John, and Bobby Kennedy are, in a real sense, the same as those among the American football quarterbacks in the Manning family: Archie, Peyton, and Eli. The father-son pairings and the elder-brother-younger-brother pairings correspond perfectly. That similarity between the two family structures illustrates the origin of the word *isomorphism* (Greek *iso-*, "same," and *-morph*, "form" or "shape"). But because the Kennedys are not the same people as the Mannings, the two genealogical structures are merely isomorphic and not equal.

Another example is more formal and more directly illustrates the motivation for distinguishing equality from isomorphism: the distinction between a finite-dimensional vector space V and its dual space $V^* = \{ \varphi: V \to \mathbf{K}\}$ of linear maps from V to its field of scalars $\mathbf{K}$. These spaces have the same dimension, and thus are isomorphic as abstract vector spaces (since algebraically, vector spaces are classified by dimension, just as sets are classified by cardinality), but there is no "natural" choice of isomorphism $V \overset{\sim}{\to} V^*$. If one chooses a basis for V, then this yields an isomorphism: For all $u. v \in V$,

$$v \overset{\sim}{\mapsto} \phi_v \in V^* \quad \text{that such} \quad \phi_v(u) = v^\mathsf{T} u$$

This corresponds to transforming a column vector (element of V) to a row vector (element of V^*) by transpose, but a different choice of basis gives a different isomorphism: the isomorphism "depends on the choice of basis". More subtly, there *is* a map from a vector space V to its double dual $V^{**} = \{\, x \colon V^* \to \mathbf{K}\,\}$ that does not depend on the choice of basis: For all $v \in V$ and $\varphi \in V^*$,

$$v \overset{\sim}{\mapsto} x_v \in V^{**} \quad \text{that such} \quad x_v(\phi) = \phi(v)$$

This leads to a third notion, that of a natural isomorphism: while V and V^{**} are different sets, there is a "natural" choice of isomorphism between them. This intuitive notion of "an isomorphism that does not depend on an arbitrary choice" is formalized in the notion of a natural transformation; briefly, that one may *consistently* identify, or more generally map from, a vector space to its double dual, $V \overset{\sim}{\to} V^{**}$, for *any* vector space in a consistent way. Formalizing this intuition is a motivation for the development of category theory.

However, there is a case where the distinction between natural isomorphism and equality is usually not made. That is for the objects that may be characterized by a universal property. In fact, there is a unique isomorphism, necessarily natural, between two objects sharing the same universal property. A typical example is the set of real numbers, which may be defined through infinite decimal expansion, infinite binary expansion, Cauchy sequences, Dedekind cuts and many other ways. Formally these constructions define different objects, which all are solutions of the same universal property. As these objects have exactly the same properties, one may forget the method of construction and considering them as equal. This is what everybody does when talking of "*the* set of the real numbers". The same occurs with quotient spaces: they are commonly constructed as sets of equivalence classes. However, talking of set of sets may be counterintuitive, and quotient spaces are commonly considered as a pair of a set of undetermined objects, often called "points", and a surjective map onto this set.

If one wishes to draw a distinction between an arbitrary isomorphism (one that depends on a choice) and a natural isomorphism (one that can be done consistently), one may write $\approx$ for an unnatural isomorphism and $\cong$ for a natural isomorphism, as in $V \approx V^*$ and $V \cong V^{**}$. This convention is not universally followed, and authors who wish to distinguish between unnatural isomorphisms and natural isomorphisms will generally explicitly state the distinction.

Generally, saying that two objects are *equal* is reserved for when there is a notion of a larger (ambient) space that these objects live in. Most often, one speaks of equality of two subsets of a given set (as in the integer set example above), but not of two objects abstractly presented. For example, the 2-dimensional unit sphere in 3-dimensional space

$$S^2 := \{(x,y,z) \in \mathbb{R}^3 \mid x^2 + y^2 + z^2 = 1\} \text{ and the Riemann sphere } \widehat{\mathbb{C}}$$

which can be presented as the one-point compactification of the complex plane $\mathbf{C} \cup \{\infty\}$ *or* as the complex projective line (a quotient space)

$$\mathbf{P}^1_{\mathbb{C}} := (\mathbb{C}^2 \setminus \{(0,0)\})/(\mathbb{C}^*)$$

are three different descriptions for a mathematical object, all of which are isomorphic, but not *equal* because they are not all subsets of a single space: the first is a subset of $\mathbf{R}^3$, the second is $\mathbf{C} \cong \mathbf{R}^{2\,[\text{note 4}]}$ plus an additional point, and the third is a subquotient of $\mathbf{C}^2$

In the context of category theory, objects are usually at most isomorphic—indeed, a motivation for the development of category theory was showing that different constructions in homology theory yielded equivalent (isomorphic) groups. Given maps between two objects X and Y, however, one asks if they are equal or not (they are both elements of the set $\mathrm{Hom}(X, Y)$, hence equality is the proper relationship), particularly in commutative diagrams.

29.5   See also

- Bisimulation

- Heap (mathematics)

- Isometry

- Isomorphism class

- Isomorphism theorem

- Universal property

29.6 Notes

[1] For clarity, by *inverse* is meant *inverse homomorphism* or *inverse morphism* respectively, not *inverse function*.

[2] The careful reader may note that *A, B, C* have a conventional order, namely alphabetical order, and similarly 1, 2, 3 have the order from the integers, and thus one particular isomorphism is "natural", namely

$$A \mapsto 1, B \mapsto 2, C \mapsto 3$$

More formally, as *sets* these are isomorphic, but not naturally isomorphic (there are multiple choices of isomorphism), while as *ordered sets* they are naturally isomorphic (there is a unique isomorphism, given above), since finite total orders are uniquely determined up to unique isomorphism by cardinality. This intuition can be formalized by saying that any two finite totally ordered sets of the same cardinality have a natural isomorphism, the one that sends the least element of the first to the least element of the second, the least element of what remains in the first to the least element of what remains in the second, and so forth, but in general, pairs of sets of a given finite cardinality are not naturally isomorphic because there is more than one choice of map—except if the cardinality is 0 or 1, where there is a unique choice.

[3] In fact, there are precisely $3! = 6$ different isomorphisms between two sets with three elements. This is equal to the number of automorphisms of a given three-element set (which in turn is equal to the order of the symmetric group on three letters), and more generally one has that the set of isomorphisms between two objects, denoted $\mathrm{Iso}(A, B)$, is a torsor for the automorphism group of *A,* $\mathrm{Aut}(A)$ and also a torsor for the automorphism group of *B.* In fact, automorphisms of an object are a key reason to be concerned with the distinction between isomorphism and equality, as demonstrated in the effect of change of basis on the identification of a vector space with its dual or with its double dual, as elaborated in the sequel.

[4] Being precise, the identification of the complex numbers with the real plane,

$$\mathbf{C} \cong \mathbf{R} \cdot 1 \oplus \mathbf{R} \cdot i = \mathbf{R}^2$$

depends on a choice of i; one can just as easily choose $(-i)$, , which yields a different identification—formally, complex conjugation is an automorphism—but in practice one often assumes that one has made such an identification.

29.7 References

[1] Awodey, Steve (2006). "Isomorphisms". *Category theory*. Oxford University Press. p. 11. ISBN 9780198568612.

[2] Vinberg, Ėrnest Borisovich (2003). *A Course in Algebra*. American Mathematical Society. p. 3. ISBN 9780821834138.

[3] Mazur 2007

29.8 Further reading

- Mazur, Barry (12 June 2007), *When is one thing equal to some other thing?* (PDF)

29.9 External links

- Hazewinkel, Michiel, ed. (2001), "Isomorphism", *Encyclopedia of Mathematics*, Springer, ISBN 978-1-55608-010-4

- Isomorphism at PlanetMath.org.

- Weisstein, Eric W., "Isomorphism", *MathWorld*.

Chapter 30

Field (mathematics)

This article is about fields in algebra. For fields in geometry, see Vector field. For other uses, see Field (disambiguation).

In abstract algebra, a **field** is a nonzero commutative division ring, or equivalently a ring whose nonzero elements form an abelian group under multiplication. As such it is an algebraic structure with notions of addition, subtraction, multiplication, and division satisfying the appropriate abelian group equations and distributive law. The most commonly used fields are the field of real numbers, the field of complex numbers, and the field of rational numbers, but there are also finite fields, fields of functions, algebraic number fields, p-adic fields, and so forth.

Any field may be used as the scalars for a vector space, which is the standard general context for linear algebra. The theory of field extensions (including Galois theory) involves the roots of polynomials with coefficients in a field; among other results, this theory leads to impossibility proofs for the classical problems of angle trisection and squaring the circle with a compass and straightedge, as well as a proof of the Abel–Ruffini theorem on the algebraic insolubility of quintic equations. In modern mathematics, the theory of fields (or **field theory**) plays an essential role in number theory and algebraic geometry.

As an algebraic structure, every field is a ring, but not every ring is a field. The most important difference is that fields allow for division (though not division by zero), while a ring need not possess multiplicative inverses; for example the integers form a ring, but $2x = 1$ has no solution in integers. Also, the multiplication operation in a field is required to be commutative. A ring in which division is possible but commutativity is not assumed (such as the quaternions) is called a *division ring* or *skew field*. (Historically, division rings were sometimes referred to as fields, while fields were called *commutative fields*.)

As a ring, a field may be classified as a specific type of integral domain, and can be characterized by the following (not exhaustive) chain of class inclusions:

> **Commutative rings ⊃ integral domains ⊃ integrally closed domains ⊃ unique factorization domains ⊃ principal ideal domains ⊃ Euclidean domains ⊃ fields ⊃ finite fields**

30.1 Definition and illustration

Intuitively, a field is a set F that is a commutative group with respect to two compatible operations, addition and multiplication (the latter excluding zero), with "compatible" being formalized by *distributivity*, and the caveat that the additive and the multiplicative identities are distinct ($0 \neq 1$).

The most common way to formalize this is by defining a *field* as a set together with two operations, usually called *addition* and *multiplication*, and denoted by + and ·, respectively, such that the following axioms hold; *subtraction* and *division* are defined in terms of the inverse operations of addition and multiplication:[note 1]

***Closure* of F under addition and multiplication** For all a, b in F, both $a + b$ and $a \cdot b$ are in F (or more formally, +

and $\cdot$ are binary operations on F).

Associativity **of addition and multiplication** For all a, b, and c in F, the following equalities hold: $a + (b + c) = (a + b) + c$ and $a \cdot (b \cdot c) = (a \cdot b) \cdot c$.

Commutativity **of addition and multiplication** For all a and b in F, the following equalities hold: $a + b = b + a$ and $a \cdot b = b \cdot a$.

Existence of additive and multiplicative *identity elements* There exists an element of F, called the *additive identity* element and denoted by 0, such that for all a in F, $a + 0 = a$. Likewise, there is an element, called the *multiplicative identity* element and denoted by 1, such that for all a in F, $a \cdot 1 = a$. To exclude the trivial ring, the additive identity and the multiplicative identity are required to be distinct.

Existence of *additive inverses* **and** *multiplicative inverses* For every a in F, there exists an element $-a$ in F, such that $a + (-a) = 0$. Similarly, for any a in F other than 0, there exists an element a^{-1} in F, such that $a \cdot a^{-1} = 1$. (The elements $a + (-b)$ and $a \cdot b^{-1}$ are also denoted $a - b$ and a/b, respectively.) In other words, *subtraction* and *division* operations exist.

Distributivity **of multiplication over addition** For all a, b and c in F, the following equality holds: $a \cdot (b + c) = (a \cdot b) + (a \cdot c)$.

A field is therefore an algebraic structure ⟨F, +, $\cdot$, $-$, $^{-1}$, 0, 1⟩; of type ⟨2, 2, 1, 1, 0, 0⟩, consisting of two abelian groups:

- F under +, $-$, and 0;

- $F \setminus \{0\}$ under $\cdot$, $^{-1}$, and 1, with $0 \neq 1$,

with $\cdot$ distributing over +.[1]

30.1.1 First example: rational numbers

A simple example of a field is the field of rational numbers, consisting of numbers which can be written as fractions a/b, where a and b are integers, and $b \neq 0$. The additive inverse of such a fraction is simply $-a/b$, and the multiplicative inverse (provided that $a \neq 0$) is b/a. To see the latter, note that

$$\frac{b}{a} \cdot \frac{a}{b} = \frac{ba}{ab} = 1.$$

The abstractly required field axioms reduce to standard properties of rational numbers, such as the law of distributivity

$$\frac{a}{b} \cdot \left(\frac{c}{d} + \frac{e}{f} \right)$$

$$= \frac{a}{b} \cdot \left(\frac{c}{d} \cdot \frac{f}{f} + \frac{e}{f} \cdot \frac{d}{d} \right)$$

$$= \frac{a}{b} \cdot \left(\frac{cf}{df} + \frac{ed}{fd} \right) = \frac{a}{b} \cdot \frac{cf + ed}{df}$$

$$= \frac{a(cf + ed)}{bdf} = \frac{acf}{bdf} + \frac{aed}{bdf} = \frac{ac}{bd} + \frac{ae}{bf}$$

$$= \frac{a}{b} \cdot \frac{c}{d} + \frac{a}{b} \cdot \frac{e}{f},$$

or the law of commutativity and law of associativity.

30.1.2   Second example: a field with four elements

In addition to familiar number systems such as the rationals, there are other, less immediate examples of fields. The following example is a field consisting of four elements called O, I, A and B. The notation is chosen such that O plays the role of the additive identity element (denoted 0 in the axioms), and I is the multiplicative identity (denoted 1 above). One can check that all field axioms are satisfied. For example:

$$A \cdot (B + A) = A \cdot I = A, \text{ which equals } A \cdot B + A \cdot A = I + B = A, \text{ as required by the distributivity.}$$

The above field is called a finite field with four elements, and can be denoted $\mathbf{F}_4$. Field theory is concerned with understanding the reasons for the existence of this field, defined in a fairly ad-hoc manner, and describing its inner structure. For example, from a glance at the multiplication table, it can be seen that any non-zero element (i.e., I, A, and B) is a power of A: $A = A^1$, $B = A^2 = A \cdot A$, and finally $I = A^3 = A \cdot A \cdot A$. This is not a coincidence, but rather one of the starting points of a deeper understanding of (finite) fields.

30.1.3   Alternative axiomatizations

As with other algebraic structures, there exist alternative axiomatizations. Because of the relations between the operations, one can alternatively axiomatize a field by explicitly assuming that there are four binary operations (add, subtract, multiply, divide) with axioms relating these, or (by functional decomposition) in terms of two binary operations (add and multiply) and two unary operations (additive inverse and multiplicative inverse), or other variants.

The usual axiomatization in terms of the two operations of addition and multiplication is brief and allows the other operations to be defined in terms of these basic ones, but in other contexts, such as topology and category theory, it is important to include all operations as explicitly given, rather than implicitly defined (compare topological group). This is because without further assumptions, the implicitly defined inverses may not be continuous (in topology), or may not be able to be defined (in category theory). Defining an inverse requires that one is working with a set, not a more general object.

For a very economical axiomatization of the field of real numbers, whose primitives are merely a set $\mathbf{R}$ with $1 \in \mathbf{R}$, addition, and a binary relation, "<". See Tarski's axiomatization of the reals.

30.2   Related algebraic structures

The axioms imposed above resemble the ones familiar from other algebraic structures. For example, the existence of the binary operation "·", together with its commutativity, associativity, (multiplicative) identity element and inverses are precisely the axioms for an abelian group. In other words, for any field, the subset of nonzero elements $F \setminus \{0\}$, also often denoted $F^\times$, is an abelian group $(F^\times, \cdot)$ usually called multiplicative group of the field. Likewise $(F, +)$ is an abelian group. The structure of a field is hence the same as specifying such two group structures (on the same set), obeying the distributivity.

Important other algebraic structures such as rings arise when requiring only part of the above axioms. For example, if the requirement of commutativity of the multiplication operation · is dropped, one gets structures usually called division rings or *skew fields*.

30.2.1   Remarks

By elementary group theory, applied to the abelian groups $(F^\times, \cdot)$, and $(F, +)$, the additive inverse $-a$ and the multiplicative inverse a^{-1} are uniquely determined by a.

Similar direct consequences from the field axioms include

$$-(a \cdot b) = (-a) \cdot b = a \cdot (-b), \text{ in particular } -a = (-1) \cdot a$$

as well as

$$a \cdot 0 = 0.$$

Both can be shown by replacing b or c with 0 in the distributive property.

30.3 History

The concept of *field* was used implicitly by Niels Henrik Abel and Évariste Galois in their work on the solvability of polynomial equations with rational coefficients of degree five or higher.

In 1857, Karl von Staudt published his Algebra of Throws which provided a geometric model satisfying the axioms of a field.[2] This construction has been frequently recalled as a contribution to the foundations of mathematics.

In 1871, Richard Dedekind introduced, for a set of real or complex numbers which is closed under the four arithmetic operations, the German word *Körper*, which means "body" or "corpus" (to suggest an organically closed entity),[3] hence the common use of the letter K to denote a field. He also defined rings (then called *order* or *order-modul*), but the term *"a ring"* (*Zahlring*) was invented by Hilbert.[4] In 1893, Eliakim Hastings Moore called the concept "field" in English.[5][6]

In 1881, Leopold Kronecker defined what he called a "domain of rationality", which is indeed a field of polynomials in modern terms. In 1893, Heinrich M. Weber gave the first clear definition of an abstract field.[7] In 1910, Ernst Steinitz published the very influential paper *Algebraische Theorie der Körper* (English: Algebraic Theory of Fields).[8] In this paper he axiomatically studies the properties of fields and defines many important field theoretic concepts like prime field, perfect field and the transcendence degree of a field extension.

Emil Artin developed the relationship between groups and fields in great detail from 1928 through 1942.

30.4 Examples

30.4.1 Rationals and algebraic numbers

The field of rational numbers **Q** has been introduced above. A related class of fields very important in number theory are algebraic number fields. We will first give an example, namely the field **Q**(ζ) consisting of numbers of the form

$$a + b\zeta$$

with $a, b \in \mathbf{Q}$, where ζ is a primitive third root of unity, i.e., a complex number satisfying $\zeta^3 = 1$, $\zeta \neq 1$. This field extension can be used to prove a special case of Fermat's last theorem, which asserts the non-existence of rational nonzero solutions to the equation

$$x^3 + y^3 = z^3.$$

In the language of field extensions detailed below, **Q**(ζ) is a field extension of degree 2. Algebraic number fields are by definition finite field extensions of **Q**, that is, fields containing **Q** having finite dimension as a **Q**-vector space.

30.4.2 Reals, complex numbers, and *p*-adic numbers

Take the real numbers **R**, under the usual operations of addition and multiplication. When the real numbers are given the usual ordering, they form a *complete ordered field*; it is this structure which provides the foundation for most formal treatments of calculus.

The complex numbers **C** consist of expressions

$$a + bi$$

where i is the imaginary unit, i.e., a (non-real) number satisfying $i^2 = -1$. Addition and multiplication of real numbers are defined in such a way that all field axioms hold for **C**. For example, the distributive law enforces

$$(a + bi) \cdot (c + di) = ac + bci + adi + bdi^2, \text{ which equals } ac - bd + (bc + ad)i.$$

The real numbers can be constructed by completing the rational numbers, i.e., filling the "gaps": for example $\sqrt{2}$ is such a gap. By a formally very similar procedure, another important class of fields, the field of p-adic numbers $\mathbf{Q}p$ is built. It is used in number theory and p-adic analysis.

Hyperreal numbers and superreal numbers extend the real numbers with the addition of infinitesimal and infinite numbers.

30.4.3 Constructible numbers

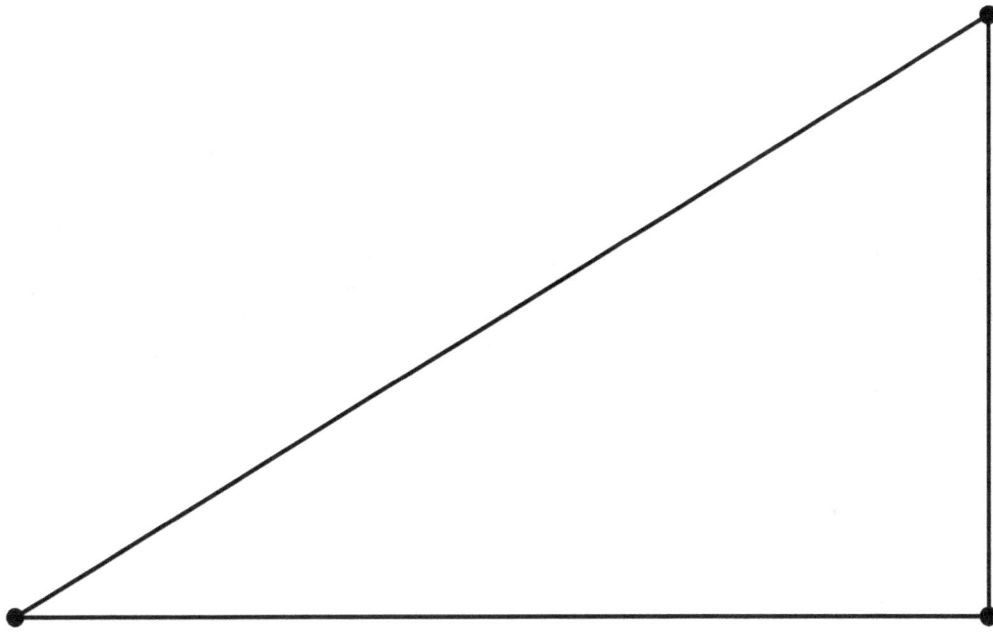

Given 0, 1, r_1 and r_2, the construction yields $r_1 \cdot r_2$

In antiquity, several geometric problems concerned the (in)feasibility of constructing certain numbers with compass and straightedge. For example it was unknown to the Greeks that it is in general impossible to trisect a given angle. Using the field notion and field theory allows these problems to be settled. To do so, the field of constructible numbers is considered. It contains, on the plane, the points 0 and 1, and all complex numbers that can be constructed from these two by a finite number of construction steps using only compass and straightedge. This set, endowed with the usual addition and multiplication of complex numbers does form a field. For example, multiplying two (real) numbers r_1 and r_2 that have already been constructed can be done using construction at the right, based on the intercept theorem. This way, the obtained field F contains all rational numbers, but is bigger than **Q**, because for any $f \in F$, the square root of f is also a constructible number.

A closely related concept is that of a Euclidean field, namely an ordered field whose positive elements are closed under square root. The real constructible numbers form the least Euclidean field, and the Euclidean fields are precisely the ordered extensions thereof.

30.4.4 Finite fields

Main article: Finite field

Finite fields (also called *Galois fields*) are fields with finitely many elements. The above introductory example $\mathbf{F}_4$ is a field with four elements. $\mathbf{F}_2$ consists of two elements, 0 and 1. This is the smallest field, because by definition a field has at least two distinct elements $1 \neq 0$. Interpreting the addition and multiplication in this latter field as XOR and AND operations, this field finds applications in computer science, especially in cryptography and coding theory.

In a finite field there is necessarily an integer n such that $1 + 1 + \cdots + 1$ (n repeated terms) equals 0. It can be shown that the smallest such n must be a prime number, called the *characteristic* of the field. If a (necessarily infinite) field has the property that $1 + 1 + \cdots + 1$ is never zero, for any number of summands, such as in $\mathbf{Q}$, for example, the characteristic is said to be zero.

A basic class of finite fields are the fields $\mathbf{F}p$ with p elements (p a prime number):

$$\mathbf{F}p = \mathbf{Z}/p\mathbf{Z} = \{0, 1, ..., p - 1\},$$

where the operations are defined by performing the operation in the set of integers $\mathbf{Z}$, dividing by p and taking the remainder; see modular arithmetic. A field K of characteristic p necessarily contains $\mathbf{F}p$,[9] and therefore may be viewed as a vector space over $\mathbf{F}p$, of finite dimension if K is finite. Thus a finite field K has prime power order, i.e., K has $q = p^n$ elements (where $n > 0$ is the number of elements in a basis of K over $\mathbf{F}p$). By developing more field theory, in particular the notion of the splitting field of a polynomial f over a field K, which is the smallest field containing K and all roots of f, one can show that two finite fields with the same number of elements are isomorphic, i.e., there is a one-to-one mapping of one field onto the other that preserves multiplication and addition. Thus we may speak of *the* finite field with q elements, usually denoted by $\mathbf{F}q$ or GF(q).

30.4.5 Archimedean fields

Main article: Archimedean field

An Archimedean field is an ordered field such that for each element there exists a finite expression $1 + 1 + \cdots + 1$ whose value is greater than that element, that is, there are no infinite elements. Equivalently, the field contains no infinitesimals; or, the field is isomorphic to a subfield of the reals. A necessary condition for an ordered field to be complete is that it be Archimedean, since in any non-Archimedean field there is neither a greatest infinitesimal nor a least positive rational, whence the sequence 1/2, 1/3, 1/4, ..., every element of which is greater than every infinitesimal, has no limit. (And since every proper subfield of the reals also contains such gaps, up to isomorphism the reals form the unique complete ordered field.)

30.4.6 Field of functions

Given a geometric object X, one can consider functions on such objects. Adding and multiplying them pointwise, i.e., $(f \cdot g)(x) = f(x) \cdot g(x)$ this leads to a field. However, for having multiplicative inverses, one has to consider partial functions, which, almost everywhere, are defined and have a non-zero value.

If X is an algebraic variety over a field F, then the rational functions $X \to F$ form a field, the function field of X. This field consists of the functions that are defined and are the quotient of two polynomial functions outside some subvariety. Likewise, if X is a Riemann surface, then the meromorphic functions $S \to \mathbf{C}$ form a field. Under certain circumstances, namely when S is compact, S can be reconstructed from this field.

30.4.7 Local and global fields

Another important distinction in the realm of fields, especially with regard to number theory, are local fields and global fields. Local fields are completions of global fields at a given place. For example, $\mathbf{Q}$ is a global field, and the attached local fields are $\mathbf{Q}p$ and $\mathbf{R}$ (Ostrowski's theorem). Algebraic number fields and function fields over $\mathbf{F}q$ are further global fields. Studying arithmetic questions in global fields may sometimes be done by looking at the corresponding questions locally—this technique is called local-global principle.

30.5 Some first theorems

- Every finite subgroup of the multiplicative group $F^{\times}$ is cyclic. This applies in particular to $\mathbf{F}q^{\times}$, it is cyclic of order $q - 1$. In the introductory example, a generator of $\mathbf{F}_4^{\times}$ is the element A.

- A integral domain is a field if and only if it has no ideals except $\{0\}$ and itself. Equivalently, an integral domain is a field if and only if its Krull dimension is 0.

- Isomorphism extension theorem

30.6 Constructing fields

30.6.1 Closure operations

Assuming the axiom of choice, for every field F, there exists a field F, called the algebraic closure of F, which contains F, is algebraic over F, which means that any element x of F satisfies a polynomial equation

$$fnx^n + fn_{-1}x^{n-1} + \cdots + f_1 x + f_0 = 0, \text{ with coefficients } fn, \dots, f_0 \in F,$$

and is algebraically closed, i.e., any such polynomial does have at least one solution in F. The algebraic closure is unique up to isomorphism inducing the identity on F. However, in many circumstances in mathematics, it is not appropriate to treat F as being uniquely determined by F, since the isomorphism above is not itself unique. In these cases, one refers to such a F as *an* algebraic closure of F. A similar concept is the separable closure, containing all roots of separable polynomials, instead of all polynomials.

For example, if $F = \mathbf{Q}$, the algebraic closure $\mathbf{Q}$ is also called *field of algebraic numbers*. The field of algebraic numbers is an example of an algebraically closed field of characteristic zero; as such it satisfies the same first-order sentences as the field of complex numbers $\mathbf{C}$.

In general, all algebraic closures of a field are isomorphic. However, there is in general no preferable isomorphism between two closures. Likewise for separable closures.

30.6.2 Subfields and field extensions

A *subfield* is, informally, a small field contained in a bigger one. Formally, a subfield E of a field F is a subset containing 0 and 1, closed under the operations $+$, $-$, $\cdot$ and multiplicative inverses and with its own operations defined by restriction. For example, the real numbers contain several interesting subfields: the real algebraic numbers, the computable numbers and the rational numbers are examples.

The notion of field extension lies at the heart of field theory, and is crucial to many other algebraic domains. A field extension F / E is simply a field F and a subfield $E \subset F$. Constructing such a field extension F / E can be done by "adding new elements" or *adjoining elements* to the field E. For example, given a field E, the set $F = E(X)$ of rational functions, i.e., equivalence classes of expressions of the kind

$$\frac{p(X)}{q(X)},$$

where $p(X)$ and $q(X)$ are polynomials with coefficients in E, and q is not the zero polynomial, forms a field. This is the simplest example of a transcendental extension of E. It also is an example of a domain (the ring of polynomials E in this case) being embedded into its field of fractions $E(X)$.

The ring of formal power series $E[[X]]$ is also a domain, and again the (equivalence classes of) fractions of the form $p(X)/q(X)$ where p and q are elements of $E[[X]]$ form the field of fractions for $E[[X]]$. This field is actually the ring of Laurent series over the field E, denoted $E((X))$.

In the above two cases, the added symbol X and its powers did not interact with elements of E. It is possible however that the adjoined symbol may interact with E. This idea will be illustrated by adjoining an element to the field of real numbers $\mathbf{R}$. As explained above, $\mathbf{C}$ is an extension of $\mathbf{R}$. $\mathbf{C}$ can be obtained from $\mathbf{R}$ by adjoining the imaginary symbol i which satisfies $i^2 = -1$. The result is that $\mathbf{R}[i]=\mathbf{C}$. This is different from adjoining the symbol X to $\mathbf{R}$, because in that case, the powers of X are all distinct objects, but here, $i^2=-1$ is actually an element of $\mathbf{R}$.

Another way to view this last example is to note that i is a zero of the polynomial $p(X) = X^2 + 1$. The quotient ring $R[X]/(X^2+1)$ can be mapped onto $\mathbf{C}$ using the map $\overline{a+bX} \rightarrow a+ib$. Since the ideal (X^2+1) is generated by a polynomial irreducible over $\mathbf{R}$, the ideal is maximal, hence the quotient ring is a field. This nonzero ring map from the quotient to $\mathbf{C}$ is necessarily an isomorphism of rings.

The above construction generalises to any irreducible polynomial in the polynomial ring $E[X]$, i.e., a polynomial $p(X)$ that cannot be written as a product of non-constant polynomials. The quotient ring $F = E[X] / (p(X))$, is again a field.

Alternatively, constructing such field extensions can also be done, if a bigger container is already given. Suppose given a field E, and a field G containing E as a subfield, for example G could be the algebraic closure of E. Let x be an element of G not in E. Then there is a smallest subfield of G containing E and x, denoted $F = E(x)$ and called *field extension F / E generated by x in G*.[10] Such extensions are also called *simple extensions*. Many extensions are of this type; see the primitive element theorem. For instance, $\mathbf{Q}(i)$ is the subfield of $\mathbf{C}$ consisting of all numbers of the form $a + bi$ where both a and b are rational numbers.

One distinguishes between extensions having various qualities. For example, an extension K of a field k is called *algebraic*, if every element of K is a root of some polynomial with coefficients in k. Otherwise, the extension is called *transcendental*. The aim of Galois theory is the study of *algebraic extensions* of a field.

30.6.3   Rings vs fields

Adding multiplicative inverses to an integral domain R yields the field of fractions of R. For example, the field of fractions of the integers $\mathbf{Z}$ is just $\mathbf{Q}$. Also, the field $F(X)$ is the quotient field of the ring of polynomials $F[X]$.

Another method to obtain a field from a commutative ring R is taking the quotient R / m, where m is any maximal ideal of R. The above construction of $F = E[X] / (p(X))$, is an example, because the irreducibility of the polynomial $p(X)$ is equivalent to the maximality of the ideal generated by this polynomial. Another example are the finite fields $\mathbf{F}p = \mathbf{Z} / p\mathbf{Z}$.

30.6.4   Ultraproducts

If I is an index set, U is an ultrafilter on I, and Fi is a field for every i in I, the ultraproduct of the Fi with respect to U is a field.

For example, a non-principal ultraproduct of finite fields is a pseudo finite field; i.e., a PAC field having exactly one extension of any degree.

30.7   Galois theory

Main article: Galois theory

Galois theory aims to study the algebraic extensions of a field by studying the symmetry in the arithmetic operations of addition and multiplication. The fundamental theorem of Galois theory shows that there is a strong relation between the structure of the symmetry group and the set of algebraic extensions.

In the case where F / E is a finite (Galois) extension, Galois theory studies the algebraic extensions of E that are subfields of F. Such fields are called intermediate extensions. Specifically, the Galois group of F over E, denoted $\mathrm{Gal}(F/E)$, is the group of field automorphisms of F that are trivial on E (i.e., the bijections $\sigma : F \to F$ that preserve addition and multiplication and that send elements of E to themselves), and the fundamental theorem of Galois theory states that there is a one-to-one correspondence between subgroups of $\mathrm{Gal}(F/E)$ and the set of intermediate extensions of the extension F/E. The theorem, in fact, gives an explicit correspondence and further properties.

To study all (separable) algebraic extensions of E at once, one must consider the absolute Galois group of E, defined as the Galois group of the separable closure, E^{sep}, of E over E (i.e., $\mathrm{Gal}(E^{\mathrm{sep}}/E)$). It is possible that the degree of this extension is infinite (as in the case of $E = \mathbf{Q}$). It is thus necessary to have a notion of Galois group for an infinite algebraic extension. The Galois group in this case is obtained as a "limit" (specifically an inverse limit) of the Galois groups of the finite Galois extensions of E. In this way, it acquires a topology.[note 2] The fundamental theorem of Galois theory can be generalized to the case of infinite Galois extensions by taking into consideration the topology of the Galois group, and in the case of E^{sep}/E it states that there this a one-to-one correspondence between *closed* subgroups of $\mathrm{Gal}(E^{\mathrm{sep}}/E)$ and the set of all separable algebraic extensions of E (technically, one only obtains those separable algebraic extensions of E that occur as subfields of the *chosen* separable closure E^{sep}, but since all separable closures of E are isomorphic, choosing a different separable closure would give the same Galois group and thus an "equivalent" set of algebraic extensions).

30.8   Generalizations

There are also proper classes with field structure, which are sometimes called **Fields**, with a capital F:

- The surreal numbers form a Field containing the reals, and would be a field except for the fact that they are a proper class, not a set.

- The nimbers form a Field. The set of nimbers with birthday smaller than 2^{2^n}, the nimbers with birthday smaller than any infinite cardinal are all examples of fields.

In a different direction, differential fields are fields equipped with a derivation. For example, the field $\mathbf{R}(X)$, together with the standard derivative of polynomials forms a differential field. These fields are central to differential Galois theory. Exponential fields, meanwhile, are fields equipped with an exponential function that provides a homomorphism between the additive and multiplicative groups within the field. The usual exponential function makes the real and complex numbers exponential fields, denoted $\mathbf{R}_{\mathrm{exp}}$ and $\mathbf{C}_{\mathrm{exp}}$ respectively.

Generalizing in a more categorical direction yields the field with one element and related objects.

30.8.1   Exponentiation

One does not in general study generalizations of fields with *three* binary operations. The familiar addition/subtraction, multiplication/division, exponentiation/root-extraction/logarithm operations from the natural numbers to the reals, each built up in terms of iteration of the last, mean that generalizing exponentiation as a binary operation is tempting, but has generally not proven fruitful; instead, an exponential field assumes a unary exponential function from the additive group to the multiplicative group, not a partially defined binary function. Note that the exponential operation of a^b is neither associative nor commutative, nor has a unique inverse (± 2 are both square roots of 4, for instance), unlike addition and multiplication, and further is not defined for many pairs—for example, $(-1)^{1/2} = \sqrt{-1}$ does not define a single number.

These all show that even for rational numbers exponentiation is not nearly as well-behaved as addition and multiplication, which is why one does not in general axiomatize exponentiation.

30.9 Applications

The concept of a field is of use, for example, in defining vectors and matrices, two structures in linear algebra whose components can be elements of an arbitrary field.

Finite fields are used in number theory, Galois theory, cryptography, coding theory and combinatorics; and again the notion of algebraic extension is an important tool.

30.10 See also

- Category of fields

- Glossary of field theory for more definitions in field theory.

- Heyting field

- Lefschetz principle

- Puiseux series

- Ring

- Vector space

- Vector spaces without fields

30.11 Notes

[1] That is, the axiom for addition only assumes a binary operation $+: F \times F \to F$, $a, b \mapsto a + b$. The axiom of inverse allows one to define a unary operation $-: F \to F$ $a \mapsto -a$ that sends an element to its negative (its additive inverse); this is not taken as given, but is implicitly defined in terms of addition as " $-a$ is the unique b such that $a + b = 0$ ", "implicitly" because it is defined in terms of solving an equation—and one then defines the binary operation of subtraction, also denoted by "−", as $-: F \times F \to F$, $a, b \mapsto a - b := a + (-b)$ in terms of addition and additive inverse. In the same way, one defines the binary operation of division $\div$ in terms of the assumed binary operation of multiplication and the implicitly defined operation of "reciprocal" (multiplicative inverse).

[2] As an inverse limit of finite discrete groups, it is equipped with the profinite topology, making it a profinite topological group

30.12 References

[1] Wallace, D A R (1998) *Groups, Rings, and Fields*, SUMS. Springer-Verlag: 151, Th. 2.

[2] Karl Georg Christian v. Staudt, *Beiträge zur Geometrie der Lage* (Contributions to the Geometry of Position), volume 2 (Nürnberg, (Germany): Bauer and Raspe, 1857). See: "Summen von Würfen" (sums of throws), pp. 166-171 ; "Produckte aus Würfen" (products of throws), pp. 171-176 ; "Potenzen von Würfen" (powers of throws), pp. 176-182.

[3] Peter Gustav Lejeune Dirichlet with R. Dedekind, *Vorlesungen über Zahlentheorie von P. G. Lejeune Dirichlet* (Lectures on Number Theory by P.G. Lejeune Dirichlet), 2nd ed., volume 1 (Braunschweig, Germany: Friedrich Vieweg und Sohn, 1871), p. 424. From page 424: *"Unter einem* Körper *wollen wir jedes System von unendlich vielen reellen oder complexen Zahlen verstehen, welches in sich so abgeschlossen und vollständig ist, dass die Addition, Subtraction, Multiplication und Division von je zwei dieser Zahlen immer wieder eine Zahl desselben Systems hervorbringt."* (By a "field" we will understand any system of

infinitely many real or complex numbers, which is so closed and complete that the addition, subtraction, multiplication, and division of any two of these numbers always again produces a number of the same system.)

[4] J J O'Connor and E F Robertson, *The development of Ring Theory*, September 2004.

[5] Moore, E. Hastings (1893), "A doubly-infinite system of simple groups", *Bulletin of the New York Mathematical Society* **3** (3): 73–78, doi:10.1090/S0002-9904-1893-00178-X, JFM 25.0198.01. From page 75: "Such a system of *s* marks [i.e., a finite field with *s* elements] we call a *field of order s*."

[6] *Earliest Known Uses of Some of the Words of Mathematics (F)*

[7] Fricke, Robert; Weber, Heinrich Martin (1924), *Lehrbuch der Algebra*, Vieweg, JFM 50.0042.03

[8] Steinitz, Ernst (1910), "Algebraische Theorie der Körper", *Journal für die reine und angewandte Mathematik* **137**: 167–309, doi:10.1515/crll.1910.137.167, ISSN 0075-4102, JFM 41.0445.03

[9] Jacobson (2009), p. 213

[10] Jacobson (2009), p. 213

30.13   Sources

- Artin, Michael (1991), *Algebra*, Prentice Hall, ISBN 978-0-13-004763-2, especially Chapter 13

- Allenby, R.B.J.T. (1991), *Rings, Fields and Groups*, Butterworth-Heinemann, ISBN 978-0-340-54440-2

- Blyth, T.S.; Robertson, E. F. (1985), *Groups, rings and fields: Algebra through practice*, Cambridge University Press. See especially Book 3 (ISBN 0-521-27288-2) and Book 6 (ISBN 0-521-27291-2).

- Jacobson, Nathan (2009), *Basic algebra* **1** (2nd ed.), Dover, ISBN 978-0-486-47189-1

- James Ax (1968), *The elementary theory of finite fields*, Ann. of Math. (2), **88**, 239–271

30.14   External links

- Hazewinkel, Michiel, ed. (2001), "Field", *Encyclopedia of Mathematics*, Springer, ISBN 978-1-55608-010-4

- Field Theory Q&A

- Fields at ProvenMath definition and basic properties.

- Field at PlanetMath.org.

30.15 Text and image sources, contributors, and licenses

30.15.1 Text

- **Lie group** *Source:* https://en.wikipedia.org/wiki/Lie_group?oldid=670427056 *Contributors:* AxelBoldt, Zundark, Josh Grosse, XJaM, Miguel~enwiki, Stevertigo, Xavic69, Michael Hardy, TakuyaMurata, GTBacchus, Looxix~enwiki, Barak~enwiki, Charles Matthews, Dysprosia, Jitse Niesen, Zoicon5, David Shay, Itai, Phys, Josh Cherry, Saaska, Tobias Bergemann, Weialawaga~enwiki, Tosha, Giftlite, JamesMLane, BenFrantzDale, Lethe, Fropuff, Wgmccallum, Jason Quinn, Bobblewik, DefLog~enwiki, Lockeownzj00, Beland, Pmanderson, Abdull, Dablaze, MuDavid, Paul August, ChrisJ, Bender235, Tompw, Rgdboer, Kwamikagami, Shanes, Cherlin, Msh210, PAR, Alex Varghese, Oleg Alexandrov, Zntrip, Joriki, Linas, Dzordzm, Isnow, SDC, AnmaFinotera, Frankie1969, Graham87, Porcher, Rjwilmsi, NatusRoma, MarSch, Salix alba, HappyCamper, R.e.b., VKokielov, BMF81, Masnevets, Chobot, Algebraist, Wavelength, Hillman, RussBot, Michael Slone, KSmrq, Archelon, Buster79, Arkapravo, Smaines, Orthografer, Ekeb, Kier07, Pred, RodVance, JDspeeder1, SmackBot, Incnis Mrsi, Tom Lougheed, FlashSheridan, Davewild, Mhss, Kmarinas86, Bluebot, Badger014, Silly rabbit, DHN-bot~enwiki, Bears16, Akriasas, KeithB, Lambiam, Ninte, Siva1979, John, Ulner, Jim.belk, Michael Kinyon, Inquisitus, Mathchem271828, Rschwieb, Krasnoludek, Yggdrasil014, CRGreathouse, CBM, Logical2u, Myasuda, Kupirijo, MotherFunctor, Dr.enh, Xantharius, Thijs!bot, Headbomb, JustAGal, RichardVeryard, RobHar, Salgueiro~enwiki, Dougher, Len Raymond, JAnDbot, Deflective, Unifey~enwiki, Homeworlds, Magioladitis, Bongwarrior, Cmelby, WhatamIdoing, Sullivan.t.j, David Eppstein, The Real Marauder, Benjamin.friedrich, David J Wilson, Jesper Carlstrom, Maproom, TomyDuby, Rocket71048576, Pidara, Fylwind, Dorftrottel, Lseixas, Borat fan, Cuzkatzimhut, Trevorgoodchild, JohnBlackburne, Ndbrian1, James.r.a.gray, Hesam7, Geometry guy, Jmath666, Eubulides, Brian Huffman, Genuine0legend, Drorata, Arcfrk, Smylei, Oscarbaltazar, YohanN7, JackSchmidt, S2000magician, Beastinwith, Mr. Stradivarius, Deciwill, Sidiropo, Leontios, Heckledpie, Cacadril, SchreiberBike, Marc van Leeuwen, MystBot, Addbot, Topology Expert, LaaknorBot, Ozob, Tanath, Tide rolls, Luckas-bot, Yobot, Ht686rg90, Niout, Amirobot, AnomieBOT, Citation bot, ArthurBot, Br77rino, Kaoru Itou, FrescoBot, Anterior1, Sławomir Biały, RedBot, Tinfoilcat, EmausBot, KbReZiE 12, Darkfight, Slawekb, Suslindisambiguator, Maschen, Zueignung, Anita5192, ClueBot NG, Mgvongoeden, Kasirbot, Helpful Pixie Bot, Daviddwd, BG19bot, CitationCleanerBot, Fraisière, NotWith, MathKnight-at-TAU, Suhagja, Brirush, CsDix, Sol1, Blackbombchu, Pwm86, Abitslow, Cbartondock, Victoryhuy, KasparBot and Anonymous: 110

- **Table of Lie groups** *Source:* https://en.wikipedia.org/wiki/Table_of_Lie_groups?oldid=670292128 *Contributors:* Zundark, TakuyaMurata, Arpingstone, Fropuff, Almit39, Rich Farmbrough, Paul August, Rgdboer, John Vandenberg, R. S. Shaw, Pschemp, Oleg Alexandrov, Linas, Salix alba, R.e.b., Mathbot, Bgwhite, KSmrq, Molinagaray, Nbarth, Vanished User 0001, Jim.belk, Sebastian Klein, Syrcatbot, Mets501, RobHar, David Eppstein, R'n'B, Pomte, Cbigorgne, Mr. Granger, Yasmar, Addbot, Matěj Grabovský, Luckas-bot, Niout, Ildeguz, Dieterich~enwiki, CsDix, Master Lenman and Anonymous: 8

- **Simple Lie group** *Source:* https://en.wikipedia.org/wiki/Simple_Lie_group?oldid=664521608 *Contributors:* Zundark, Nonenmac, Michael Hardy, TakuyaMurata, Charles Matthews, Phys, Giftlite, Fropuff, Cambyses, Sigfpe, Tomruen, MIT Trekkie, Oleg Alexandrov, GregorB, Rjwilmsi, Salix alba, R.e.b., Buster79, Kier07, SmackBot, Bluebot, Nbarth, Jim.belk, MOBle, CRGreathouse, Myasuda, Secular mind~enwiki, Thijs!bot, Escarbot, Ludvikus, R'n'B, TomyDuby, Gill110951, Red Act, Michael H 34, Arcfrk, Shadrack-dva, Addbot, Discrepancy, Niout, AnomieBOT, Omnipaedista, Erik9bot, GoingBatty, Helpful Pixie Bot, CsDix and Anonymous: 16

- **Lie algebra** *Source:* https://en.wikipedia.org/wiki/Lie_algebra?oldid=668578572 *Contributors:* AxelBoldt, Zundark, Miguel~enwiki, Michael Hardy, Wshun, Joel Koerwer, TakuyaMurata, Suisui, Kragen, Rossami, Iorsh, Loren Rosen, Charles Matthews, Dysprosia, Michael Larsen, Grendelkhan, Phys, Tobias Bergemann, David Gerard, Weialawaga~enwiki, Tosha, Giftlite, BenFrantzDale, Lethe, Fropuff, Curps, Jeremy Henty, Jason Quinn, Python eggs, Chameleon, DefLog~enwiki, CryptoDerk, CSTAR, Pyrop, Guanabot, Pj.de.bruin, Vsmith, Gauge, Pt, Kwamikagami, Wood Thrush, Reinyday, Foobaz, Msh210, Arthena, Spangineer, Dirac1933, Drbreznjev, Oleg Alexandrov, Linas, Isnow, BD2412, NatusRoma, MarSch, Mathbot, Margosbot~enwiki, RexNL, Masnevets, YurikBot, Wavelength, Hairy Dude, Michael Slone, Lenthe, Stephenb, Grubber, Trovatore, Asimy, Crasshopper, Curpsbot-unicodify, Sbyrnes321, SmackBot, Incnis Mrsi, Grokmoo, Kmarinas86, Bluebot, Silly rabbit, Nbarth, Thomas Bliem, Chlewbot, BlackFingolfin, Noegenesis, Rschwieb, AlainD, Harold f, CmdrObot, Shirulashem, Headbomb, Second Quantization, Dachande, RobHar, B-80, Jrw@pobox.com, Deflective, Englebert, Vanish2, R'n'B, Bogey97, Maurice Carbonaro, Supermanifold, Policron, Fylwind, Cuzkatzimhut, VolkovBot, JohnBlackburne, LokiClock, Ndbrian1, Hesam7, Geometry guy, Drorata, Arcfrk, StevenJohnston, YohanN7, SieBot, Stca74, Jenny Lam, Paolo.dL, JackSchmidt, Mr. Stradivarius, Fatchat, Veromies, JP.Martin-Flatin, Count Truthstein, Addbot, Roentgenium111, Lightbot, Legobot, Luckas-bot, Yobot, Niout, Jason Recliner, Esq., Delilahblue, AnomieBOT, Twri, SassoBot, Kaoru Itou, D'ohBot, Darij, Juniuswikiae, Prtmrz, Rausch, Jkock, Adam cohenus, TobeBot, Lotje, Doctor Zook, Slawekb, Quondum, Mikhail Ryazanov, ClueBot NG, Dd314, Teika kazura, Walterpfeifer, Pfeiferwalter, IkamusumeFan, Flbsimas, Deltahedron, Saung Tadashi, Mark L MacDonald, Danielbrice, Enyokoyama, CsDix, 314Username, Forgetfulfunctor00, CaptainLama, KasparBot, Texnico and Anonymous: 91

- **Differentiable manifold** *Source:* https://en.wikipedia.org/wiki/Differentiable_manifold?oldid=669483065 *Contributors:* Toby Bartels, Michael Hardy, Ixfd64, Nikai, Rmilson, Charles Matthews, Modulatum, Aetheling, Tobias Bergemann, Giftlite, MFNickster, Rich Farmbrough, Ben Standeven, Gauge, MBisanz, Bobo192, C S, Pearle, Tsirel, Atlant, PAR, Culix, SteinbDJ, Gene Nygaard, GiovanniS, Oleg Alexandrov, Joriki, Linas, Guardian of Light, SeventyThree, BD2412, Jmh2o, Rjwilmsi, MarSch, Salix alba, Strobilomyces, Mathbot, BradBeattie, YurikBot, Ugha, Hairy Dude, 4C~enwiki, RussBot, Gaius Cornelius, Gwaihir, SmackBot, Selfworm, David Farris, Silly rabbit, Nbarth, Drphilharmonic, Vina-iwbot~enwiki, Lambiam, Ao1977, JMK, Jambaugh, Yggdrasil014, Geomprof, CmdrObot, Bobblehead, RobHar, Komponisto, Turgidson, Danielegrandini, Sullivan.t.j, David Eppstein, Lizhuoru, Maurice Carbonaro, TomyDuby, Policron, GregWoodhouse, Geometry guy, Arcfrk, Rybu, Tcamps42, Julie.larson, Bananastalktome, Paolo.dL, Dmoskovich, Iknowyourider, Martarius, MABadger, UKoch, Sun Creator, Brews ohare, SchreiberBike, Addbot, DOI bot, Topology Expert, Cuaxdon, Loupeter, Jarble, Yobot, Ht686rg90, TaBOT-zerem, AnomieBOT, Citation bot, LilHelpa, Point-set topologist, Sławomir Biały, Citation bot 1, Lost-n-translation, Tkuvho, IhorLviv, Rausch, Tinfoilcat, Le Docteur, Neurotip, Set theorist, Slawekb, Suslindisambiguator, Quondum, D.Lazard, J Dan Christensen, Crown Prince, Helpful Pixie Bot, HMSSolent, Frank.manus, Alesak23, Brad7777, Qetuth, Comfr, Saung Tadashi, Mark L MacDonald, Spectral sequence, Amenlight, Stomatapoll, Mark viking, YiFeiBot, Pwm86, Michael Lee Baker, Stewart.M.Nash and Anonymous: 63

- **Differential structure** *Source:* https://en.wikipedia.org/wiki/Differential_structure?oldid=599737802 *Contributors:* Michael Hardy, Jimfbleak, Susurrus, Charles Matthews, Giftlite, BenFrantzDale, Lethe, Dratman, Mboverload, Elroch, TedPavlic, Don Reba, Oleg Alexandrov,

Joriki, Rjwilmsi, R.e.b., John Baez, KSmrq, Helge Rosé, Orthografer, ABehrens, SmackBot, Silly rabbit, Nbarth, Daqu, CmdrObot, Ranicki, Ben pcc, Aizenr, Turgidson, Torsten Asselmeyer-Maluga, STBot, PerezTerron, Rybu, Ideal gas equation, Addbot, Uncia, Drpickem, Yobot, ⁇⁇, MetaplecticGroup, DSisyphBot, FrescoBot, Sławomir Biały, SepIHw, Magmalex, Quondum, Helpful Pixie Bot, Gilgoldm, Kodip, Karl2828, Tentinator and Anonymous: 31

- **Glossary of group theory** *Source:* https://en.wikipedia.org/wiki/Glossary_of_group_theory?oldid=648705156 *Contributors:* AxelBoldt, Zundark, Michael Hardy, Wshun, Dcljr, Loren Rosen, Charles Matthews, Dysprosia, Zoicon5, Robbot, Tobias Bergemann, Weialawaga~enwiki, Tosha, Giftlite, Fropuff, Waltpohl, D6, Oleg Alexandrov, Ruud Koot, Marudubshinki, Salix alba, Penumbra2000, Grubber, Xyzzyplugh, Jim.belk, Cydebot, RobHar, MikeLynch, Magioladitis, Jakob.scholbach, David Eppstein, CopyToWiktionaryBot, R'n'B, Jesper Carlstrom, Synthebot, JackSchmidt, Mr. Stradivarius, Niceguyedc, TimothyRias, Addbot, Yobot, Vroo, Ilikebeansandweiners, Citation bot, El Caro, Citation bot 1, Quondum, Spectral sequence, CsDix and Anonymous: 15

- **Group theory** *Source:* https://en.wikipedia.org/wiki/Group_theory?oldid=668380925 *Contributors:* AxelBoldt, Zundark, The Anome, KF, Cwitty, Edward, Michael Hardy, Wshun, Dcljr, Ellywa, JWSchmidt, Bogdangiusca, Poor Yorick, Rossami, Jordi Burguet Castell, Charles Matthews, Lfh, Dysprosia, Jitse Niesen, Hyacinth, Fibonacci, Phys, Bevo, Kwantus, Finlay McWalter, PuzzletChung, Gromlakh, Romanm, Mayooranathan, Gandalf61, MathMartin, Rursus, Papadopc, ComplexZeta, Giftlite, Graeme Bartlett, Recentchanges, Dratman, Doshell, Li-Daobing, Alberto da Calvairate~enwiki, Karl-Henner, Rich Farmbrough, FT2, Luqui, ArnoldReinhold, H00kwurm, Paul August, Tompw, Jaimedv, Adan, Obradovic Goran, Friviere, Ranveig, Masv~enwiki, HenryLi, Oleg Alexandrov, Tbsmith, Archie Paulson, OdedSchramm, Kmg90, PeterPearson, V8rik, BD2412, Chun-hian, Josh Parris, Rjwilmsi, Dennis Estenson II, Salix alba, Ligulem, R.e.b., Brighterorange, FlaBot, Chris Pressey, Mathbot, Margosbot~enwiki, Rune.welsh, MTC, Chobot, YurikBot, Hairy Dude, Hillman, Michael Slone, Grubber, Cate, Merlincooper, Petter Strandmark, DYLAN LENNON~enwiki, Crasshopper, Googl, Tigershrike, Willtron, GrinBot~enwiki, RonnieBrown, Palapa, SmackBot, Reedy, Melchoir, Scullin, Natebarney, Cessator, BiT, GBL, Bluebot, Pieter Kuiper, MalafayaBot, Ligulembot, Pilotguy, Davipo, Christopherodonovan, Lambiam, Richard L. Peterson, Utopianheaven, Mike Fikes, Tawkerbot2, Chetvorno, CRGreathouse, Ale jrb, Gregbard, Rifleman 82, Tyskis, Mungomba, Headbomb, WVhybrid, Nadav1, RobHar, NERIUM, Escarbot, Seaphoto, M cuffa, VictorAnyakin, JAnDbot, Bongwarrior, Jakob.scholbach, CountingPine, Baccyak4H, Gabriel Kielland, David Eppstein, MaEr, David Callan, J.delanoy, Cmbankester, Indeed123, Gombang, Treisijs, Useight, Lemonaftertaste, VolkovBot, JohnBlackburne, EchoBravo, Philip Trueman, Eakirkman, Magmi, Eubulides, ArzelaAscoli, Arcfrk, Andreas Carter, Peter Stalin, Drschawrz, SieBot, Ivan Štambuk, WereSpielChequers, Viskonsas, Messagetolove, Lightmouse, JackSchmidt, NobillyT, StaticGull, Alpha Beta Epsilon, Justin W Smith, Alksentrs, Padicgroup, Bhuna71, Mspraveen, Avouac, Watchduck, Edwinconnell, Xylthixlm, Hans Adler, Vegetator, Johnuniq, TimothyRias, XLinkBot, Jin-Jian, CàlculIntegral, Addbot, Manuel Trujillo Berges, SpellingBot, Fluffernutter, Kristine8~enwiki, Favonian, Tide rolls, Luckas-bot, Yobot, TaBOT-zerem, Julia W, Eamonster, AnomieBOT, DemocraticLuntz, Rubinbot, Μυρμηγκάκι, WinoWeritas, Citation bot, Calcio33, Auclairde, FrescoBot, Lothar von Richthofen, Orhanghazi, Sławomir Biały, Citation bot 1, Boulaur, Hard Sin, Hamtechperson, Ngyikp, D stankov, Jauhienij, Debator of mathematics, Lightlowemon, FoxBot, Yger, SomeRandomPerson23, EmausBot, Fly by Night, Tommy2010, Shishir332, D15724C710N, Quondum, Kranix, Adgjdghjdety, Gottlob Gödel, ClueBot NG, Lord Roem, Ciro.santilli, HMSSolent, BG19bot, Ijgt, CimanyD, Meclee, Brad7777, Jochen Burghardt, Brirush, CsDix, Laxfan1977, Chetan bagora, Edmundthe, KasparBot and Anonymous: 136

- **General linear group** *Source:* https://en.wikipedia.org/wiki/General_linear_group?oldid=666798863 *Contributors:* AxelBoldt, Zundark, Patrick, Chas zzz brown, Michael Hardy, Zhaoway~enwiki, A5, Charles Matthews, Dysprosia, Jitse Niesen, Shizhao, Huppybanny, Weialawaga~enwiki, Giftlite, MSGJ, Fropuff, Dratman, Paul August, Gauge, EmilJ, Msh210, Oleg Alexandrov, Linas, Salix alba, HappyCamper, R.e.b., Goudzovski, YurikBot, Dmharvey, RussBot, Michael Slone, KSmrq, Gaius Cornelius, Gwaihir, Cullinane, KnightRider~enwiki, Llanowan, Mhss, Bluebot, Silly rabbit, Nbarth, Harryboyles, Jim.belk, Zero sharp, Dycedarg, RobHar, Albmont, Spvo, Sullivan.t.j, Franp9am, Ixionid, Jeepday, Policron, Pleasantville, Drschawrz, YohanN7, YonaBot, JackSchmidt, Anchor Link Bot, ClueBot, Watchduck, Addbot, Roentgenium111, Topology Expert, Luckas-bot, Yobot, Ht686rg90, Ptbotgourou, Niout, Kilom691, AnomieBOT, Marconet, MondalorBot, Trappist the monk, Greenfernglade, John of Reading, ZéroBot, AvicAWB, Quondum, Emc2fred83, ClueBot NG, Brad7777, Moritorium, Mogism, Spectral sequence, Mark viking, CsDix, Diademodon, Some1Redirects4You and Anonymous: 35

- **Algebraic group** *Source:* https://en.wikipedia.org/wiki/Algebraic_group?oldid=670511317 *Contributors:* AxelBoldt, Michael Hardy, Takuya-Murata, Charles Matthews, Dysprosia, Giftlite, Fropuff, Waltpohl, Vivacissamamente, Paul August, Zaslav, Linas, R.e.b., Wavelength, Crasshopper, Ppntori, Nbarth, Lesnail, Joerg Winkelmann~enwiki, Cronholm144, Jim.belk, Michael Kinyon, Krasnoludek, Bprsolt Qaoddz, Thijs!bot, Headbomb, Turgidson, Jakob.scholbach, David Eppstein, LokiClock, Hesam7, JackSchmidt, Mr. Stradivarius, DeaconJohnFairfax, Alexbot, Addbot, Luckas-bot, FrescoBot, LucienBOT, Artem M. Pelenitsyn, ChuispastonBot, Mgvongoeden, Yasinzaehringer, Mark viking, CsDix, Jodosma, Danneks, K9re11 and Anonymous: 12

- **Discrete group** *Source:* https://en.wikipedia.org/wiki/Discrete_group?oldid=668598567 *Contributors:* Zundark, Patrick, Michael Hardy, Charles Matthews, Giftlite, Fropuff, Cambyses, Rich Farmbrough, Linas, Wavelength, Mosher, SmackBot, RDBury, Maksim-e~enwiki, Commander Keane bot, Mhss, Dreadstar, Jim.belk, Dr Greg, Noleander, Jakob.scholbach, JoergenB, Kyle the bot, Arcfrk, Mr. Stradivarius, DragonBot, Addbot, Topology Expert, Luckas-bot, Amirobot, Erik9bot, Foobarnix, Armando-Martin, 28bot, Brirush, CsDix, Blackbombchu and Anonymous: 12

- **Finite group** *Source:* https://en.wikipedia.org/wiki/Finite_group?oldid=662606326 *Contributors:* AxelBoldt, Zundark, Patrick, Michael Hardy, TakuyaMurata, Silverfish, Schneelocke, Loren Rosen, Charles Matthews, Phys, Schutz, DHN, Tobias Bergemann, Giftlite, D3, Alberto da Calvairate~enwiki, Rgdboer, Andi5, Vipul, ABCD, HenryLi, Oleg Alexandrov, R.e.b., Ysangkok, Wavelength, Cullinane, RDBrown, Mhym, Dreadstar, Cydebot, Kilva, Baccyak4H, STBot, SparsityProblem, TXiKiBoT, Geometry guy, Radagast3, GirasoleDE, Messagetolove, Thehotelambush, JackSchmidt, Mr. Stradivarius, Sfan00 IMG, SilvonenBot, Good Olfactory, Addbot, Lightbot, Luckas-bot, LGB, AnomieBOT, Ciphers, JackieBot, Omnipaedista, Wpiechowski, Quondum, Wikfr, ChuispastonBot, Rezabot, KLBot2, Fraqtive42, Luizpuodzius, ChrisGualtieri, AHusain314, Brirush, CsDix, KasparBot and Anonymous: 22

- **Group action** *Source:* https://en.wikipedia.org/wiki/Group_action?oldid=670376278 *Contributors:* AxelBoldt, LC~enwiki, Bryan Derksen, Zundark, The Anome, Toby Bartels, Patrick, Chas zzz brown, Michael Hardy, Dominus, Chinju, TakuyaMurata, BenKovitz, Charles Matthews, Dysprosia, Jitse Niesen, Michael Larsen, Phys, Schutz, Rvollmert, MathMartin, Tobias Bergemann, Tosha, Giftlite, Rs2, BenFrantzDale, Lethe, MSGJ, Fropuff, Piotrus, Elroch, Pyrop, Guanabot, ArnoldReinhold, Zaslav, El C, 3mta3, Haham hanuka, Eric Kvaalen, Ultramarine,

JackSchmidt, Svick, Razimantv, Virginia-American, Addbot, Мыша, Glane23, AHbot, Luckas-bot, Calle, AnomieBOT, Nishantjr, GrouchoBot, Ringspectrum, FrescoBot, RedAcer, Jesse V., EmausBot, Chaohuang, Netheril96, Quondum, Git2010, ClueBot NG, XDorksiclex, Infodonor, Jmdx, Brirush, CsDix, K9re11, Airwoz, Lor, Iwilsonp, Some1Redirects4You and Anonymous: 65

- **Lattice (discrete subgroup)** *Source:* https://en.wikipedia.org/wiki/Lattice_(discrete_subgroup)?oldid=648705147 *Contributors:* Zundark, Michael Hardy, Charles Matthews, Giftlite, BD2412, Headbomb, Arcfrk, Mr. Stradivarius, Addbot, Yobot, EmausBot, Eransoko, Handsofftibet, Brirush, CsDix and Anonymous: 4

- **Continuous symmetry** *Source:* https://en.wikipedia.org/wiki/Continuous_symmetry?oldid=635098217 *Contributors:* Patrick, Charles Matthews, BD2412, SmackBot, Ealdent, Cronholm144, Special-T, Kilva, Addbot, Erik9bot, Brirush and Anonymous: 2

- **Mathematical object** *Source:* https://en.wikipedia.org/wiki/Mathematical_object?oldid=668632459 *Contributors:* Michael Hardy, Andycjp, Rich Farmbrough, SixWingedSeraph, Reverendgraham, SmackBot, Byelf2007, Bn, Vaughan Pratt, Gregbard, EagleFan, Robin S, Maurice Carbonaro, Tautologist, Certes, Addbot, OlEnglish, JEN9841, Luckas-bot, Yobot, LGB, JRB-Europe, Twri, ArthurBot, FrescoBot, Tkuvho, GreenGrammarian, TobeBot, Slawekb, ZéroBot, Super-real dance, ClueBot NG, Rjs.swarnkar, Brad7777, Bg9989, Loraof and Anonymous: 21

- **Semisimple Lie algebra** *Source:* https://en.wikipedia.org/wiki/Semisimple_Lie_algebra?oldid=648705137 *Contributors:* TakuyaMurata, Charles Matthews, Giftlite, Fropuff, VivaEmilyDavies, Rjwilmsi, R.e.b., Masnevets, YurikBot, Grafen, SmackBot, Nbarth, TenPoundHammer, Mathsci, Yggdrasil014, RobHar, David Eppstein, R'n'B, VolkovBot, JackSchmidt, Mr. Stradivarius, ELLinng, Addbot, Yobot, Niout, Omnipaedista, Sławomir Biały, Night Jaguar, ClueBot NG, Echsecutor, 吴先生, Foursided Triangle, The Disambiguator and Anonymous: 20

- **Homogeneous space** *Source:* https://en.wikipedia.org/wiki/Homogeneous_space?oldid=660895840 *Contributors:* Michael Hardy, TakuyaMurata, Charles Matthews, Dcoetzee, Dysprosia, Phys, Choni, Tobias Bergemann, Giftlite, Fropuff, Fleminra, Tomruen, Paul August, Gauge, Killing Vector, Oleg Alexandrov, Joriki, Linas, MarSch, Mathbot, Chobot, Eienmaru, Siddhant, YurikBot, Archelon, Silly rabbit, Nbarth, YK Times, Apon, Ixionid, Lantonov, Squids and Chips, Trigamma, YoungFrog, LokiClock, TXiKiBoT, Mr. Stradivarius, Alexbot, Nilradical, SilvonenBot, Addbot, Topology Expert, Fluffernutter, Point-set topologist, Jschnur, Fly by Night, Dewritech, Quondum, D.Lazard, Helpful Pixie Bot, Brad7777, Qetuth, Brirush, Vskrin and Anonymous: 25

- **Representation theory** *Source:* https://en.wikipedia.org/wiki/Representation_theory?oldid=666160673 *Contributors:* Zundark, TakuyaMurata, Tobias Bergemann, Unfree, Giftlite, BenFrantzDale, Frau Holle, Linas, BD2412, Wavelength, RussBot, Pred, SmackBot, Melchoir, Colonies Chris, John, Cesium 133, CmdrObot, CBM, Myasuda, RobHar, David Eppstein, R'n'B, Policron, JohnBlackburne, PaulTanenbaum, Geometry guy, Falcon8765, YohanN7, Hugh16, KathrynLybarger, Cyfal, Mr. Stradivarius, The Thing That Should Not Be, Mild Bill Hiccup, Rhubbarb, Alexbot, Addbot, Lightbot, Legobot, Andresswift, Rubinbot, Citation bot, Xqbot, Citation bot 1, Darij, Kiefer.Wolfowitz, Trappist the monk, EmausBot, ZéroBot, Quondum, Helpful Pixie Bot, Beaumont877, Nosuchforever, AdventurousSquirrel, Brad7777, Majesty of Knowledge, Enyokoyama, Paritto, Jochen Burghardt, CsDix, Hamoudafg, Sol1, Liz, Melcous and Anonymous: 34

- **Vector space** *Source:* https://en.wikipedia.org/wiki/Vector_space?oldid=670326684 *Contributors:* AxelBoldt, Bryan Derksen, Zundark, The Anome, Taw, Awaterl, Youandme, N8chz, Olivier, Tomo, Patrick, Michael Hardy, Tim Starling, Wshun, Nixdorf, Kku, Gabbe, Wapcaplet, TakuyaMurata, Pcb21, Iulianu, Glenn, Ciphergoth, Dysprosia, Jitse Niesen, Jogloran, Phys, Kwantus, Aenar, Robbot, Romanm, P0lyglut, Tobias Bergemann, Giftlite, BenFrantzDale, Lethe, MathKnight, Fropuff, Waltpohl, Andris, Daniel Brockman, Python eggs, Chowbok, Sreyan, Lockeownzj00, MarkSweep, Profvk, Maximaximax, Barnaby dawson, Mh, Klaas van Aarsen, TedPavlic, Rama, Smyth, Notinasnaid, Paul August, Bender235, Rgdboer, Shoujun, Army1987, Cmdrjameson, Stephen Bain, Tsirel, Msh210, Orimosenzon, ChrisUK, Ncik~enwiki, Eric Kvaalen, ABCD, Sligocki, Jheald, Eddie Dealtry, Dirac1933, Woodstone, Kbolino, Oleg Alexandrov, Woohookitty, Mindmatrix, ^demon, Hfarmer, Mpatel, MFH, Graham87, Ilya, Rjwilmsi, Koavf, MarSch, Omnieiunium, Salix alba, Titoxd, FlaBot, VKokielov, Therearenospoons, Nihiltres, Ssafarik, Srleffler, Kri, R160K, Chobot, Gwernol, Algebraist, YurikBot, Wavelength, Spacepotato, Hairy Dude, RussBot, Michael Slone, CambridgeBayWeather, Rick Norwood, Kinser, Guruparan, Trovatore, Vanished user 1029384756, Nick, Bota47, BraneJ, Martinwilke1980, Antiduh, Lonerville, Netrapt, Curpsbot-unicodify, Cjfsyntropy, Paul D. Anderson, GrinBot~enwiki, SmackBot, RDBury, InverseHypercube, KocjoBot~enwiki, Davidsiegel, Chris the speller, Geneb1955, SMP, Silly rabbit, Complexica, Nbarth, DHNbot~enwiki, Colonies Chris, Chlewbot, Vanished User 0001, Cícero, Cybercobra, Daqu, Mattpat, James084, Lambiam, Tbjw, Breno, Terry Bollinger, Michael Kinyon, Lim Wei Quan, Rcowlagi, SandyGeorgia, Whackawhackawoo, Inquisitus, Rschwieb, Levineps, Madmath789, Markan~enwiki, Tawkerbot2, Igni, CRGreathouse, Mct mht, Cydebot, Danman3459, Guitardemon666, Mikewax, Thijs!bot, Headbomb, RobHar, CharlotteWebb, Urdutext, Escarbot, JAnDbot, Thenub314, Englebert, Magioladitis, Jakob.scholbach, Kookas, SwiftBot, WhatamIdoing, David Eppstein, Cpl Syx, Charitwo, Akhil999in, Infovarius, Frenchef, TechnoFaye, CommonsDelinker, Paranomia, Michaelp7, Mitsuruaoyama, Trumpet marietta 45750, Daniele.tampieri, Gombang, Policron, Fylwind, Cartiod, Camrn86, AlnoktaBOT, Hakankösem~enwiki, TXiKiBoT, Hlevkin, Gwib, Anonymous Dissident, Imasleepviking, Hrrr, Mechakucha, Geometry guy, Terabyte06, Tommyinla, Wikithesource, Staka, AlleborgoBot, Deconstructhis, Newbyguesses, YohanN7, SieBot, Ivan Štambuk, Portalian, ToePeu.bot, Lucasbfrbot, Tiptoety, Paolo.dL, Henry Delforn (old), Thehotelambush, JackSchmidt, Jorgen W, AlanUS, Randomblue, Jludwig, ClueBot, Alksentrs, Nsk92, JP.Martin-Flatin, FractalFusion, Niceguyedc, DifferCake, Auntof6, 0ladne, PixelBot, Brews ohare, Jotterbot, Hans Adler, SchreiberBike, Jasanas~enwiki, Humanengr, TimothyRias, BodhisattvaBot, SilvonenBot, Jaan Vajakas, Addbot, Gabriele ricci, AndrewHarvey4, Topology Expert, NjardarBot, Looie496, Uncia, ChenzwBot, Ozob, Wikomidia, TeH nOmInAtOr, Jarble, CountryBot, Yobot, Kan8eDie, THEN WHO WAS PHONE?, AnomieBOT, ^musaz, Götz, Citation bot, Xqbot, Txebixev, GeometryGirl, Point-set topologist, RibotBOT, Charvest, Quartl, Lisp21, FrescoBot, Nageh, Rckrone, Sławomir Biały, Citation bot 1, Kiefer.Wolfowitz, Jonesey95, MarcelB612, Stpasha, Mathstudent3000, Jujutacular, Dashed, TobeBot, Javierito92, January, Setitup, TjBot, EmausBot, WikitanvirBot, Brydustin, Fly by Night, Slawekb, Chricho, Ldboer, Quondum, D.Lazard, Milad pourrahmani, RaptureBot, Cloudmichael, ClueBot NG, Wcherowi, Chitransh Gaurav, Jiri 1984, Joel B. Lewis, Widr, Helpful Pixie Bot, Ma snx, David815, Alesak23, Probability0001, JOSmithIII, Duxwing, PsiEpsilon, IkamusumeFan, Սասա Ցսհիսյսս, IPWAI, JYBot, Dexbot, Catclock, Tch3n93, Fycafterpro, CsDix, Hella.chillz, Jose Brox, François Robere, Loganfalco, Newestcastleman, K9re11, Monkbot, AntiqueReader, KurtHeckman, Isambard Kingdom, Shivakrishna .Srinivas. Dasari and Anonymous: 213

- **Group homomorphism** *Source:* https://en.wikipedia.org/wiki/Group_homomorphism?oldid=670432256 *Contributors:* AxelBoldt, Bryan Derksen, XJaM, Toby Bartels, Edward, Michael Hardy, TakuyaMurata, Revolver, Charles Matthews, Dysprosia, Mattblack82, Altenmann, Hemanshu, Giftlite, Gene Ward Smith, Dratman, Cambyses, TheObtuseAngleOfDoom, Guanabot, Luqui, Goochelaar, Elwikipedista~enwiki, Adan, Rgdboer, Andi5, Obradovic Goran, Msh210, Arthena, Oleg Alexandrov, Graham87, Chenxlee, FlaBot, Chobot, Algebraist, Michael Slone,

Grubber, Pred, KocjoBot~enwiki, Mhss, Bird of paradox, Cronholm144, Noleander, Michael C Price, Dragonflare82, Konradek, RobHar, Albmont, Ling.Nut, Jakob.scholbach, Kedlav, STBot, Inquam, LordAnubisBOT, Hiwk, STBotD, VolkovBot, Hesam7, Cgwaldman, SieBot, ToePeu.bot, Thehotelambush, Mr. Stradivarius, Razimantv, Alexbot, MystBot, Addbot, Download, PV=nRT, Ptbotgourou, JRB-Europe, ArthurBot, Rsepahi, RibotBOT, Kaoru Itou, Sjcjoosten, Quondum, Sahimrobot, Bezik, ChrisGualtieri, Brirush, CsDix, Nigellwh and Anonymous: 26

- **Automorphism** *Source:* https://en.wikipedia.org/wiki/Automorphism?oldid=639114467 *Contributors:* AxelBoldt, LC~enwiki, Tarquin, Jan Hidders, Youssefsan, Hephaestos, Edward, Bdesham, Patrick, Chas zzz brown, Michael Hardy, William M. Connolley, AugPi, MatrixFrog, Dysprosia, Phys, SirJective, Robbot, Huppybanny, Altenmann, Tosha, Giftlite, MSGJ, Fropuff, Peruvianllama, Kaldari, Rdsmith4, Sam Hocevar, Noisy, Rich Farmbrough, Xezbeth, Zaslav, Elwikipedista~enwiki, Gauge, Rgdboer, Crisófilax, Obradovic Goran, Keenan Pepper, Mcmillin24, Oleg Alexandrov, Pixeltoo, Qwertyus, Amire80, VKokielov, Mathbot, Algebraist, YurikBot, Archelon, TechnoGuyRob, BOT-Superzerocool, Cbogart2, Reyk, Pred, Nbarth, Henning Makholm, Mets501, Dreftymac, Happy-melon, Gregbard, Gogo Dodo, Dogaroon, MishaMisha, Hannes Eder, Salgueiro~enwiki, Ksanyi, David Eppstein, Policron, Caiodnh, TXiKiBoT, Mskalak13, Thehotelambush, JackSchmidt, Mild Bill Hiccup, Brews ohare, MelonBot, TimothyRias, Marc van Leeuwen, DOI bot, Topology Expert, PV=nRT, Legobot, JRB-Europe, Citation bot, ArthurBot, Nishantjr, Omnipaedista, SassoBot, Shadowjams, FrescoBot, Citation bot 1, Orenburg1, Quondum, Tommy Jantarek, Helpful Pixie Bot, Vagobot, MathKnight-at-TAU, Valentinovna, Bg9989, Monkbot, Dyott and Anonymous: 24

- **Isomorphism** *Source:* https://en.wikipedia.org/wiki/Isomorphism?oldid=670285447 *Contributors:* AxelBoldt, Zundark, Andre Engels, Youssefsan, Ghakko, Edemaine, Ryguasu, Youandme, Stevertigo, Patrick, Michael Hardy, Isomorphic, TakuyaMurata, Glenn, Netsnipe, Mxn, Revolver, Charles Matthews, Reddi, Dysprosia, Andrewman327, Zero0000, Phys, Robbot, Bkell, Marc Venot, Tosha, Giftlite, MathKnight, Peruvianllama, MarkSweep, PhotoBox, Wrp103, Paul August, Bender235, Elwikipedista~enwiki, Nabla, Rgdboer, EmilJ, Army1987, Msh210, Philip Cross, Oleg Alexandrov, LOL, BD2412, Yurik, Zbxgscqf, Mattmacf, Mathbot, Chobot, YurikBot, Wavelength, Spacepotato, Jlittlet, Michael Slone, KSmrq, Mathwiz777, Grubber, Vanished user 1029384756, Stuhacking, Banus, SmackBot, Rljacobson, The Rhymesmith, Kmarinas86, Lubos, Nbarth, Chlewbot, Maksim-bot, Bryanmcdonald, Spinality, Nick Green, Fantomdrives, Cronholm144, Jim.belk, 16@r, Mets501, Rschwieb, Yuide, CRGreathouse, Krauss, Sam Staton, Rlupsa, QuiteUnusual, Hannes Eder, TK-925, JAnDbot, Avaya1, Bahar, Coolhandscot, Magioladitis, Koberozendaal, JamesBWatson, Albmont, Uncle Dick, Smite-Meister, Cpiral, Trumpet marietta 45750, Policron, Bigdumbdinosaur, STBotD, Mcole13, Michael Angelkovich, Borhan0, LokiClock, Thaddeus Slamp, Anonymous Dissident, JhsBot, PaulTanenbaum, Spinningspark, Mike4ty4, SieBot, Ivan Štambuk, Ssavelan, Iamthedeus, Taemyr, Thehotelambush, Richard Molnár-Szipai, Anchor Link Bot, DixonD, Superbatfish, Martarius, Alexbot, Bender2k14, Ioannis.Demetriou, Brews ohare, Subversive.sound, Addbot, Omnipedian, Debresser, Ozob, PV=nRT, Jarble, Luckas-bot, Ezequiels.90, Nallimbot, AnomieBOT, Xqbot, XZeroBot, Omnipaedista, AllCluesKey, Sławomir Biały, Ebony Jackson, RedBot, Lars Washington, Pokus9999, Gryllida, عقیل فشاك, EmausBot, Fly by Night, Racerx11, GoingBatty, Slawekb, Zell08v, Quondum, D.Lazard, SporkBot, YnnusOiramo, ClueBot NG, Frietjes, Mesoderm, HMSSolent, BG19bot, Канеюку, M hariprasad, Brad7777, Pratyya Ghosh, Pariefracture, Mark viking, Eyesnore and Anonymous: 86

- **Field (mathematics)** *Source:* https://en.wikipedia.org/wiki/Field_(mathematics)?oldid=670186165 *Contributors:* AxelBoldt, Bryan Derksen, Zundark, The Anome, Andre Engels, Josh Grosse, XJaM, Toby Bartels, Miguel~enwiki, Lir, Patrick, Michael Hardy, Wshun, DIG~enwiki, TakuyaMurata, Karada, Looxix~enwiki, Rossami, Andres, Loren Rosen, Revolver, RodC, Dysprosia, Jitse Niesen, Prumpf, Tero~enwiki, Phys, Philipp, R3m0t, Jmabel, Mattblack82, MathMartin, P0lyglut, Wikibot, Tobias Bergemann, Unfree, Marc Venot, Giftlite, Highlandwolf, Gene Ward Smith, Lethe, Zigger, Fropuff, Millerc, Waltpohl, Python eggs, Gubbubu, CSTAR, Pmanderson, Barnaby dawson, PhotoBox, Mormegil, Jørgen Friis Bak, Discospinster, Guanabot, Sperling, Paul August, Zaslav, Elwikipedista~enwiki, El C, Rgdboer, EmilJ, Touriste, Army1987, Giraffedata, Obradovic Goran, OoberMick, Msh210, Mlm42, Olegalexandrov, RJFJR, Oleg Alexandrov, Woohookitty, Linas, Arneth, Bkkbrad, Hypercube~enwiki, MarkTempeit, Damicatz, MFH, Isnow, Palica, Graham87, FreplySpang, Chenxlee, Josh Parris, Rjwilmsi, Hiberniantears, Salix alba, R.e.b., FlaBot, Codazzi~enwiki, Jrtayloriv, R160K, Chobot, Abu Amaal, Algebraist, Wavelength, Dmesg, Eraserhead1, Hairy Dude, KSmrq, Grubber, Archelon, Rintrah, Rat144, Rick Norwood, Trovatore, DYLAN LENNON~enwiki, Crasshopper, RaSten, DavidHouse~enwiki, Mgnbar, Children of the dragon, SmackBot, Mmernex, Melchoir, Gilliam, Nbarth, Charlotte Hobbs, Lesnail, Cybercobra, Acepectif, Slawekb, Bidabadi~enwiki, Lambiam, Jim.belk, Schildt.a, Mets501, DabMachine, Rschwieb, WAREL, Newone, Vaughan Pratt, CRGreathouse, Kupirijo, Tiphareth, DEWEY, Eulerianpath, Pedro Fonini, Goldencako, BobNiichel, Xantharius, KLIP~enwiki, JLISP, Headbomb, RobHar, Nick Number, Turgidson, Kprateek88, Martinkunev, Magioladitis, Bongwarrior, VoABot II, JamesBWatson, Jakob.scholbach, SwiftBot, Catgut, Lukeaw, MORI, Cpiral, Maproom, Gombang, Policron, Barylior, Umarekawari, LokiClock, Red Act, Anonymous Dissident, Hesam7, Joeldl, Dave703, Zermalo, Shellgirl, Cwkmail, Soler97, JackSchmidt, Jorgen W, Anchor Link Bot, Willy, your mate, Oekaki, UKe-CH, ClueBot, Mild Bill Hiccup, Tcklein, Niceguyedc, He7d3r, Bender2k14, Squirreljml, Palnot, ZooFari, Addbot, Gabriele ricci, Download, Unzerlegbarkeit, Cesiumfrog, Yobot, Ht686rg90, TaBOT-zerem, Zagothal, AnomieBOT, DSisyphBot, Depassp, Danielschreiber, MegaMouthBolt123, Point-set topologist, Charvest, KirarinSnow, FrescoBot, Mjmarkowitz, RandomDSdevel, Ebony Jackson, D stankov, Girish.ponkiya2007, Kunle102, DASHBot, Sedrikov, Tom.kemp90, Tommy2010, Wikipelli, Shishir332, Lfrazier11, Quondum, D.Lazard, JimMeiss, ClueBot NG, Ankur1vi, Wcherowi, Frietjes, MerlIwBot, Helpful Pixie Bot, !mcbloobyenstein!!, Or elharar, Fabio.nsantos, Rjs.swarnkar, Topgraph28, Deltahedron, Sanipriya, GigaGerard, CsDix, YiFeiBot, Teddyktchan, GeoffreyT2000, Charlotte Aryanne, Vluczkow and Anonymous: 133

30.15.2 Images

- **File:Affine_subspace.svg** *Source:* https://upload.wikimedia.org/wikipedia/commons/8/8c/Affine_subspace.svg *License:* CC BY-SA 3.0 *Contributors:* Own work *Original artist:* Jakob.scholbach

- **File:Alternating_group_4;_Cayley_table;_numbers.svg** *Source:* https://upload.wikimedia.org/wikipedia/commons/8/8b/Alternating_group_4%3B_Cayley_table%3B_numbers.svg *License:* CC BY 3.0 *Contributors:* Own work *Original artist:* Watchduck (a.k.a. Tilman Piesk)

- **File:Caesar3.svg** *Source:* https://upload.wikimedia.org/wikipedia/commons/2/2b/Caesar3.svg *License:* Public domain *Contributors:* Own work *Original artist:* Cepheus

- **File:Cayley_graph_of_F2.svg** *Source:* https://upload.wikimedia.org/wikipedia/commons/d/d2/Cayley_graph_of_F2.svg *License:* Public domain *Contributors:* ? *Original artist:* ?

- **File:Split-arrows.svg** *Source:* https://upload.wikimedia.org/wikipedia/commons/a/a7/Split-arrows.svg *License:* Public domain *Contributors:* ? *Original artist:* ?

- **File:Symmetric_group_3;_Cayley_table;_GL(2,2).svg** *Source:* https://upload.wikimedia.org/wikipedia/commons/2/27/Symmetric_group_3%3B_Cayley_table%3B_GL%282%2C2%29.svg *License:* Public domain *Contributors:* Own work *Original artist:* Watchduck (a.k.a. Tilman Piesk)

- **File:Symmetric_group_3;_Cayley_table;_subgroup_of_S4_(elements_0,1,14,15,20,21).svg** *Source:* https://upload.wikimedia.org/wikipedia/commons/2/26/Symmetric_group_3%3B_Cayley_table%3B_subgroup_of_S4_%28elements_0%2C1%2C14%2C15%2C20%2C21%29.svg *License:* CC BY 3.0 *Contributors:* Own work *Original artist:* Watchduck (a.k.a. Tilman Piesk)

- **File:Symmetric_group_3;_Cayley_table;_subgroup_of_S4_(elements_0,1,2,3,4,5).svg** *Source:* https://upload.wikimedia.org/wikipedia/commons/8/8e/Symmetric_group_3%3B_Cayley_table%3B_subgroup_of_S4_%28elements_0%2C1%2C2%2C3%2C4%2C5%29.svg *License:* CC BY 3.0 *Contributors:* Own work *Original artist:* Watchduck (a.k.a. Tilman Piesk)

- **File:Symmetric_group_3;_Cayley_table;_subgroup_of_S4_(elements_0,2,6,8,12,14).svg** *Source:* https://upload.wikimedia.org/wikipedia/commons/e/e6/Symmetric_group_3%3B_Cayley_table%3B_subgroup_of_S4_%28elements_0%2C2%2C6%2C8%2C12%2C14%29.svg *License:* CC BY 3.0 *Contributors:* Own work *Original artist:* Watchduck (a.k.a. Tilman Piesk)

- **File:Symmetric_group_3;_Cayley_table;_subgroup_of_S4_(elements_0,5,6,11,19,21).svg** *Source:* https://upload.wikimedia.org/wikipedia/commons/3/35/Symmetric_group_3%3B_Cayley_table%3B_subgroup_of_S4_%28elements_0%2C5%2C6%2C11%2C19%2C21%29.svg *License:* CC BY 3.0 *Contributors:* Own work *Original artist:* Watchduck (a.k.a. Tilman Piesk)

- **File:Symmetric_group_4;_Cayley_graph_4,9.svg** *Source:* https://upload.wikimedia.org/wikipedia/commons/b/b5/Symmetric_group_4%3B_Cayley_graph_4%2C9.svg *License:* Public domain *Contributors:*

- GrapheCayley-S4-Plan.svg *Original artist:* GrapheCayley-S4-Plan.svg: Fool (talk)

- **File:Symmetric_group_4;_Cayley_table;_numbers.svg** *Source:* https://upload.wikimedia.org/wikipedia/commons/d/d1/Symmetric_group_4%3B_Cayley_table%3B_numbers.svg *License:* CC BY 3.0 *Contributors:* Own work *Original artist:* Watchduck (a.k.a. Tilman Piesk)

- **File:Symmetric_group_4;_Lattice_of_subgroups_Hasse_diagram.svg** *Source:* https://upload.wikimedia.org/wikipedia/commons/3/32/Symmetric_group_4%3B_Lattice_of_subgroups_Hasse_diagram.svg *License:* CC BY 3.0 *Contributors:* ? *Original artist:* ?

- **File:Text_document_with_red_question_mark.svg** *Source:* https://upload.wikimedia.org/wikipedia/commons/a/a4/Text_document_with_red_question_mark.svg *License:* Public domain *Contributors:* Created by bdesham with Inkscape; based upon Text-x-generic.svg from the Tango project. *Original artist:* Benjamin D. Esham (bdesham)

- **File:Torus.png** *Source:* https://upload.wikimedia.org/wikipedia/commons/1/17/Torus.png *License:* Public domain *Contributors:* ? *Original artist:* ?

- **File:Two_coordinate_charts_on_a_manifold.svg** *Source:* https://upload.wikimedia.org/wikipedia/commons/0/06/Two_coordinate_charts_on_a_manifold.svg *License:* CC BY-SA 3.0 *Contributors:* Own work *Original artist:* Stomatapoll

- **File:Universal_tensor_prod.svg** *Source:* https://upload.wikimedia.org/wikipedia/commons/d/d3/Universal_tensor_prod.svg *License:* CC BY-SA 4.0 *Contributors:* Own work *Original artist:* IkamusumeFan

- **File:Vector_addition3.svg** *Source:* https://upload.wikimedia.org/wikipedia/commons/e/e6/Vector_addition3.svg *License:* CC BY-SA 3.0 *Contributors:* Own work *Original artist:* Jakob.scholbach

- **File:Vector_addition_ans_scaling.png** *Source:* https://upload.wikimedia.org/wikipedia/commons/7/72/Vector_addition_ans_scaling.png *License:* CC-BY-SA-3.0 *Contributors:* Own work *Original artist:* Jakob.scholbach

- **File:Vector_components.svg** *Source:* https://upload.wikimedia.org/wikipedia/commons/8/87/Vector_components.svg *License:* CC BY-SA 3.0 *Contributors:* Own work *Original artist:* Jakob.scholbach

- **File:Vector_components_and_base_change.svg** *Source:* https://upload.wikimedia.org/wikipedia/commons/a/a6/Vector_components_and_base_change.svg *License:* CC BY-SA 3.0 *Contributors:* Own work *Original artist:* Jakob.scholbach

- **File:Vector_norms2.svg** *Source:* https://upload.wikimedia.org/wikipedia/commons/0/02/Vector_norms2.svg *License:* CC BY-SA 3.0 *Contributors:* Own work (Original text: *I created this work entirely by myself.*) *Original artist:* Jakob.scholbach (talk)

- **File:Wiki_letter_w_cropped.svg** *Source:* https://upload.wikimedia.org/wikipedia/commons/1/1c/Wiki_letter_w_cropped.svg *License:* CC-BY-SA-3.0 *Contributors:*

- Wiki_letter_w.svg *Original artist:* Wiki_letter_w.svg: Jarkko Piiroinen

- **File:Wikibooks-logo-en-noslogan.svg** *Source:* https://upload.wikimedia.org/wikipedia/commons/d/df/Wikibooks-logo-en-noslogan.svg *License:* CC BY-SA 3.0 *Contributors:* Own work *Original artist:* User:Bastique, User:Ramac et al.

- **File:Wiktionary-logo-en.svg** *Source:* https://upload.wikimedia.org/wikipedia/commons/f/f8/Wiktionary-logo-en.svg *License:* Public domain *Contributors:* Vector version of Image:Wiktionary-logo-en.png. *Original artist:* Vectorized by Fvasconcellos (talk · contribs), based on original logo tossed together by Brion Vibber

30.15.3 Content license

- Creative Commons Attribution-Share Alike 3.0

www.ingramcontent.com/pod-product-compliance
Lightning Source LLC
Chambersburg PA
CBHW080803180526
45168CB00006B/2310